Immunopharmacology of the Heart

THE HANDBOOK OF IMMUNOPHARMACOLOGY

Series Editor: Clive Page
King's College London, UK

Titles in this series

Cells and Mediators

Immunopharmacology of Eosinophils
(edited by H. Smith and R. Cook)

Immunopharmacology of Mast
Cells and Basophils
(edited by J.C. Foreman,
forthcoming)

Adhesion Molecules
(edited by C.D. Wegner,
forthcoming)

Lipid Mediators
(edited by F. Cunningham,
forthcoming)

Immunopharmacology of
Lymphocytes
(edited by M. Rola-Pleszczynski,
forthcoming)

Immunopharmacology of Platelets
(edited by M. Joseph, forthcoming)

Immunopharmacology of
Neutrophils
(edited by P.G. Hellewell and
T.J. Williams)

Immunopharmacology of
Macrophages and Other
Antigen-Presenting Cells
(edited by C.A.F.M. Bruijnzeel-
Koomen, forthcoming)

Systems

Immunopharmacology of the
Gastrointestinal System
(edited by J.L. Wallace)

Immunopharmacology of the Heart
(edited by M.J. Curtis)

Immunopharmacology of Joints
and Connective Tissue
(edited by J. Dingle and M.E.
Davies, forthcoming)

Immunopharmacology of Epithelial
Barriers
(edited by R. Goldie, forthcoming)

Immunopharmacology of the Renal
System
(edited by C. Tetta, forthcoming)

Immunopharmacology of
Microcirculation
(edited by S. Brain, forthcoming)

Drugs

Immunotherapy for Immune-
related Diseases
(edited by W.J. Metzger,
forthcoming)

Immunopharmacology of AIDS
(forthcoming)

Immunosuppressive Drugs
(forthcoming)

Glucocorticosteroids
(forthcoming)

Angiogenesis
(forthcoming)

Immunopharmacology of Free
Radical Species
(forthcoming)

Immunopharmacology
of
the Heart

edited by

Michael J. Curtis
King's College, London, UK

ACADEMIC PRESS
Harcourt Brace and Company, Publishers
London San Diego New York
Boston Sydney Tokyo Toronto

ACADEMIC PRESS LIMITED
24/28 Oval Road
London NW1 7DX

United States Edition published by
ACADEMIC PRESS INC.
San Diego, CA 92101

A catalogue record for this book
is available from the British Library

ISBN 0-12-200245-8

Typeset by Mathematical Composition Setters Ltd, Salisbury, Wiltshire
Printed and bound in Great Britain by The Bath Press, Avon

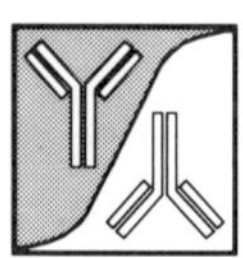

Contents

6. *Complement Activation in Cardiac Disease* 75
Roger D. Rossen

7. *Prevention of Sudden Cardiac Death by Immunopharmacological Intervention* 87
Michael J. Curtis

8. *Inflammatory Mediators and the Stunned Myocardium* 101
Garrett J. Gross

9. *Immunotherapy and Reperfusion Injury in the Heart* 109
Tsutomu Yamazaki, Yoshinori Seko, Ryozo Nagai *and* Yoshio Yazaki

10. *Immunopharmacology of Heart Transplantation* 117

Š. Nyulassy, J. Slezák *and* T. Ravingerová

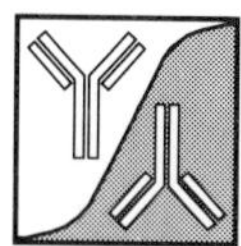

Contributors

A. Albertini
Università degli Studi di Brescia,
Facoltà di Medicina e Chirurgia Cettedre di
 Cardiologia,
P le Spedali Civili 1,
Spedali Civili 15123 Brescia,
Italy

C.M Ballantyne
Section of Cardiovascular Sciences,
Department of Medicine,
Baylor College of Medicine,
One Baylor Plaza,
Houston,
TX 77221,
USA

C. Ceconi
Università degli Studi di Brescia,
Facoltà di Medicine e Chirurgia Cettedra di
 Cardiologia,
P le Spedali Civili 1,
Spedali Civili 25123 Brescia,
Italy

L. Comini
Università degli Studi di Brescia,
Facoltà di Medicina e Chirurgia Cettedra di
 Cardiologia,
P le Spedali Civili 1,
Spedali Civili 25123 Brescia,
Italy

M.J. Curtis
Cardiovascular Research Laboratories,
Pharmacology Group,
King's College London,
London SW3 6LX,
UK

V.M. Darley-Usmar
The Wellcome Research Laboratories,
Langley Court,
South Eden Park Road,
Beckenham,
Kent BR3 3BS,
UK

M.L Entman
Section of Cardiovascular Sciences,
Department of Medicine,
Baylor College of Medicine
One Baylor Plaza
Houston,
TX 77211,
USA

R. Ferrari
Università degli Studi di Brescia,
Facoltà di Medicina e Chirurgia Cettedra di
 Cardiologia,
P le Spedali Civili 1,
Spedali Civili 25123 Brescia,
Italy

G.J. Gross
Department of Pharmacology,
Medical College of Wisconsin,
8701 Watertown Plank Road,
Milwaukee,
WI 53226,
USA

D.G. Hassall
The Wellcome Research Laboratories,
Langley Court,
South Eden Park Road,
Beckenham,
Kent BR3 3BS,
UK

G.L. Kukielka
Section of Cardiovascular Sciences,
Department of Medicine,
Baylor College of Medicine,
One Baylor Plaza,
Houston,
TX 77211
USA

A.M. Lefer
Department of Physiology,
Jefferson Medical College,
Thomas Jefferson University,
1020 Locust Street,
Philadelphia,
PA 19107,
USA

R. Nagai
The Third Department of Internal Medicine,
Faculty of Medicine,
University of Tokyo,
7-3-1 Hongo,
Bunkyo-ku
Tokyo 113,
Japan

Š. Nyulassy
Slovak Academy of Sciences,
Institute for Heart Research,
842 33 Bratislava,
Dúbravská cesta 9,
Republic of Slovakia

M.K. Pugsley
Department of Pharmacology and Therapeutics,
University of British Columbia,
Vancouver, B.C.,
Canada V6T 1W5

T. Ravingerová
Slovak Academy of Sciences,
Institute for Heart Research,
842 33 Bratislava,
Dúbravská cesta 9,
Republic of Slovakia

R.D. Rossen
Immunology Research Laboratory,
Building 211, Room 203,
VA Medical Center,
2002 Holcome Boulevard,
Houston,
TX 77211,
USA

Y. Seko
The Third Department of Internal Medicine,
Faculty of Medicine,
University of Tokyo,
7-3-1 Hongo,
Bunkyo-ku,
Tokyo 113,
Japan

J. Slezák
Slovak Academy of Sciences,
Institure for Heart Research,
842 33 Bratislava,
Dúbravská cesta 9,
Republic of Slovakia

C.W. Smith
Section of Cardiovascular Sciences,
Department of Medicine,
Baylor College of Medicine,
One Baylor Plaza,
Houston,
TX 77211,
USA

O. Visioli
Università degli Studi di Brescia,
Facoltà di Medicina e Chirurgia Cettedre di
   Cardiologia,
P le Spedali Civili 1,
Spedali Civili 25123 Brescia,
Italy

M.J.A Walker
Department of Pharmacology and Therapeutics,
University of British Columbia,
Vancouver, B.C.,
Canada V6T 1W5

T. Yamazaki
The Third Department of Internal Medicine,
Faculty of Medicine,
University of Tokyo,
7-3-1 Hongo,
Bunkyo-ku,
Tokyo 113,
Japan

Y. Yazaki
The Third Department of Internal Medicine,
Faculty of Medicine,
University of Tokyo,
7-3-1 Hongo,
Bunkyo-ku,
Tokyo 113,
Japan.

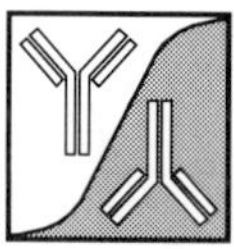

Series Preface

The consequences of diseases involving the immune system such as AIDS, and chronic inflammatory diseases such as bronchial asthma, rheumatoid arthritis and atherosclerosis, now account for a considerable economic burden to governments worldwide. In response to this, there has been a massive research effort investigating the basic mechanisms underlying such diseases, and a tremendous drive to identify novel therapeutic applications for the prevention and treatment of such diseases. Despite this effort, however, much of it within the pharmaceutical industries, this area of medical research has not gained the prominence of cardiovascular pharmacology or neuropharmacology. Over the last decade there has been a plethora of research papers and publications on immunology, but comparatively little written about the implications of such research for drug development. There is also no focal information source for pharmacologists with an interest in diseases affecting the immune system or the inflammatory response to consult, whether as a teaching aid or as a research reference. The main impetus behind the creation of this series was to provide such a source by commissioning a comprehensive collection of volumes on all aspects of immunopharmacology. It has been a deliberate policy to seek editors for each volume who are not only active in their respective areas of expertise, but who also have a distinctly *pharmacological* bias to their research. My hope is that *The Handbook of Immunopharmacology* will become indispensable to researchers and teachers for many years to come, with volumes being regularly updated.

The series follows three main themes, each theme represented by volumes on individual component topics.

The first covers each of the major cell types and classes of inflammatory mediators. The second covers each of the major organ systems and the diseases involving the immune and inflammatory responses that can affect them. The series will thus include clinical aspects along with basic science. The third covers different classes of drugs that are currently being used to treat inflammatory disease or diseases involving the immune system, as well as novel classes of drugs under development for the treatment of such diseases.

To enhance the usefulness of the series as a reference and teaching aid, a standardized artwork policy has been adopted. A particular cell type, for instance, is represented identically throughout the series. An appendix of these standard drawings is published in each volume. Likewise, a standardized system of abbreviations of terms has been implemented and will be developed by the editors involved in individual volumes as the series grows. A glossary of abbreviated terms is also published in each volume. This should facilitate cross-referencing between volumes. In time, it is hoped that the glossary will be regarded as a source of standard terms.

While the series has been developed to be an integrated whole, each volume is complete in itself and may be used as an authoritative review of its designated topic.

I am extremely grateful to the officers of Academic Press, and in particular to Dr Carey Chapman, for their vision in agreeing to collaborate on such a venture, and greatly hope that the series does indeed prove to be invaluable to the medical and scientific community.

C.P. Page

Preface

The objective of this volum is to consider a variety of cardiac diseases for which drugs may play a therapeutic role by virture of their effects on aspects of the immune system. Since many readers will be oriented more towards immunopharmacology in general than towards cardiology specifically, the first two chapters are designed to set the stage by reviewing diseases of the heart which may involve an immunological component, and methods and techniques for the study of physiological and biochemical function in the heart. The heart's own vessels are a poignant focus for immunological consideration, and chapters have been written which consider the immunopharmacology of the coronary vascular endothelium and the role of cellular and biochemical components of the immune system in the pathogenesis of atherosclerosis. Leukocytes have a broad potential involvement in ischaemic heart disease, and a substantial chapter is devoted to this topic. Complement activation is relevant to several types of cardiac disease, and is likewise and separately addressed. Sudden cardiac death, the stunned myocardium and reperfusion injury each have a multifactorial pathophysiological basis and may respond to diverse therapies; in each case a chapter is devoted to the potential for immunopharmacological intervention. Finally, and perhaps most obviously, the use of immunologically relevant agents in the setting of cardiac transplantation has been reviewed from the clinical perspective.

Immunotherapy has a definite role to play in cardiology which is greater or lesser than other types of intervention according to the specific type of disease. I would hope that this volume serves to identify where and clarify how this role is met, and perhaps shed some light on likely future developments. I am indebted, therefore, to many colleagues from the international scientific and medical community, each of whom have contributed a chapter, and to Clive Page who gave me the privilege of putting the volume together.

Michael J. Curtis

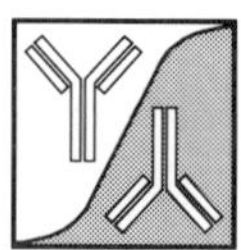

1. Diseases of the Heart: Overview and Identification of Possible Targets for Immunopharmacological Intervention

R. Ferrari, C. Ceconi, L. Comini, A. Albertini and O. Visioli

1. Introduction

The modern cardiologist can no longer understand the pathological processes of the cardiovascular system without a knowledge of immunology. During the last ten years there has been a tremendous growth in our understanding of the immune system and in the technology related to monoclonal antibodies.

As diagnostic tools, monoclonal antibodies have been used as imaging agents to diagnose disorders such as myocardial infarction (Khaw and Haber, 1989), myocarditis (Dec *et al.*, 1988) and cardiac transplant rejection (Frist *et al.*, 1987). As therapeutic agents to a selected target, monoclonal antibodies are very promising for treating digoxin toxicity (Smith *et al.*, 1982) and as adjuncts in the treatment of coronary artery disease (Runge and Haber, 1992) and cardiac transplantation rejection (Swinnen *et al.*, 1990).

Immunological injury, which in the early days was confined to the pathogenesis of rheumatic fever, now seems to be implicated in a number of clinical syndromes affecting the myocardium and the blood vessels. The precise mechanisms of immunological injury are still not completely identified. Several hypotheses have been considered such as autoimmune reactions, cross-reaction between unrelated antigens, immune complex deposition and immune responses to the site of injury when non-recognized epitopes are exposed to the immune system.

In this chapter we consider clinical conditions which are clearly due to immunological injury as well as other conditions in which such types of injury may play a role, although which are not necessarily treated by immunopharmacological interventions.

2. Acute Rheumatic Fever

Clinical, prophylactic, epidemiological and immunological evidence supports a streptococcal etiology of rheumatic fever. However, the link between group A streptococci throat infection and the development of rheumatic fever is still unknown. Certainly, there is no streptococcal colonization in the affected organs or a reaction against toxins hibernated by the organism. Probably the disease is due to a complex interrelationship between the genetics of the host's immune system and the streptococcus bacillus (Zabriskie, 1985). An immunological origin has been suspected because rheumatic fever occurs in those individuals who have a significant antibody response to streptococcal proteins (Seigal *et al.*, 1961). It was postulated that streptococci contain

Immunopharmacology of the Heart
ISBN 0-12-200245-8

components that are antigenically cross-reactive with human tissues and streptococcal antigens have been identified to cross-react with myocardium, cardiac valves and with the cytoplasm of the cells of subthalamic and caudate nuclei. Patients with acute rheumatic fever show a serum concentration of antibodies to the myocardium at greater concentration than that observed in patients with uncomplicated streptoccocal infections (Kaplan, 1963; Kaplan *et al.*, 1964). These antibodies remain elevated for a period of 3–5 years after the acute attack (Zabriskie *et al.*, 1970).

There have been many attempts to identify the antigen cross-reacting with cardiac tissue (Kaplan *et al.*, 1964; Kaplan, 1965, 1967; Zabriskie and Fremier, 1966; Zabriskie *et al.*, 1970). An antigen which completely blocks the binding of heart-reactive antibodies from patients with acute rheumatic fever to normal myocardium has been purified from group A streptococcal membrane (Van de Rijn *et al.*, 1977). This antigen cross-reacts with the sarcolemma of cardiac myocytes. In addition, it has been demonstrated that M proteins from different serotypes of group A streptococci share immunological epitopes on cardiac myosin heavy chain (Cavelti, 1955; Stollerman, 1975; DiSciascio and Taranta, 1980; Dale and Beachey, 1985). Therefore streptococcal proteins which have common antigenic determinants with human cardiac tissue have been identified. This explains why an immunological reaction to a group A streptococcus would result in an antibody being able to attack the myocardium.

Similarly an immunological cross-reaction exists between the structural glycoprotein components of the human heart valves and a polysaccharide of the group A streptococcus which would explain the valvular lesions.

It is likely that many of the streptococci, as result of a long-term evolutionary process, have evolved to be similar to the antigens of normal human tissue in order to elude the immunological response. In general the organism can colonize humans in the absence of an immunological response targeted to its eradication but, in a minority of cases having acute streptococcal pharyngitis, an immune response is evoked. At this stage the antibodies cannot discriminate between the host and the invading organism resulting in the destruction of each.

3.   *Myocarditis*

The hypothesis that viral infections may induce immunologically mediated chronic damage which causes symptomatic congestive heart failure has achieved widespread acceptance despite the lack of absolute proof in man (Huber, 1987).

The use of endomyocardial biopsy allows a greater specificity in the diagnosis of myocarditis and it has implicated immunological injury in the progression of the disease (Fenoglio *et al.*, 1983; Midei *et al.*, 1990). Histology shows a classifical pattern: in the early stages there is predominance of monocyte infiltration and interstitial oedema. Then areas of necrosis with decreasing numbers of inflammatory cells become evident, culminating in a late stage when there are only large areas of necrosis with no evidence of active inflammation (Wynne and Braunwald, 1988; Wenger *et al.*, 1990).

Thus, immunologically mediated damage to the heart following viral myocarditis is thought to cause the idiopathic dilated cardiomyopathy which follows the myocardial infection.

At the experimental level, the murine model of enteroviral myocarditis may have anologies to the human disease. Infection of mice with cardiotropic coxsackie virus B or encephalomyocarditis virus causes a short-term viral replication in the myocardium, followed by a chronic phase of the illness characterized by progressive myocyte damage due to a genetically predetermined humoral or cell-mediated autoimmune response by cytotoxic T lymphocytes (which is, in turn) driven by intrinsic myocardial antigens such as myosin or a fibroblast neoantigen. Deposition of gamma globulin is observed in the sarcolemma, and serum antibody to the virus and autoantibody to the myocytes can be measured. In the long term, congestive heart failure due to a pathophysiological state analogous to human dilated cardiomyopathy develops (Reyes *et al.*, 1981; Huber and Lodge, 1986; Herskowitz *et al.*, 1987; Wenger *et al.*, 1990).

The evidence supporting similar pathogenetic mechanisms in man is, at the best, scanty. Equally, the mechanisms by which myocarditis or pericarditis occur occasionally after a systemic infection have not been elucidated. Most viral illnesses and infections of the myocardium by a cardiotropic virus are self-limiting and, typically, asymptomatic. It follows that biopsy or viral culture are not taken and these subjects do not present to their physician until late in the illness, by which time symptoms of congestive heart failure have developed (O'Connell, 1983; Grady and Costanzo-Nordin, 1989). At this stage the viremic phase will have passed as well as the opportunities to obtain sera for immunological study.

A variety of noninfectious agents may damage the myocardium and cause histopathologic changes similar to myocarditis. These include chemotherapeutic agents, radiation, psychotropic drugs and antibiotics. In addition, insect stings and illicit drugs such as cocaine may cause necrosis with histopathological changes consistent with immunologically mediated damage.

Studies in which the incidence of dilated cardiomyopathy is calculated from follow-up analysis of patients who have recovered from acute viral myocarditis proven by biopsy show that 25–50% develop a dilated cardiomyopathy (Torp, 1978; O'Connell and Mason, 1987). These data may overestimate the incidence of the disease because the subjects considered in these investigations represent the extreme of the spectrum of severity of acute viral myocarditis, i.e. those with clinically significant cardiac symptoms.

The reason that only a proportion of subjects with myocarditis progress to dilated cardiomyopathy is unclear, but an immunoregulatory defect in suppressor cell function may be present, which, hypothetically, will exaggerate immunoresponses and result in persistent myocardial inflammation and, ultimately, congestive heart failure (Fowles *et al.*, 1982).

This hypothesis has prompted the use of immunosuppressive therapy in biopsy-proven myocarditis (Mason *et al.*, 1980; Daly *et al.*, 1984; Ledford and Espinoza, 1987; Yacoub *et al.*, 1987; Hardesty, 1988; Duquesnoy and Cramer, 1990). The data from early uncontrolled studies (on 97 patients) show clinical improvement in 56% of cases. This might appear convincing evidence for a role of immunosuppressive therapy. However, other studies show a spontaneous improvement in 48% of patients who did not receive immunosuppression. Thus, a better estimation of the real value of immunosuppressive therapy in myocarditis must await completion of ongoing multicenter, randomized, prospective trials.

4. *Pericarditis*

The accessibility of the pericardial space and the possibility of analysis of the pericardial effusion has improved diagnosis of pericarditis (Mason *et al.*, 1980). However, in clinical practice, as in the case of myocarditis, the specific etiology of pericarditis is not determined and in general terms the infection is attributed to a virus or to an immunological process which follows the viral infection, or to a trauma, usually as a consequence of cardiectomy (Mason *et al.*, 1980; Ledford and Espinoza, 1987). As in the case of myocarditis, immunological treatment of this condition awaits appropriate confirmation.

5. *Cardiac Transplant Rejection*

Management of cardiac transplant rejection is the area in which immunotherapy causes an enhanced clinical outcome (see Chapter 10). Over the past decade, immunosuppression for cardiac transplantation has undergone a series of modifications, the most remarkable of which was the introduction of cyclosporin A in 1980. This marked the beginning of new era with a considerable increase in both numbers of transplants and institutions involved.

During the last 10 years much effort has been focused on understanding the several immune mechanisms that participate in transplant rejection. It is now clear that the early hyperacute rejection of the graft is due to recipient antibodies directed against foreign antigens. The major histocompatibility antigens which arise from a linked group of genes on the short arm of chromosome six are the most important mediators of transplant rejection. The cellular rejection of the graft is mediated by T lymphocytes and, in animal models, it has been shown to be related to class II antigen (HLA) mismatching (Daly *et al.*, 1984).

The chronic rejection, manifested by severe atherosclerosis in the coronary arteries of the transplanted heart, is a more complex process influenced not only by histocompatibility antigen mismatching but, probably, by other unknown factors. Several proposals have been put forward: (a) damage to the coronary endothelium by uncontrolled viruses because of an immunocompromised host; (b) immune damage as a result of low-grade chronic rejection; (c) treatment with immunosuppressive therapy causing injury to the vascular endothelium (Yacoub *et al.*, 1987; Duquesnoy and Cramer, 1990).

Without doubt, development of coronary artery disease in the transplanted heart is a challenge for the future and the reason for the greatest concern at the present time. Within three years almost 60% of the patients who have undergone cardiac transplantation have significant coronary artery disease and the proportion of multivessel disease increases with the time. This complication is not influenced by azathioprine, high-dose steroids or cyclosporin-based immunosuppression (Hardesty, 1988). Its prevention clearly awaits elucidation of the basic mechanisms involved in its pathogenesis. If histocompatibility proves to be important in long-term development of coronary artery disease, there will be a need for better donor matching as well as intervention with more tissue target-selective therapy. A murine monoclonal antibody (OKT3) has been used as a major part of the immunosuppressive prophylaxis in 102 patients (Gay *et al.*, 1989). OKT3 is specific for the human CD3 surface antigen found on all T lymphocytes (Reinherz and Schlossman, 1980). Therefore, it interferes with the ability of the human lymphocyte to recognize a foreign antigen, thus delaying or ablating the initiation of cell-mediated rejection. OKT3 has been found to be more effective than conventional high-dose steroids in reversing rejection in renal transplant recipients (Cosimi *et al.*, 1981; Kreis *et al.*, 1985). Encouraging results have been reported for the treatment of refractory rejection and for immunosuppressive prophylaxis in cardiac transplantation with OKT3 (Gilbert *et al.*, 1987a,b; Sweeney *et al.*, 1987). This therapy, however, has also been questioned because it may cause lymphoproliferative disorders (Swinnen *et al.*, 1990).

6. *Vasculitis*

The vasculitis associated with cardiac transplant rejection is not the only immune-mediated damage to vascular endothelium (Lindsay *et al.*, 1990; Smith and Perdue, 1990). There are several other forms of vasculitis characterized by inflammation and necrosis of the blood vessel with eventual ischaemia of the related tissue. These are: systemic necrotizing vasculitis such as polyarteritis nodosa; allergic angiitis; and hypersensitivity vasculitis such as serum sickness and Henoch–Schönlein purpura. These syndromes are caused by immune complex deposition and the affected vessels, either arteries, venules or capillaries, show necrosis and infiltration of polymorphonuclear lymphocytes and eosinophils. In addition, vasculitis may be caused by a cell-mediated mechanism, such as granulomatous arteritis.

It is interesting to recall here that several studies have demonstrated the efficacy of intravenous gamma globulin in the prevention of coronary artery abnormalities and giant aneurysms in Kawasaki disease (Furusho *et al.*, 1984, 1987; Rowley *et al.*, 1988).

7.   Acquired Immunodeficiency Syndrome

Cardiovascular disorders are often associated with an acquired immunodeficiency syndrome. The most common findings are myocarditis, endocarditis, pericarditis and vascular lesions. However, at present, cardiac death is not a common event in patients with acquired immunodeficiency syndrome, although cardiac dysfunction is likely to become an important clinical problem when more effective treatment for the syndrome becomes available.

The incidence of myocarditis with ventricular dilatation and heart failure is high (25–50%) at autopsy (Cammarosano and Lewis, 1985; Lewis *et al.*, 1985; Anderson *et al.*, 1988). Usually, it is characterized by infiltration of the myocardium by Kaposi's sarcoma. Interestingly, despite the presence of ventricular dilatation at biopsy, clinical symptoms of heart failure or arrhythmias are rare in patients with acquired immunodeficiency syndrome. Pericarditis is often (10%) associated with mycobacterium tuberculosis or cryptococcus infection.

Non-bacterial thrombotic or "marantic" endocarditis is not rare in patients with acquired immunodeficiency syndrome (Coplan and Bruno, 1989; Goldfarb *et al.*, 1989). Bacterial endocarditis, however, is unusual and typical only of patients affected by the syndrome with a clear history of drug abuse (Nagoshi and Fukuyama, 1988; Lewis, 1989).

Finally, vascular lesions have been observed in these patients, including aneurism of the coronary arteries. These are more frequent in the right coronary artery than in the left and preliminary reports suggest that these lesions differ morphologically and immunologically from the fibrocalcific arteriopathy typical of Kawasaki's disease (Acierno, 1989).

8.   Immune Cytokines and Cardiac Disease in General

Recently, it has been proposed that in conditions such as idiopathic dilated congestive cardiomyopathy, cardiac allograft rejection, and post-ischaemic reperfusion (see Chapter 9), the immune system can reversibly impair cardiac function. Interestingly, severe myocardial dysfunction might occur in conjunction with leukocytic infiltrates in the absence of significant myocyte necrosis. The link between lymphocyte and macrophage infiltrates and cardiac dysfunction could be the cytokines interleukin I and tumor necrosis factor (see Chapters 3–5).

Cytokines are produced by lymphocytes and macrophages and exert their effects only on other leukocytes, such as neutrophils, B-lymphocytes, macrophages and T-cell subset (Golub and Green, 1991). These are normally called interleukins, whose accepted number is 10. Macrophages release a variety of factors that can virtually affect all cells in the body (Nathan, 1987). They include tumor necrosis factor and a variety of growth factors. Interestingly, both interleukin I and tumor necrosis factor are able to modulate myocardial contractility by reversibly modulating cyclic-AMP metabolism in cardiac myocytes. These effects appear to be mediated by alterations in G1 protein with uncoupling of the β receptor from adenylate cyclase (Gulik *et al.*, 1988, 1991; Chung *et al.*, 1990). Therefore attendant disturbances in contractility can potentially produce a failing myocardium, which is unresponsive to β adrenergic stimulation.

Clinically, reversible changes in contractility are associated with conditions characterized by leukocytic infiltrates such as cardiac transplant rejection and idiopathic dilated congestive cardiomyopathy associated with lymphocytic myocarditis, some cases of post-partum cardiomyopathy as well as some cases of familial cardiomyopathy.

Leukocytes and neutrophils do accumulate also after an acute ischaemic insult (see Chapter 5) and they might contribute to left ventricular dysfunction or stunning (Woodruff, 1980; Fuster *et al.*, 1981; Johnson and Palacios, 1982; Marboc *et al.*, 1984).

In conclusion, although it is too early for drawing a firm conclusion, it is likely that *in situ* immune effects are important in the generation and maintenance of clinically relevant conditions associated with reversible myocardial dysfunction and heart failure. Elucidation of the immunomodulation of cardiac function provides a number of opportunities for both diagnostic and therapeutic intervention in conditions in which inflammatory cells invade the heart. As the following chapters show, exploitation of the immune system as a target for therapeutic intervention is receiving considerable attention at present by basic cardiac scientists and clinicians alike, in relation to many of the cardiac diseases mentioned above.

9. References

Acierno, L.J. (1989). Cardiac complications in acquired immune deficiency syndrome (AIDS): a review. J. Am. Coll. Cardiol. 13, 1144–1154.

Anderson, D.W., Virmani, R., Reilly, J.M. *et al.* (1988). Prevalent myocarditis at necroscopy in the acquired immune deficiency syndrome. J. Am. Coll. Cardiol. 11, 792–799.

Cammarosano, C. and Lewis, W. (1985). Cardiac lesions in acquired immune deficieny syndrome (AIDS). J. Am. Coll. Cardiol. 5, 703–706.

Cavelti, P.A. (1955). Autoimmunologic disease. J. Allergy, 26, 95–101.

Chung, M.K., Gulick, T.S., Rotondo, R.E., Schreiner, G.F. and Lange, L.G. (1990). Mechanism of cytokine inhibition of β-adrenergic agonist stimulation of cyclic AMP in rat cardiac myocytes: impairment of signal transduction. Circ. Res. 67, 753–753.

Coplan, N.L. and Bruno, M.S. (1989). Acquired immune deficiency syndrome and heart disease: the present and the future. Am. Heart J. 117, 1175–1177.

Cosimi, A.B., Burton, R.C. and Colvin, R.B. (1981). Treatment of acute renal allograft rejection with OKT3 monoclonal antibody. Transplantation 32, 535–539.

Dale, J.B. and Beachey, E.H. (1985). Epitopes of streptococcal M proteins shared with cardiac myosin. J. Exp. Med. 162, 583–588.

Daly, K., Richardson, P.J., Olsen, E.G.J., Morgan-Capner, P., McSorley, C., Jackson, G. and Jewitt, D.E. (1984). Acute myocarditis, role of histologic and virological examination in the diagnosis and assessment of immunosuppressive treatment. Br. Heart J. 51, 30–35.

Dec, G.W., Fallon, J.T., Southern, J.F. and Palacios, I.F. (1988). Relation between histological findings on early repeat right ventricular biopsy and ventricular function in patients with myocarditis. Br. Heart J. 60, 332–337.

DiSciascio, G. and Taranta, A. (1980). Rheumatic fever in children. Am. Heart J. 99, 635–643.

Duquesnoy, R.J. and Cramer, D.V. (1990). Immunologic mechanisms of cardiac transplant rejection. Cardiovasc. Clin. 20, 87–103.

Fenoglio, J.J., Jr, Ursell, P.C., Kellogg, C.F., Drusin, R.E. and Weiss, M.B. (1983). Diagnosis and classification of myocarditis by endomyocardial biopsy. N. Engl. J. Med. 308, 12–18.

Fowles, R.E., Bieber, C.P. and Stinson, E.B. (1982). Defective in vitro suppressor cell function in idiopathic congestive cardiomyopathy. Circulation 65, 1224–1229.

Frist, W., Yasuda, T., Segall, G., Khaw, B.A., Strauss, H.W., Gold, H., Sinson, E., Oyer, P., Baldwin, H., Billingham, M., McDougall, I.R. and Haber, E. (1987). Noninvasive detection of human cardiac transplant rejection with indium-111 antimyosin (Fab) imaging. Circulation 76, V-81–V-85.

Furusho, K., Kamiya, T., Nakano, H. *et al.* (1984). High-dose intravenous gammaglobulin therapy for Kawasaki disease. Lancet 2, 1055–1058.

Furusho, K., Kamiya, T., Nakano, H. *et al.* (1987). In "Japanese Gamma Globulin Trials for Kawasaki Disease" (ed S.T. Shulman). pp. 425–432. Alan R. Liss, New York.

Fuster, V., Geish, B.J., Glullani, E.R., Tavik, A.J., Brandenburg, R.O. and Frye, R.L. (1981). The national history of idiopathic dilated cardiomyopathy. Am. J. Cardiol. 47, 525–531.

Gay, W.A., O'Connell, J.G., Burton, N.A., Karwande, S.V., Renlund, D.G. and Bristow, M.R. (1989). OKT3 monoclonal antibody in cardiac transplantation. From the Utah Cardiac Transplant Program, Salt Lake City, Utah.

Gilbert, E.M., Eiswirth, C.C., Renlund, D.G. *et al.* (1987a). Use of orthoclome OKT3 monoclonal antibody in cardiac transplantation: early experience with rejection prophylaxis and treatment of refractory rejection. Transplant. Proc. 19(Suppl 1), 45–53.

Gilbert, E.M., DeWitt, C.W., Eiswirth, C.C. *et al.* (1987b). Treatment of refractory cardiac allograft rejection with OKT3 monoclonal antibody. Am. J. Med. 82, 202–206.

Goldfarb, A., King, C.L., Rosenzweig, B.P., Feit, F., Kamat, B.R., Rumancik, W.M. and Kronzon, I. (1989). Cardiac lymphoma in the acquired immunodeficiency syndrome. Am. Heart J. 118, 1340–1344.

Golub, E. and Green, D. (1991). "Immunology: A Synthesis", pp. 444–461. Sunderland, MA, Sinauer Associates.

Grady, K.L. and Costanzo-Nordin, M.R. (1989). Myocarditis: Review of a clinical enigma. Heart Lung 18, 347–354.

Gulik, T., Chung, M.K., Pieper, S.J., Schreiner, G.F. and Lange, L.G. (1988). Immune cytokine inhibition of β-adrenergic agonist stimulated cyclic AMP generation in cardiac myocytes. Biochem. Biophys. Res. Commun. 150, 1–9.

Gulick, T., Pieper, S.J., Murphy, M.A., Lange, L.G. and Schreiner, G.F. (1991). A new method for assessment of cultured cardiac myocyte contractility detects immune factor-mediated inhibition of β-adrenergic responses. Circulation 84, 313–321.

Hardesty, R.L. (1988). Heart transplantation at the University of Pittsburgh: 1980–1987. Transplantation Proc. 20, 737–740.

Herskowitz, A., Wolfgram, L.J., Rose, N.R. and Beisel, K.W. (1987). Coxsackievirus B3 murine myocarditis: a pathologic spectrum of myocarditis in genetically defined inbred strains. J. Am. Coll. Cardiol. 9, 1311–1319.

Huber, S.A. (1987). In "Immunology and Molecular Biology of Cardiovascular Diseases" (ed Spry), pp. 143–159. MTP Press, Boston.

Humber, S.A. and Lodge, P.A. (1986). Coxsackie B-3 myocarditis. Identification of different pathogenic mechanisms in DBA/2 and BALB/c mice. Am. J. Path. 122, 284–291.

Johnson, R.A. and Palacios, I. (1982). Dilated cardiomyopathies of the adult. N. Engl. J. Med. 308, 1051–1057.

Kaplan, M.H. (1963). Immunologic relation of streptococcal and tissue antigens: I. properties of an antigen in certain strains of group A streptococci exhibiting an immulogic cross-reaction with human heart tissue. J. Immunol. 90, 595–603.

Kaplan, M.H. (1965). Autoantibodies to heart and rheumatic fever: the induction of autoimmunity to heart by streptococcal antigen cross-reactive with heart. Ann. N.Y. Acad. Sci. 124, 904–1009.

Kaplan, M.H. (1967). In "Cross-reacting and Neoantigens" (ed J.J. Trentin), pp. 375–384. Williams & Wilkins, Baltimore.

Kaplan, M.H., Bolande, R. and Rakita, L. (1964). Presence of bound immunoglobulins and complement in the myocardium in acute rheumatic fever. N. Engl. J. Med. 271, 637–641.

Kaplan, M.H. and Svec, K.H. (1964). Immunologic relation of streptococcal and tissue antigens. J. Exp. Med. 119, 651–661.

Khaw, B.A. and Haber, E. (1989). Imaging necrotic myocardium: detection with 99mTc-pyrophosphate and radiolabeled antimyosin. Cardiol. Clin 7, 577–588.

Kreis, H., Chkoff, N., Vigeral, P.H. *et al.* (1985). Therapeutic use of monoclonal antibodies in kidney transplantation. Adv. Nephrol. 14, 389–407.

Ledford, D.K. and Espinoza, L.R. (1987). Immunologic aspects of cardiovascular disease. JAMA 258, 2974–2982.

Lewis, W. (1989). AIDS: cardiac findings from 115 autopsies. Prog. Cardiovasc. Dis. XXXII, 207–215.

Lewis, W., Lipsick, J. and Cammarosano, C. (1985). Cryptococcal myocarditis in acquired immune deficiency syndrome. Am. J. Cardiol. 55, 1240.

Lindsay, J., Jr, DeBakey, M.E. and Beall, A.C., Jr (1990). In "The Heart, Arteries and Veins" (ed J.W. Hurst), pp. 1408–1422. McGraw-Hill, New York.

Marboc, C.C., Schlerman, S.W., Rose, E., Reemtsma, K. and Fenoglio, J.J., Jr (1984). Characterization of monocuclear cell infiltrates in human cardiac allografts. Transplant Proc. 16, 1598–1599.

Mason, J.W., Billingham, M.E. and Ricci, D.R. (1980). Treatment of acute inflammatory myocarditis assisted by endomyocardial biopsy. Am. J. Cardiol. 45, 1037–1044.

Midei, M.G., DeMent, S.H., Feldman, A.M., Hutchins, G.M. and Baughman, K.L. (1990). Peripartum myocarditis and cardiomyopathy. Circulation 81, 922–928.

Nagoshi, M.H. and Fukuyama, O. (1988). Cardiac disease associated with the acquired immunodeficiency syndrome: a case report and review of the literature. Hawaii Med. J. 47, 531–532.

Nathan, C.F. (1987). Secretory products of macrophages. J. Clin. Invest. 79, 319–326.

O'Connell, J.B. (1983). In "Myocarditis: Precursor of Cardiomyopathy" (eds Robinson, A. and O'Connell, J.B.), pp. 93–105. Macmillan, New York.

O'Connell, J.B. and Mason, J.W. (1987). In "Pathogenesis of Myocarditis and Cardiomyopathy (eds I. Kawai, R. Abelmann, and A. Matsumori), pp. 281–292. University of Tokyo Press, Tokyo.

Reinherz, E.L. and Schlossman, S.F. (1980). The differentiation and function of human T-lymphocytes. Cell 19, 821–827.

Reyes, M.P., Ho, K., Smith, F. and Lerner, A.M. (1981). A mouse model of dilated-type cardiomyopathy due to coxsackie virus B-3. J. Infect. Dis. 144, 232–236.

Rowley, A.H., Duffy, E. and Shulman, S.T. (1988). Prevention of giant coronary artery aneurysms in Kawasaki disease by intravenous gamma globulin therapy. J. Pediatr. 113, 290–294.

Runge, M.S. and Haber, E. (1992). In "Immunological Principles in Cardiovascular disease" (ed H.A. Fozzard), pp. 355–365. Raven Press, New York.

Seigal, A.C., Johnson, E.E. and Stollerman, G.H. (1961). Controlled studies of streptococcal pharyngitis in a pediatric population. N. Engl. J. Med. 265, 559–565.

Smith, R.B. and Perdue, G.D. (1990). In "The Heart, Arteries and Veins" (ed J.W. Hurst), pp. 1423–1445. McGraw-Hill, New York.

Smith, T.W. , Butler, V.P., Jr and Haber, E. (1982). Treatment of life-threatening digitalis intoxication with digoxin-specific Fab antibody fragments: experience in 26 cases. N. Engl. J. Med. 307, 1357–1362.

Stollerman, G.H. (1975). "Rheumatic Fever and Streptococcal Infection". Grune & Stratton, New York.

Sweeney, M.S., Sinnott, J.T., Cullison, J.P. and Weinstein, S.S. (1987). The use of OKT3 for stubborn heart allograft rejection: an advance in clinical immunotherapy? J. Heart Transplant. 6, 324–328.

Swinnen, L.J., Costanzo-Nordin, M.R., Fisher, S.G., O'Sullivan, E.J., Maryl, H.D., Johnson, M.G., Heroux, A.L., Dizikes, G.J., Pifarre, R. and Fisher, R. (1990). Increasing incidence of lymphoproliferative disorder after immunosuppression with monoclonal antibody OKT3 in cardiac transplant recipients. N. Engl. J. Med. 323, 1723–1728.

Torp, A. (1978). Incidence of congestive cardiomyopathy. Postgrad. Med. J. 54, 435–437.

Van de Rijn, I., Zabriskie, J.B. and McCarty, M. (1977). Group A streptococcal antigen cross-reactive with myocardium. Purification of heart-reactive antibody and isolation and characterization of the streptococcal antigen. J. Exp. Med. 146, 579–586.

Wenger, N.K., Abelmann, W.H. and Roberts, W.C. (1990). In "The Heart, Arteries and Veins" (ed J.W. Hurst), pp. 1256–1277. McGraw-Hill, New York.

Woodruff, J.F. (1980). Viral myocarditis: a review. Am. J. Pathol. 101, 426–479.

Wynne, J. and Braunwald, E. (1988). In "Heart Disease: A Textbook of Cardiovascular Medicine" (ed E. Braunwald), pp. 1410–1469. W.B. Saunders, Philadelphia.

Yacoub, M., Festenstein, H. and Doyle, P. (1987). The influence of HLA matching in cardiac allograft recipients receiving cyclosporine and azathioprine. Transplant. Proc. 19, 2487–2489.

Zabriskie, J.B. (1985). Rheumatic fever: The interplay between host, genetics, and microbe. Circulation 71, 1077–1086.

Zabriskie, J.B. and Fremier, E.H. (1966). An immunological relationship between the group A streptococcus and mammalian muscle. J. Exp. Med. 124, 661–668.

Zabriskie, J.B., Hsu, K.C. and Seegal, B.C. (1970). Heart reactive antibody associated with rheumatic fever: characterization and diagnostic significance. Clin. Exp. Immunol. 7, 147–151.

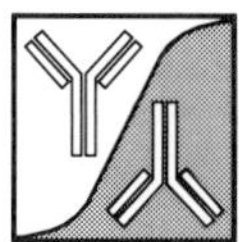

2. Methods for Evaluating Cardiac Function

Michael K. Pugsley *and* M.J.A. Walker

1. Introduction

1.1 IMMUNOLOGY AND CARDIAC FUNCTION

Although the heart is not an organ which is highly susceptible, immunological damage can occur as a result of: (1) immediate hypersensitivity reactions; (2) cytotoxic effects of antibodies; (3) deposition of immune complexes; and (4) delayed hypersensitivity reactions (see Chapter 1, and Williams and Rampart, 1987). Immunological mechanisms are also at play in a number of cardiovascular diseases; even the most common of these diseases, atherosclerosis, may have an immunological component (see Chapter 4, and Gero *et al.*, 1981). There are a number of myocarditides with an immunological component, e.g. Coxsackie and Echo viruses of the piconavirus family (Fletcher and Brennan, 1957; Grist & Bell, 1969; Russell & Bell, 1970) or chronic myocarditis associated with Chagas' disease (Kass Wenger, 1974). Immune complexes can directly damage cardiac valves as one of the sequelae of rheumatic fever following streptococcal infections (Williams, 1981). In addition, activation of components of the immune system can release cardiotoxic substances (Jacob, 1981).

As indicated above, in even the most common of the cardiovascular diseases, atherosclerosis, an immunological component has been implicated (Beaumont and Beaumont, 1978). There is growing clinical and experimental evidence which suggests a link between infection and atherosclerosis (Valtonen, 1991). It has been suggested that infection plays a role in the atherogenic process by inducing damage and inflammation in the presence of hypercholesterolaemia. Viral components appear to be involved in the genesis of some types of experimental and possibly clinical atherosclerosis (Denman, 1981). Immunological processes involving

Immunopharmacology of the Heart
ISBN 0−12−200245−8

viruses may damage the endothelium of coronary vessels thereby initiating the process of plaque formation (see Chapter 3). There is also epidemiological evidence of an increased incidence of myocardial infarction following epidemics of influenza (Bainton *et al.*, 1978).

In man, the degree of importance of different immunological processes which can influence cardiac function is dependent upon economic and geographic factors. Thus in South America the occurrence of trypanosomal infections (Chagas' disease) assumes a significant role in cardiac disorders whereas the same disease is virtually unknown in the Northern hemisphere (Hanson, 1977; Miles 1979, 1981; Hudson and Ribeiro-dos-Santos, 1981). On the other hand, atherosclerosis and myocardial infarction are important throughout the world, although not equally so. At one time in Western Europe and North America, streptococcal infection and the resulting rheumatic fever caused myocardial valvular damage in large numbers of people but such damage is not so prevalent nowadays (UK and US Joint Report, 1955; Stollerman, 1975). The precise mechanism of interaction between host, and the cardiopathic bacteria involved in rheumatic heart disease, is not known although cross-reactivity between antigenic constituents of the bacterial cell wall and the glycoproteins of heart valves has been demonstrated by Kaplan (1963). The incidence of rheumatic heart damage has dropped remarkably in the last few decades, partly as a consequence of antibiotic treatment of the precipitating infection (Thamlikitkul *et al.*, 1992). In richer countries the growing number and success of cardiac transplantation has given rise to several new challenges relating to cardiac rejection and the involvement of the immune system.

Given that it is necessary to examine the cardiac function of hearts in the presence of, or after, an immunological event it is important that appropriate techniques are used for this purpose. There are techniques in which cardiac function can be examined while the immunological event is occurring, and other techniques which can be utilized to assess the impairment of cardiac function following previous immunological damage.

1.2 Evaluation of Cardiac Function

Cardiac function, in its narrowest sense, refers to the ability of the heart to pump blood. The best measure of this is cardiac output corrected for factors such as preload, afterload, heart rate and autonomic tone which influence output. However, in a broader sense cardiac function can be assessed in a number of different ways that only relate indirectly to measurement of cardiac output. Thus the functioning of the heart can be assessed in terms of mechanical function of the whole, or components, of the heart, heart rate, electrophysiological behaviour (ECG and transmembrane potentials),

coronary flow, biochemical and histological markers. Such studies can be performed *in vitro* (using cardiac tissue or hearts isolated from the body) as well as in intact animals. Some of the methods for determining function are invasive, others are not. In order to follow the time course of changes in cardiac function in whole animals it is clear that noninvasive methods are to be preferred. However, as a generalization, the most invasive of procedures are generally the most precise and accurate.

The overwhelmingly important function of the heart is to pump blood. Provided venous return is adequate, the maintenance of a satisfactory cardiac output (i.e. cardiac function) depends upon intrinsic factors such as:

(a) mechanics;
(b) electrophysiology and rhythm;
(c) coronary flow;
(d) biochemical status;
(e) anatomy (macroscopic and microscopic).

The above factors ensure adequate cardiac function independently of neuronal influences. In the following we will discuss methods for assessing cardiac function under the headings, mechanics, electrophysiology, coronary flow, biochemical status and anatomy (gross and microscopic) as outlined in Fig. 2.1.

1.2.1 Mechanics

Ultimately the mechanical function of the heart results in the pumping of blood. The flow-pressure functions of the heart therefore define cardiac function. Flow-pressure functions can be assessed both *in vivo* and *in vitro* from the following: cardiac output, chamber (primarily ventricular) pressures, heart and chamber sizes as well as length and tension changes in various parts of the heart. All of the foregoing can be measured in a direct physical manner, or indirectly by a variety of visualization and other techniques.

1.2.2 Electrophysiology

The electrical behaviour of the heart can be examined from the level of the whole heart down to the level of single ion channels located in the cardiac sarcolemma. The ECG, or its main derivatives, gives useful insights into the electrical functioning of the whole heart as well as its component parts such as sinus and atrioventricular nodes and conducting systems (Einthoven, 1912). Regional electrograms can give further regional definition down to the level of monophasic action potentials. Intracellular potential techniques have advanced to the degree that, at least *in vitro*, all of the different cell types found in the heart can be studied. Further variants of the intracellular technique include voltage and patch clamp techniques. The latter also allow the behaviour of single ion channels in cardiac muscle, such as the sodium channel, to be investigated (Aldrich *et al.*, 1983; Vandenberg and Horn, 1984; Ju *et al.*, 1992). To some extent it is now possible to produce a coherent

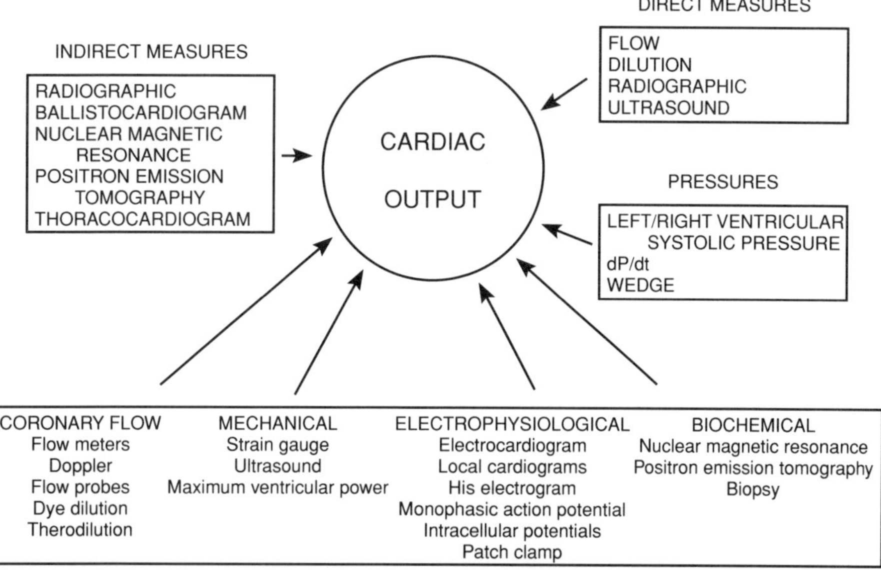

Figure 2.1 Schematic of the range of methods available for determining cardiac function. These methods yield direct and indirect estimates of the ultimate function of the heart, i.e. maintenance of cardiac output.

explanation of cardiac electrical activity from single channels to the gross ECG. Unfortunately, except in the case of cardiac arrhythmias, the link between electrical and mechanical function is such that the former cannot be used to explain or predict changes in the latter.

1.2.3   Flow

Flow through myocardial vasculature is best assessed experimentally by means of suitably implanted flow probes (Terry, 1980) although it can be assessed less directly by angiographic and allied procedures (Pitt *et al.*, 1969; Zeitler and Grosse-Vorholt, 1980). The ECG also gives indirect measures of the quality of blood flow through the myocardium in that partial or complete ischaemia produces characteristic changes in the S-T segment of the ECG (Samson *et al.*, 1960; Scherlag *et al.*, 1970; Kjekshus *et al.*, 1972).

1.2.4   Biochemistry

Many biochemical determinations require that tissue samples are removed from the heart and analysed, i.e. a method which requires destruction of the source material. This approach denies the possibility of continuous assessment with time. Nuclear magnetic resonance spectroscopy, however, allows for the continuous assay of myocardial concentrations of biologically important phosphates such as ATP, ADP, etc (Mogelvang *et al.*, 1982; Haselgrove *et al.*, 1983; Nunnally, 1986). On the other hand, it is possible to detect substances in the bloodstream as a result of their release from stressed or damaged hearts.

1.2.5   Anatomy

The anatomical structure of the heart, whether macroscopic or microscopic, has to be assessed from tissue samples, or the whole dissected heart. It is possible, however, to biopsy the heart on a recurring basis in order to provide samples for histological assessment.

2.   *Choice of Cardiac Tissue for Study*

In clinical situations there are limitations as to what can be studied since cardiac function has to be evaluated in the presence of immunological processes that are underway, or have already occurred. However, in experimental situations various paradigms are available for studying the effects of immunological processes on cardiac function. Thus one can study hearts in animals exposed to an immunological challenge imposed in the form of an infection, or to injection of the products of immunological reactions. Alternatively one can directly examine the immunological process in heterotopically transplanted hearts (Novitzky *et al.*, 1984; Scheutz *et al.*, 1992). It is also possible to expose essentially normal cardiac tissue isolated from the body to various immunological challenges, or to the products of immunological reactions.

2.1   IMMUNOLOGICALLY CHALLENGED CARDIAC TISSUE

Experimental animals have been exposed to a variety of immunological challenges, either acutely, or chronically, and the resulting effects on cardiac function observed. Many types of immunological challenge can influence the heart. For example, a murine model of virally induced myocarditis (Woodruff, 1980) has allowed the elucidation of ultrastructural alterations of the cardiac conducting system as well as a quantitative analysis of the characteristics of persistent viral infection of heart muscle (Terasaki *et al.*, 1992; Klingel *et al.*, 1992). The induction of atherosclerosis by high cholesterol diets is facilitated by concomitant viral infections, possibly as a result of impaired host defence systems (Klurfeld *et al.*, 1979; Hajjar, 1991).

2.2 TRANSPLANTED CARDIAC TISSUE

One of the most important procedures that can be used to study the immunological process of rejection is transplantation of the heart. Transplantation initiates a classic host versus graft immunological reaction (Bildsoe, 1972; Muller, 1990). Cardiac transplantation was first reported by Carrel and Guthrie in dogs in 1905 (Carrel and Guthrie, 1905) and over the intervening years a variety of improvements were made prior to the first human heart transplant by Barnard in 1967 (Barnard, 1967). The number of human heart transplants is growing. In 1988 alone 2500 cardiac transplants were performed in 173 centres (Heck *et al.*, 1989). The specific issue of the immunopharmacology of cardiac transplantation is addressed in detail in Chapter 10. In the present chapter the use of transplantation as a means of providing cardiac tissue for investigational purposes is considered.

Although transplantation has often been performed in larger species it is most often performed experimentally in small laboratory species such as the rat. Cardiac transplantation studies in the dog have often been performed with particular relevance to human cardiac transplantation, but in rats the subject of investigation is often the rejection phenomenon itself, rather than a particular clinical problem. Many techniques have been developed for transplanting hearts in rats, beginning in 1964 with Abbott's studies (Abbott, 1964). Improvements have been made such as those due to Ono and Lindsay (1969), Lee *et al.* (1982) and others. One can use rats genetically close to one another (including syngeneic animals) or genetically dissimilar animals (Klempnauer *et al.*, 1985). For the sake of convenience, neonatal hearts may be grafted, without a need for microsurgery, into highly vascular tissue such as the plantar space or the pinna of the ear (Wotherspoon and Dorsch, 1986).

A large number of surgical procedures are available for cardiac transplantation in rats (Konertz *et al.*, 1985). These include transplantation made in the chest (orthotopic) with or without co-transplantation of lungs, and in the neck region or in the abdomen (heterotopic) (Jara *et al.*, 1979; Lee *et al.*, 1982; Lee, 1990; Rao and Lisitza,

1985). The most popular site for transplantation is the abdomen. Typically the recipient aorta is anastomosed to donor aorta, as in the Fox and Montorsi (1980) method. In addition to using the abdominal aorta, others have used a carotid artery (e.g. Timmermann, 1985; Muller, 1990). The necessary surgical techniques have been extensively discussed by Engemann (1985). Essentially all of the required skills can be learnt quite easily by the average experimenter and such techniques require a minimum of instruments and facilities.

Many of the older preparations of heterotopically transplanted heart were not working heart preparations inasmuch as the transplanted heart did not eject blood. However, in preparations in which the superior and inferior vena cava are ligated with a suture, and a lobe of lung left intact, coronary flow in the transplanted heart is directed into the right atrium, through the right ventricle into the donor pulmonary lobe through which it circulates before entering the left atrium and ventricle to be pumped into the aorta (Lee *et al.*, 1982). This type of heterotopic transplantation results in a heart which performs work, but has a low cardiac output. The coronary flow constitutes the volume of blood which is pumped through the heart, i.e. approximately 20 per cent of the usual cardiac output (Korecky and Masika, 1990). Transplanted hearts can be used for a variety of different studies, including electrophysiological, mechanical and pathohistological (Chisholm, 1987). A major disadvantage, however, is that the interdependence of cardiac output and coronary flow makes it difficult to separate these two variables in analysing experimental outcomes.

It is easy to monitor the process of rejection in transplanted hearts in rats. The transplanted heart can be removed at any time and examined histologically for the presence of rejection as shown by hallmarks such as neutrophil and macrophage infiltration. Histological changes induced by immunological processes can, nevertheless, be readily correlated with the results of functional studies utilizing transplanted hearts (Ueda *et al.*, 1991). Thus transplanted hearts have great utility in a large number of studies related to well-known immunological problems (see Chapter 10).

The immunogenetic basis of heart transplantation is illustrated by the work of Klempnauer *et al.* (1985). Using appropriate donor and recipient strain combinations, it was possible to analyse the relative contributions of Class I and Class II major histocompatibility complex alloantigens as well as non-MHC histocompatibilities (Katz *et al.*, 1983; Klempnauer *et al.*, 1985).

Galinanes and Hearse (1991) have recently used heterotopically transplanted rat heart to assess recovery of heart function, changes in metabolism (loss of high-energy phosphates) and histological changes in hearts subject to global ischaemia. Their model allowed assessment of functional, metabolic and histological changes in hearts at up to one week post-transplantation.

2.3 NORMAL CARDIAC TISSUE

In addition to the above paradigms involving whole animals (*in vivo* preparations), it is possible to use isolated cardiac tissues (*in vitro* preparations) and expose them to immunological challenges, or to cellular and chemical mediators of the immune response. An example of this is our own studies with isolated hearts in which macrophage cultures were immunologically challenged and the cardiac effects of this challenge detected by perfusing the released materials directly into the coronary arteries of isolated hearts. Effects on cardiac function in hearts perfused in this manner were assessed in terms of left ventricular pressures, coronary flow and ECG changes (Salari & Walker, 1989). Similar experiments can be performed on isolated segments of heart such as ventricles, Purkinje fibres, atria, or even in cultured heart cells.

3. Methods Used to Determine Cardiac Function in Intact Animals and Intact Hearts

3.1 CARDIAC OUTPUT, CARDIAC CHAMBER PRESSURES AND CORONARY FLOW (INVASIVE AND NON-INVASIVE)

The following is a description of methods that are generally applicable to the examination of cardiac function in many conditions; none are particular to problems in immunopharmacology. These methods (stylized in Fig. 2.2) can be divided into those used *in vivo*, and those used *in vitro*. *In vivo* methods are used in whole animals where the heart is normally connected to the lungs and the rest of the circulation. Such methods are basically concerned with measuring the electrical or mechanical performance of the heart. The mechanical performance of the heart *in vivo* can be assessed in terms of cardiac output, ventricular pressures, as well as movements of chamber walls.

A large number of methods exist for determining cardiac output. Some of these methods are direct in that they measure the volume of blood per unit time either going to, or being ejected from the heart. Thus cardiac output can be measured in terms of either venous return or the output through the pulmonary artery or aortic root (Braunwald, 1971).

Direct measurement of cardiac output, normally reported in millilitres or litres per minute depending upon the size of the animal, is best accomplished by flow probes. Flow probes can be positioned around the major arteries in such a way that the volume of blood flowing through them can be measured. The disadvantage of

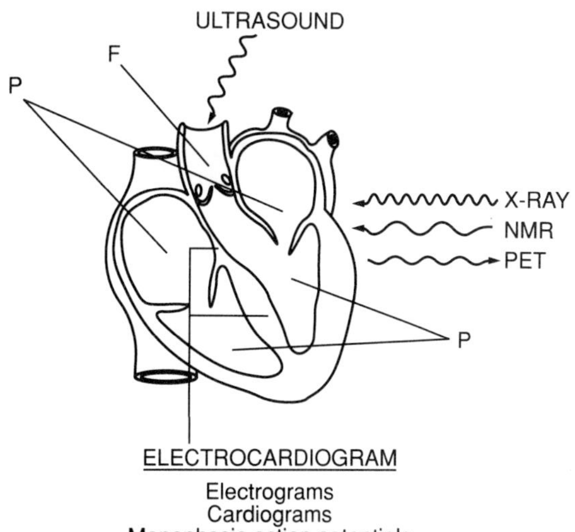

Figure 2.2 A diagrammatic overview of some of the physical techniques used to assess cardiac function. P = pressure; F = flow.

these direct measures of flow is the need to use surgery to implant flow probes around arteries, although one of the available Doppler ultrasonic methods avoids this (Grayburn *et al.*, 1992).

Various types of flow probes are used. The most common are electromagnetic and ultrasonic (see review of topic by Harper *et al.*, 1974). The function of the former depends upon "distortion" of a magnetic field caused by moving red cells or, more specifically, by movement of the iron in haemoglobin. The electric field induced in the probe is perpendicular to the direction of flow and its magnitude is proportional to the applied magnetic field and average velocity of the blood (Wyatt, 1960; Doring and Dehnert, 1988). Electromagnetic flow probes give fairly direct measures of blood flow in an artery and can be positioned around the vessel as a cuff, or are even used as a catheter-tip probe. In large vessels a U- or C-shaped magnet, in a suitable cuff, provides a homogeneous magnetic field and accurate flow measurements. Smaller probes are less accurate (Harper *et al.*, 1974) while catheter-tip probes have to be small enough not to disturb flow.

Ultrasonic flow probes are inherently less accurate than magnetic flow probes (Vatner *et al.*, 1985). They either rely on the change in ultrasound between emitter and receiver, or on the Doppler principle whereby the velocity of red blood cells beneath the flow probe changes the frequency of the returned sound and thereby gives an indirect measure of flow. Since ultrasonic flow probes only measure velocity they are useful for measuring patterns of changes in flow rather than absolute flow itself.

In addition to measuring cardiac output by probes or flow meters, one can also use indirect dye dilution techniques (reviewed by Lund-Johansen, 1990, and Fagard and Conway, 1990) according to the Fick principle. The Fick principle relies on the fact that the rate at which a known amount of rapidly injected tracer material (originally Evan's blue and later indocyanine green) is diluted in blood gives a measure of cardiac output. From the characteristic graph of concentration in blood, versus time, one can calculate cardiac output.

More recently, instead of using dye injection, boluses of cold saline coupled to precise measurement of temperature are used in a procedure known as thermodilution (see Conway and Lund-Johansen, 1990; Kadota, 1985) for measurement of cardiac output. This method is less cumbersome than dye dilution. A bolus injection of a known volume of saline or glucose solution at a particular temperature is injected and changes in temperature at a distant site monitored (Fegler, 1954). Cardiac output is determined by changes in temperature (thermodilution) using appropriate computer software. This essentially clinical technique can readily be applied to animal experiments. However, the method is liable to gross errors unless certain conditions are adhered to. Of particular importance is the care taken with injectates. The validity of data used for estimating cardiac output needs to be thoroughly checked; the thermodilution curve should show a smooth and rapid ascent phase as well as a monoexponential decline (Vliers *et al.*, 1973). The thermodilution procedure can be used repeatedly to give estimates of cardiac output which vary by as little as 7%.

Inherently, dye or thermodilution procedures cannot give a beat to beat determination of cardiac output while circulatory shunts, low flow rates and lack of stability with exercise, all cause errors. In terms of the possible utility of such a method in immunopharmacology, where one is less interested in beat to beat changes of cardiac output, dilution techniques have advantages over flow probes which have to be implanted by surgical techniques.

Variations on the basic dilution technique include procedures for using radio-opaque dyes or radioactive materials with subsequent detection of radioactivity (Fouad-Tarazi and McIntyre, 1990). Technetium-labelled serum albumen, or red blood cells, are used for dilution techniques. Radionuclides (Baur, 1989) can also be detected by collimated scintillation counters so as to allow for scintigraphic representations. Thus technetium and thallium have been used to show areas of damaged myocardium. Thallium uptake in the heart is dependent upon factors such as regional blood flow, hypoxia and drug treatments (Aizawa *et al.*, 1980).

Myocardial imaging has been routinely used for assessing infarct size (Williams *et al.*, 1992). Many radioactive agents have been used including various technetium salts which accumulate within the developing infarct. On the other hand radionuclides such as ^{129}Cs,

^{43}K, ^{81}Rb and ^{201}Tl concentrate in normal, but not infarcted, or infarcting, myocardium (Ritchie, 1978).

A further method which relies on the dilution principle is rebreathing of carbon dioxide. Carbon dioxide rebreathing methods (Reybrouck and Fagard, 1990) use one of two methods for calculating cardiac output, either equilibration according to the Collier procedure, or the exponential method utilizing the Fick principle.

Measurement of cardiac output, in a semantic sense, is the ultimate procedure for determining the functionality of the heart since the role of the heart is to pump an adequate supply of blood to the rest of the body. However, in addition to cardiac output, there are other physical indices of cardiac function which are of value. According to the well-known Frank–Starling principle, the performance (contractility) of the heart depends upon ventricular filling pressure, or diastolic fibre length, since the strength of contraction in myocardial tissue depends upon its initial length or tension (Starling, 1918). It is important, therefore, in assessing variations in cardiac output to determine the associated pressures within right or left ventricle, particularly the left ventricle.

One of the first signs of cardiac output failure is a rise in end-diastolic pressures in left or right ventricles, or in both (Parker *et al.*, 1969; Henderson, 1988). It is a fairly simple matter to measure end diastolic pressures in the right ventricle *in vivo*. A catheter is inserted through a vein and its tip positioned in the right ventricle so as to record pressures. Measurement of left ventricular pressure is made more difficult by the obstructive nature of the aortic valves as well as the higher pressures encountered in the aorta and left ventricle. In animals these factors are of lesser importance than in man. An alternative to cannulating the left ventricle is to measure pulmonary wedge pressures by floating a catheter in the blood stream through the right atrium into the lung vasculature (Geddes, 1991). Once the catheter tip is located in a branch of the pulmonary arterial tree a balloon at the tip of the catheter is inflated to occlude the vessel. Thus the open tip pressure of the catheter approximates to filling pressure of the left side of the heart, a so-called wedge pressure measurement since the catheter is wedged into an arterial branch.

While cardiac output and associated ventricular pressures are the best indices of the functional state of the myocardium it is possible to obtain other measures of the contractility state of myocardial chambers. One such measure is the first differential of ventricular pressure with time, i.e. dP/dt. A positive dP/dt is generated by the increasing ventricular pressure with systole while the negative dP/dt occurs upon relaxation. They are indirect measures of contraction and relaxation process, respectively. The value of dP/dt measurements depends to various extents on the state of the myocardium. Inaccurate values are liable to occur in the presence of an infarct (Guth *et al.*, 1990) or other cardiac damage.

Contraction and relaxation can be assessed more directly by measuring the length or tension changes which develop in cardiac muscle with each contraction. These measurements can be obtained directly by suturing strain gauges onto the surface of cardiac chambers, usually the ventricles, in variations of a procedure originally developed by Brodie (Boniface *et al.*, 1953). Two ends of a strain gauge are sutured onto the myocardium and the tension generated between them measured electrically (Boniface *et al.*, 1953). As reviewed by Sandler *et al.* (1982) there are many ways of measuring changes in cardiac dimensions. Thus in experimental animals strain and inductance gauges can be used, as can ultrasonic crystals, accelerometers and even diffracto-meters (Theroux *et al.*, 1974). Less invasive techniques, including those used in man, include X-ray techniques (with clip or bead detectors), ultrasound, radioisotope, or sensors on catheter tips (Carlsson and Milne, 1967; Mitchell *et al.*, 1969; Bodenheimer *et al.*, 1978; Meier *et al.*, 1980; Sandler *et al.*, 1982).

Many variations on electromechanical strain gauges are used. Micro strain gauges can be implanted directly onto the endocardium of the ventricles to allow for the calculation of chamber dimensions as well as tensions developed. Strain gauges can be implanted onto, or into, any cardiac chamber but implantation of these devices requires surgical skill and the necessity for "open-chest" surgical procedures. A derivative of strain measurements, maximum ventricular power, gives a measure of contractile function that is little influenced by the loading (pressure) conditions of the ventricle (Kass and Beyar, 1991). This power function is the product of ventricular pressure and flow.

The latest strain gauges are technologically sophisti-cated and often involve the utilization of ultrasound or even laser principles although the procedures for their actual use in animals are the same as with older devices. The fibre optic strain gauge has been used, in con-junction with an NMR probe, to study the mechanics of human atrial trabeculae in studies of tissue preservation for human transplantation (Lareau *et al.*, 1991).

An alternative method for determining the rate of development of contraction or relaxation involves the use of ultrasound probes or radiographic procedures. Doppler echocardiography (Mark *et al.*, 1986; Labovitz and Williams, 1992) uses two modes in examining cardiac dimensions, either the continuous wave mode using side-by-side transducers, or the pulsed wave technique. By suitable timing ("gating"), using the ECG or other triggering signals, it is possible to localize structures at different depths within the heart. M-mode 2D echo-cardiography (Wallerson *et al.*, 1990) is the oldest and simplest technique.

Doppler probes can also be used to directly measure aortic flow via a transducer suitably placed at the sternal notch and pointing into the aorta. This approach gives an ultrasonic view of blood being ejected from ventricles (Takayama *et al.*, 1984). Simultaneous ECG recording gives a wealth of further information with respect to the timing of intervals and other cardiac events (Teague, 1986). In summary, noninvasive Doppler ultrasound can be very useful in assessing ventricular function *in vivo* in terms of cardiac output or ejection-phase indices such as acceleration or velocity of blood ejection, and their timing in the cardiac cycle. Acceleration is the most sensitive measure of cardiac function. Thus, at least at a clinical level, the accurate and noninvasive aspects of Doppler measurements makes them very useful in the assessment of ventricular function for a wide variety of situations.

The "gated angiogram" gives an instantaneous view of the extent of filling and emptying of cardiac chambers. It shows, on a moment to moment basis, the speed and pattern of entry and ejection of blood from the ventricles (Gerlock *et al.*, 1988). On the other hand, ultrasound procedures allow for the location of the borders of cardiac chambers. By relating the movement of these borders to the stage of the cardiac cycle it is easy to monitor contraction and/or relaxation process. In addition to measuring the velocity and direction of wall movement it is also possible to calculate ejection fractions, i.e. how much of the total chamber volume is ejected from the heart. Ejection fraction is a good clinical index of ventricular function (Parmley and Sonneblick, 1971; Berman *et al.*, 1975).

Ballisto-cardiography, which began with Starr in 1936, relies on the principle that as the heart expels blood it acts within the heart cavity as a ballistic "missile" with kinetic properties (see review by Ty Smith, 1974). In ballisto-cardiography the patient lies on a freely suspended table which is attached to accelerometers so as to measure the extent of acceleration of the table with each heart beat and so provide information on maximal aortic blood flow and acceleration. The technique is technically difficult to perform and has been superseded to various extents by other techniques. It is the oldest noninvasive method of cardiac evaluation and is less commonly used today although recent improvements have been made (Jansen *et al.*, 1991), such as the use of static-charge sensitive beds.

A related technique, the thoracocardiograph, measures body wall movements as they relate to heart movements and to respiration (Sackner *et al.*, 1991a); it is currently under clinical and experimental development. Elastic inductance plethysmographs are placed transversely at, or near, the xiphoid process to detect chest wall movement due to both respiration and the beating heart. Using an ensemble-averaging technique triggered by the ECG, the cardiac waveform can be extracted from the recorded data; this waveform has the appearance of a ventricular volume curve. It appears that this new technique is capable of noninvasively, and continuously, monitoring stroke volume and cardiac output (Sackner, 1991b). In addition, it appears capable of detecting ischaemic or scarred myocardium (Sackner *et al.*, 1991a).

Increasing clinical use is being made of the non-hazardous technique of magnetic resonance imaging (MRI) (Longmore, 1989) for "viewing" the heart. MRI is a synonym for the older term nuclear magnetic resonance (NMR) (Chance, 1989). NMR has been used for many years experimentally in assessing the biochemical and structural status of the heart (Mogelvang *et al.*, 1982). For imaging purposes MRI (Longmore, 1989; Pohost and Canby, 1987) utilizes the properties of hydrogen nuclei in the free water of the heart. In the experimental situation it can also be used to measure cardiac levels of high energy phosphate compounds. It is possible by MRI to measure myocardial concentrations of ATP, ADP, CP and inorganic phosphate, potentially in a three-dimensional format. The measurement of high energy phosphate compounds gives a good indication of the energy status of the heart. Secondly, by measuring the inorganic phosphate peak relative to the position of the peaks for other phosphates, it is possible to determine intracellular pH within heart cells (Radda and Seeley, 1979; Nunnally, 1986).

In terms of these (Craige, 1974) and similar imaging techniques, Omvik (1990) has considered the differences and similarities between computerized tomography and MRI. Both methods provide accurate and highly reproducible measures of stroke volume and cardiac output at rest. They are less applicable during physical exercise, while at the moment, neither method is very practical in terms of usability and expense. However, both techniques have potential value in the measurement of cardiac output as well as being applicable to measurement of wall motion, myocardial structure, and tissue perfusion, although the value of their use for the latter has still to be fully evaluated. MRI tomography gives a true image with a resolution theoretically close to that of a low-powered microscope (Longmore, 1989).

Another highly technical procedure which will be increasingly used in assessing biochemical, if not mechanical, function of the heart is PET. PET relies on the use of positron-emitting substances which are produced as needed in a cyclotron (Wolf and Jones, 1983). PET images reflect actual tissue concentrations of radionuclides and therefore can be noninvasively quantified in the form of a high-contrast image. Over 200 radiotracers are available for PET. These are classified according to their suitability for examining metabolism, membrane function, blood flow or neurons (Schelbert and Mody, 1991). However, clinical experience with the PET technique is still limited.

Effective contraction of cardiac muscle can only occur *in vivo* if adequate oxygen is supplied to the muscle and the products of metabolism washed out. Thus it is always important to measure coronary blood flow both globally for the whole heart as well as regionally. Flow can be measured directly by means of flow probes located around a coronary artery or secondarily by means of microspheres or angiography.

There are problems with coronary angiograms when they are used to assess the severity of artery stenosis (White *et al.*, 1984). This has resulted in studies into novel and more physiological methods for measuring coronary flow and coronary flow reserve. The latter is maximal coronary flow rate divided by resting flow rate and can best be assessed by digital subtraction analyses (Hangiandreou *et al.*, 1991). Recently Wilson (1991) has developed a Doppler catheter for use in the diagnosis of microvascular dysfunction caused by coronary stenosis.

Optical methods can be used to assess contractile function in cardiac tissue and measure electrical activity. A microscopic optical method for determining the contractile state that has been used in cardiac transplantation studies is birefringence, a technique originally used in stained giant axons in an attempt to monitor membrane potential (Ross *et al.*, 1977). In cardiac tissue, changes in birefringence occur upon exposure to ATP and $CaCl_2$ as a result of contractions of myofilaments. Such birefringence thereby gives an index of the contractile status of the heart.

Wijngaard *et al.* (1990) used quantitative birefringence to assess and diagnose acute rejection after heart transplantation. Acute rejection was assessed by morphological inspection of mononuclear cell infiltration while birefringence was used to indicate myocardial cell contractile function.

3.2 Electrophysiological Methods

The electrical function of the heart can be assessed at various levels. The electrical activity of the whole heart can be recorded in terms of the ECG, whether as a conventional ECG, or some derivative such as vector, polarcardiograph, or signal-averaged high resolution ECG. In addition to examining the electrical activity of the whole heart it is possible to examine the electrical activity of different chambers and the individual structures within the heart. With respect to this, electrodes can be inserted into any of the chambers of the heart so as to examine the electrical activity of discrete regions such as sino-atrial or atrioventricular nodes, the His-Purkinje system, atria and ventricles (Winslow, 1984). Stimulation electrodes can be placed as required in any part of the heart. It is possible using such electrodes and appropriate stimulation to electrically "challenge" any of the different tissues or parts of the heart. Electrical stimulation makes it possible to assess the electrical status of cardiac tissue (i.e. ionic channel status) as well other factors such as vulnerability to arrhythmias.

With suitably constructed electrodes one can record extracellular monophasic action potentials (MAPs) from almost any part of the endocardium (see review by Franz, 1991), or epicardium, of the intact heart, ventricles or atria. The MAP technique has been further developed in recent times since first being described by Jochim *et al.* (1935). Recent advances have included the development of better contact electrodes, particularly hemispherical ones (1.5 mm diameter). The electrodes for recording monophasic action potentials from the endocardium are anchored against the muscle wall and are co-located with a suitable indifferent or ground electrode. Stable monophasic action potentials can be recorded in this manner. These action potentials have many of the characteristics of a good extracellular recordings inasmuch as signals of 40 or 50 mV can be faithfully recorded. The monophasic action potential is an extracellular recording technique which accurately reproduces the time course of repolarization.

The contact pressure between electrode and tissue (10–30 g) has to be sufficient to produce local depolarization, and hence the MAP, without damaging the underlying tissue. As a result, the potential recorded is a type of "injury" potential which only approximates the intracellular potential. However, the MAP can be used to record phases 1–3 of the action potential in a reasonably faithful manner. The multicellular component of the potential means, of course, that features seen in individual cells are averaged. Accurate recording of phase 0 or phase 4 is not possible. The dV/dt maximum recorded by this technique is much smaller than that recorded intracellularly, i.e. approximately 5 V/s versus the 200–300 V/s found intracellularly. Furthermore, the MAP recording technique is sensitive to motion artifacts.

It is possible, in situations where the heart is already exposed, to record from multiple extracellular sites across the epicardial surface. Routinely a grid of extracellular electrodes is used to criss-cross the heart (El-Sherif *et al.*, 1981; Mehra *et al.*, 1983).

A variety of optical techniques have been used in attempts to quantify membrane potentials. These include the use of voltage-sensitive dyes, membrane potential-dependent optical probes which give a measure of action potentials in neurons and cardiac myocytes (Ross *et al.*, 1974; Salama and Morad, 1976; Waggoner *et al.*, 1977; Kamino *et al.*, 1981). They are largely classified according to chemical structure (Waggoner *et al.*, 1977). Their optical activity resides predominantly within cyanine, merocyanine-thiobarbital, merocyanine-rhodanine, oxonol and styryl molecules.

A good dye gives changes in fluorescence during an action potential as well as good signal : noise ratios (Ross *et al.*, 1974). Of all dyes available for use, M-540 has been most extensively studied. It gives a large fluorescence as well as a large absorption signal, and possesses a relatively high signal : noise ratio (10 : 1) (Dragsten and Webb, 1978). Fluorescent action potentials can be produced which correspond to the time scale of action potentials recorded by microelectrodes (Dragsten and Webb, 1978; Kamino *et al.*, 1981). Waggoner *et al.* (1977) have discussed potential-dependent changes in light absorption across lipid bilayers in the presence of cyanine and oxonol dyes.

Salama and Morad (1976) have shown that M-540 has 20–40% higher fluorescence in atrial tissue versus ventricles, owing to the presence of oxymyoglobin in the latter. In addition, the signal is unchanged by changes in other variables (Salama and Morad, 1976) and thus is specific for monitoring membrane potential.

The precise mechanism by which these dyes work at the level of the plasmalemmal bilayer is not known. Many theories have been proposed based upon biophysical and optical properties of both membrane and dye molecules, and their interaction. Ross *et al.* (1974) initially proposed that changes in the optical properties of a dye are due to changes in its microenvironment. They proposed that depolarization of the membrane shifts the dynamic membrane-bound dye equilibrium from a slightly fluorescent dimeric state to a highly fluorescent monomeric state. Waggoner *et al.* (1977) expanded upon this "on–off" hypothesis.

Thus, voltage-sensitive dyes provide an alternative means to microelectrode techniques by which to examine spontaneous electrical signals and propagation pathways in the heart.

3.3 INDUCTION OF ARRHYTHMIAS

The successful function of a mechanically sound heart can be seriously impaired in the presence of arrhythmias, especially those of a ventricular origin. Thus in the presence of ventricular tachycardia even an otherwise normal myocardium will suffer impaired cardiac output. In the presence of ventricular fibrillation there is no cardiac output. Clinically and experimentally it is easy to detect arrhythmias on the basis of ECG and blood pressure changes.

The experimental methods for inducing arrhythmias are complex since arrhythmias can be induced chemically, electrically or as a result of myocardial ischaemia and infarction. These methods have been reviewed many times (e.g. Antoni, 1971; Szekeres, 1971; Winslow, 1984); the following is a summary.

The chemical induction of arrhythmias is best performed in small animals although local applications of aconitine have been used to induce atrial fibrillation in dogs. The standard chemical methods for inducing arrhythmias in whole animals include administration of catecholamines (especially adrenaline in the dog), chloroform in mice, aconitine in rats and cardiac glycosides in guinea-pigs. The type and mechanisms of arrhythmias induced by these chemical agents are specific for the agent and species used. Such arrhythmias also show characteristic responses to antiarrhythmic drugs (Winslow, 1984).

Less commonly, arrhythmias are induced by chemical means in isolated hearts. One particularly novel way to accomplish this is the dual perfusion method of Curtis (1991). In a modification of this technique, the major coronary beds can be selectively perfused with arrhythmogens without the necessity for cannulation of individual arteries (Avkiran and Curtis, 1991).

Arrhythmias can be readily induced in hearts of any size if appropriate electrical stimulation is used. Stimulation is applied through suitable electrodes implanted, as required, into any part of the heart (Szekeres, 1979; Winslow, 1984). Responses to defined patterns of stimulation reflect the status of ion channels. Thus threshold currents and pulse widths needed to induce paced activity reflect sodium channel availability. Similarly, refractory periods reflect sodium and potassium channel activity.

Electrical stimulation can be used to induce extra beats, tachycardias or fibrillation. Such stimulation can be performed in normal hearts, or in hearts previously subjected to damage, such as myocardial infarction (Buchanan *et al.*, 1992). Thus electrical stimulation of arrhythmias is used to test for arrhythmia propensity.

The most clinically relevant way of inducing arrhythmias is by pathological means. Pathological changes can be induced in hearts by surgical and other techniques. The easiest way to induce clinically relevant pathological changes is to occlude a coronary artery and thereby produce myocardial ischaemia (Tillmans *et al.*, 1975). If the ischaemia is maintained for a sufficient length of time, infarction results. The extent of infarction depends on the duration and intensity of the ischaemia. These factors determine the severity of arrhythmias (Botting *et al.*, 1985). Sudden reversal of the ischaemia by reperfusion also produces arrhythmias (Tennant and Wiggers, 1935). After consolidation of the infarct, arrhythmias can continue for weeks and even months (Ritchie *et al.*, 1979; Botting *et al.*, 1985).

Most arrhythmias, especially tachycardia and fibrillation, occur at critical times after the onset of ischaemia (Walker *et al.*, 1991) and upon reperfusion after a critical preceding period of ischaemia (Manning and Hearse, 1984), or when a new episode of ischaemia is imposed on an old infarct (Patterson and Lucchesi, 1981; Patterson *et al.*, 1983). Techniques such as those described above can be employed in almost any species providing the heart is of sufficient size, i.e. rat, or larger species.

The presence of coronary collateralization is important since arrhythmias depend on the size of the ischaemic zone in models of regional ischaemia (Austin *et al.*, 1982; Johnston *et al.*, 1983). Guinea-pigs have many coronary collaterals and as a result have few, if any, arrhythmias with regional ischaemia (Johns and Olson, 1954). Arrhythmic responses are also erratic in dogs due to variations between individual dogs in the extent of their preexisting coronary collateralization (Schaper, 1971, 1979; Selye *et al.*, 1960; Winbury, 1975). Pigs, rats, primates and rabbits have few, if any, coronary collaterals and consequently have more predictable responses to ischaemia and reperfusion (Johns and Olson 1954; Maxwell *et al.*, 1984; Winkler *et al.*, 1984; Schaper *et al.*, 1986). In order to ensure reproducibility in ischaemia and infarction experiments, a series of design and methodological conventions for such studies have been published (Walker *et al.*, 1988).

3.4 INDUCTION OF MYOCARDIAL ISCHAEMIA AND INFARCTION

In addition to its value as a model of human disease, the induction of myocardial ischaemia and infarction can be used generally as a pathological insult to test the status of cardiovascular function. Thus infarcts are often produced in animals concomitantly subjected to another pathological condition.

Ischaemia and infarction can be produced by a variety of methods including coronary occlusion or thrombosis. Whether an infarct is produced experimentally, or occurs naturally, it is important to be able to estimate its size. In view of the large volume of experimental and clinical data available on techniques for estimating infarct size, this chapter will present only a brief overview of some better known methods (for a concise overview see Wagner, 1982).

Direct anatomical estimate of myocardial infarct size is the standard against which all methods of determining infarct size should be made (Ideker *et al.*, 1982). Gross anatomic quantification includes cross-sectioning the ventricles into slices by cutting perpendicular to the long axis and estimating the area of infarct per slice. Summation of such estimates gives the total infarct size. Such a method has been greatly improved by using stains which are chemically altered by normal tissue in such a manner as to demarcate the infarct.

The dye most frequently used to quantify myocardial necrosis is triphenyltetrazolium chloride (Sandritter and Jestadt, 1957). When this dye is exposed to living tissue its reaction with succinic dehydrogenase produces a coloured formazan dye. Since dead (infarcted) tissue lacks this enzyme it remains uncoloured (Fallon, 1982). Staining procedures include slice incubation (Jestadt and Sandritter, 1959) and postmortem perfusion (Lichtig *et al.*, 1973; Feldman *et al.*, 1976). The latter is preferred by most researchers. The dye methods provide an accurate estimate of infarct size easily controlled by visual inspection.

Noninvasive techniques can be used to approximate or reflect the size of an infarct or occluded zone (zone-at-risk). These depend upon the characteristic changes which occur in the heart during infarction. Electrical (ECG changes), mechanical (reduced cardiac performance) and enzymatic (increased levels of cardiac enzymes such as creatinine kinase) methods have been developed (Goldberg and Winfield, 1972; Kostuk *et al.*, 1973; Selvester *et al.*, 1982). However, postmortem anatomic estimates of infarct size remain the most accurate (Ideker *et al.*, 1982).

Occlusion of a coronary artery results in ischaemia which progresses to an infarct if the occlusion is maintained. As ischaemic tissue becomes infarcted, characteristic changes occur in the ECG and these changes can be used to monitor the pathobiology of ischaemia and infarction. Using a 12-lead ECG, it is possible to locate infarcted tissue within the left ventricle and predict its size (Selvester *et al.*, 1982). Studies involving infarct geometry and conduction defects led to the development of comprehensive computer simulations of left ventricular infarcts ranging in size between 1 and 50 per cent of the left ventricle (Selvester *et al.*, 1982). This simulation data, together with angiographic and pathologic data about infarct size in various coronary artery beds, has allowed for a detailed set of criteria involving QRS changes which accurately represent infarct size and location (Selvester, 1975).

Infarction in the left ventricle results in changes in ventricular distensibility, compliance and reduced pump performance (Rogers *et al.*, 1982). Changes in left ventricular function include an elevated left ventricular end-diastolic pressure, reduced systolic emptying and stroke volume (Swan *et al.*, 1972); in addition, the ejection fraction decreases (Kostuk *et al.*, 1973). All these changes can be measured by catheterization (Swan *et al.*, 1970) and each provide an indirect measure of infarct size. The limitations to such measurements, in addition to the need for catheterization, include the fact that changes may reflect a cumulative effect of acute ischaemia and any established infarct. Also, small infarcts depending upon anatomic location, may produce disproportionately large haemodynamic changes (Rogers *et al.*, 1982).

Contrast ventriculography can reveal changes in left ventricular function produced by infarction at a global or regional level (Beher, 1982). Biplanar radiographic techniques can determine pressure–volume curves (Dodge *et al.*, 1962) in single cardiac cycles.

Radionuclide ventriculography provides a more accurate means of assessing cardiac performance. It is a simple procedure which provides information on the ejection fractions of either ventricle as well as regional left ventricular wall motion. A radionuclide tracer (15–20 mCi of technetium-99m pertechnetate) is used to generate cardiac images so as to provide assessment of performance. In the presence of an infarct the decrease in left or right ventricular performance can be measured by this technique (Rackley and Russell, 1972). It is, however, only an indirect measure of the severity of damage to the myocardium and thereby provides only an approximation of infarct size.

Similar limitations apply to radionuclide angiocardiography. However, the method is sufficiently sensitive to detect imbalances in myocardial perfusion before ischaemia is apparent on the ECG (Wackers *et al.*, 1982). Many suitable tracers are available (Okada *et al.*, 1982). Substances which are distributed in tissues in a manner equivalent to that of potassium such as ^{131}Cs, ^{201}Tl and ^{81}Rb, are frequently used. ^{201}Tl is most often used because it has a high myocardial extraction ratio (Lebowitz *et al.*, 1975).

Radioactive particles (10–50 μm) consisting of ^{99m}Tc, ^{133}In, or ^{131}I-labelled albumin aggregates, or human albumin microspheres, can be injected. They distribute throughout the myocardium in direct proportion to regional myocardial blood flow (Ueda *et al.*, 1965).

Diffusible, chemically and physiologically inert gases such as ^{133}Xe and ^{81m}Kr can be used to monitor myocardial blood flow (Herd *et al.*, 1962; Ross *et al.*, 1964). The short biological half-lives of such tracers allows for repetitive angiograms at frequent intervals so as to quantify changes in blood flow patterns and ischaemia as they occur with time. It is also possible to use radiolabelled metabolically active substrates such as ^{13}N-labelled ammonia or ^{11}C-palmitic acid (Weiss *et al.*, 1977) to detect regions of ischaemia and sites of infarction.

Technetium (99m) pyrophosphate is the current radio-pharmaceutical of choice for infarct imaging (Marcus *et al.*, 1982). This agent, which avidly binds to infarct tissue, gives the relative size of an infarct and discriminates between infarct and ischaemia in myocardial tissue. It concentrates in infarcts by virtue of complexing with the calcium which accumulates in infarcted tissue (Buja *et al.*, 1977). While this agent detects necrosis in individual cells (Marcus *et al.*, 1976) it cannot be easily used to assess the extent of an infarct. This lack of resolution is partly due to the complex geometry of an infarct. Since scintogram resolution deteriorates with distance from the scintillator the determination of inferior and posterior infarct size is inherently less accurate than that for anterior infarcts (Marcus *et al.*, 1982).

Cell death results in the release of cellular contents. These include creatinine kinase (CK) an enzyme which is uniformly distributed in the myocardium and released by myocytes during infarction. The usual cardiac : blood gradient for creatinine kinase is 1600 IU/1 g myocardium : 0.05 IU/ml blood (Roberts, 1982). Although total creatinine kinase lacks diagnostic specificity, its isoenzyme, CKMB, is located exclusively in cardiac tissue and therefore serves as a specific marker for myocardial infarction (Roberts *et al.*, 1975).

4. *Isolated Cardiac Tissue Preparations*

The types of *in vitro* cardiac preparations which can be used to study cardiac function are extensive. They range from whole perfused hearts to single isolated myocardial cells.

4.1   ISOLATED AND CULTURED HEART CELLS

Single adult myocardial cells can be isolated from the myocardia of a number of species, including man (Powell *et al.*, 1981). The principal species used are rat and guinea-pig. Single myocardial cells are harvested from isolated hearts by perfusing with a dispersing solution of collagenase dissolved in calcium-free physiological salt solution. Calcium is usually excluded in the dispersing solution so as to avoid excessive damage occurring in cells as they are isolated. However, it is possible to obtain cells which will tolerate normal calcium after dispersion (Powell, 1988). Brady (1991) describes the isolation of cells in perfusates containing the usual amounts of calcium. In addition Haworth *et al.* (1989) have shown that a critical calcium concentration, $20\ \mu M$, must be present in the digestion medium during the first few minutes of perfusion with the dispensing solution in order to successfully obtain single cells.

Under strictly controlled conditions, the heart is perfused with dispersing solution before it is removed for further treatment. Alternatively, single cells are collected as they are released from the heart. Often a process of teasing and physical dispersion is used to help harvest intact isolated cells (Nag and Zak, 1979).

Intact cells, isolated in the above manner, have many of the characteristics of cells in intact hearts: their electrophysiology is similar, as is their morphology, whether assessed by light or electron microscopy (Dow *et al.*, 1981; Schwarzfeld and Jacobson, 1981; Moses and Claycombe, 1982). However, it must be recognized that during the process of cell separation most of the cells in the heart die as a direct result of damage caused by the process and only a small percentage of the total number of heart cells survive (Gould and Powell, 1972). As cells die they release proteolytic and lysosomal enzymes. This means that viable intact isolated cells are subjected, to one degree or another, to exposure to these cytotoxic enzymes. Therefore cells successfully harvested may possess a degree of sarcolemmal disruption sufficient to modify their function in comparison with the intact cell *in vivo*.

Despite the above drawbacks, it is obvious that isolated single cells can be used in a variety of useful ways. The greatest use of single isolated heart cells is in electrophysiological studies.

Isolated cells can also be used for biochemical studies and even for contractile function studies. Contraction is measured in such cells by determining either the movement of the Z-bands or the border of the cell. Tsujioka *et al.* (1986) used laser diffraction techniques to overcome problems inherent in the fact that sarcomere dynamics are not uniform. Such a technique gives a better index of contraction. The problem of attaching cells for subsequent analysis of mechanical activity is quite complex. Attachment has been obtained with poly-L-lysine-coated glass beads (Tarr *et al.*, 1979). Other techniques include the use of glues, fibrin (Copelas *et al.*, 1987) and suction micropipettes (Brady *et al.*, 1979).

The main advantage of the use of single isolated cells is that immunological processes suspected of acting directly upon myocardial cells can be studied independently of the other cells which also constitute the whole heart, i.e. endothelial cells, smooth muscle cells and fibroblasts.

Unlike single isolated myocardial cells, cultured heart cells are capable of undergoing cell division (Piper *et al.*, 1982) and thus are best obtained from hearts in which myocardial cell division still occurs. The classic source tissue for cultured heart cells is chicken embryo, but this has now been replaced by neonatal rat. With the latter as a source, cultured heart cells can readily be prepared from hearts for up to 7 days after birth, but not from the hearts of older animals (Harary and Farley, 1960). Other sources include mouse embryos, postnatal hamsters and human foetuses (Pinson *et al.*, 1987).

Unfortunately, after preparation from source material, cultured heart cells often become dedifferentiated (Harary and Farley, 1960). Because of this it can be difficult to distinguish between myocardial cells and other cells that are co-isolated (smooth muscle cells and, more importantly, fibroblasts). The usual harvest of cells for culture has approximately 50% myoblasts and 50% fibroblasts. Fortunately, selective adhesion techniques have proven to be a most useful technique for separating these two types of cell. Fibroblasts adhere readily to substratum and grow across the surface whereas myoblasts take longer to adhere while retaining their characteristic shape (Pinson *et al.*, 1987). Fibroblasts may be selectively removed by growing the cells in serum free medium and treating the culture with cholesterol derivatives or with the calcium ionophore A23187 (Kaneko and Goshima, 1982). However, such procedures carry the risk of damaging the myoblasts.

As previously mentioned, cultured heart cells are generally grossly dedifferentiated. In the first few days of culture, it is not possible to see the morphology expected of cardiac cells. Cultured cells do not have cross-striations, Z-lines, or obvious contractile material (Jacobson and Piper, 1986; Piper *et al.*, 1982). However, morphology improves with time in culture although it never fully takes on the appearance seen in intact hearts, or even in single isolated heart cells. Thus cultured cells have a flat plate-like shape unlike the branched rectangular shape of adult cardiac ventricular cells. Such cells easily adhere to a suitable substrate whereas isolated cells adhere with difficulty and are often left in free suspension.

Cultured heart cells have been used for many years for electrophysiological, mechanical and biochemical studies as well as for studying the possible toxic effects of immune mediators. A recent use of cultured heart cells has involved examining cell conditions where volume and oxygen restrictions were made, i.e. conditions which mimic ischaemia at the cellular level (Musters *et al.*, 1991). The great advantage of cultured cells is that they can be co-cultured in the presence of cells important to the immunological system, such as macrophages or leukocytes, in a controlled manner.

4.2 ISOLATED CARDIAC TISSUE *IN VITRO*

For many years, isolated cardiac tissues have been used to study myocardial function. The preparation of isolated myocardial tissue is extremely simple; the easiest tissue to use is atrial tissue. Both right and left atria can be easily excised and studied under superfusion conditions in a bath of suitable physiological salt solution (see Table 2.1). Characteristically, the right atrium continues to beat at a rate dominated by the sinus node. The left atrium will beat only if damaged (and injured cells are exhibiting abnormal automaticity).

Atria survive well when superfused with suitable physiological salt solutions. On the other hand the successful functioning of ventricles requires perfusion since the thickness of ventricular tissue constitutes a diffusion barrier, and superfused ventricle will become hypoxic and starved of metabolic substrate. Preparations in which perfusion has been described include whole ventricles and the ventricular septum (Nadeau and Amir-Jahed, 1965; Levy, 1971; Carmeleit and Zaman, 1979; Grupp, 1984; Grupp and Grupp, 1984) and even isolated perfused atrioventricular nodes (Wit and Cranefield, 1974; Satch *et al.*, 1979, 1981). Perfused preparations are characteristically used for assessment of mechanical function and for electrophysiological studies.

In addition to perfusion it is also possible to use bathed preparations of selected tissue. Thus, preparations have been made of Purkinje fibres, atrial trabeculae, septum and other types of myocardial tissue. In these isolated preparations it is possible to measure mechanical and electrical function as well as biochemistry (Levy, 1971; Grupp and Grupp, 1984).

When determining the mechanical function of isolated tissue a variety of physiological factors have to be taken into consideration. In the first place inadequate oxygenation results in disruption of the centre of a tissue sample if the thickness of the sample is greater than a few cells. This is particularly relevant in the case of ventricles (Grupp and Grupp, 1984). Perfused tissue does not suffer from this drawback since it receives an adequate supply of oxygen as well as an efficient washout of metabolites, provided flow is adequate. The longevity of isolated perfused hearts can be greatly improved by modifying the perfusion fluid (e.g. by addition of EDTA) so as to chelate the heavy metals which are present as trace material in the salts used for preparing the fluid (Neely *et al.*, 1967; MacConaill, 1985).

Some typical solutions used as physiological solutions for cardiac tissue are shown in Table 2.1. In the table a comparison is made between the concentration of ions (total and free forms) found in typical mammalian serum (in this case, rat) and the concentrations used in the various solutions. There is considerable divergence between the *in vivo* concentrations and those used in the solutions. This has arisen for a number of reasons. In the first place many of the solutions were devised empirically, and decades ago. Later there was confusion regarding concentration of ions versus their chemical activity, i.e. free ion form. Furthermore, plasma concentrations are not the same as those in interstitial fluids. While it is relatively easy to mimic the ionic concentrations found in plasma, or in interstitial fluid, it is more difficult to ensure that osmolarity, pH and oxygen contents are appropriate.

Various attempts have been made to ensure correct osmolarity. Thus in the synthetic interstitial fluid described by Bretag (1969), sodium gluconate (9.64 mM) substitutes for organic ions, including protein. Urea has been used for the same purpose. Other substances used to generate osmotic pressure include fluorosol-43 and hydroxymethyl starch (Segel and Rendig, 1982). Doring and Dehnert (1988) discussed the problems associated with osmolarity and describe how the simple technique of immersing hearts 1–2 cm deep in bathing fluid reduces oedema and maintains optimal contractile force for 6–7 hours.

Another major problem with physiological salt solutions is the supply of energy and oxygen. A number of substances other than the ubiquitous glucose (with or without insulin) have been used, including acetoacetate. The problem of supplying adequate oxygen to tissue has never been completely resolved. Doring and Dehnert (1988) list a variety of procedures for improving oxygen delivery including the addition of bovine erythrocytes. The use of synthetic organic fluorine compounds readily improves oxygen availability although such compounds are very expensive.

The common technique used to obtain appropriate pH values involves either bicarbonate or phosphate buffer systems. The former requires the use of carbon dioxide in the gassing solution as shown in Table 2.1.

When contractile activity in isolated cardiac tissue is recorded it is under either isometric or isotonic conditions, i.e. either the fibres are kept at a constant length, or are allowed to shorten under conditions of constant load. In the working heart preparation and *in vivo*, muscle contracts under mixed conditions in that there is both shortening and tension development, i.e. auxotonic conditions (Blattner *et al.*, 1978). Very often the mechanical activity of cardiac tissue preparations is studied under isometric conditions and the resulting developed tension or pressure recorded. The force developed in tissue preparations can be normalized for different sizes, or shapes, of tissue. Thus the muscle can be weighed, and its mass determined, or its physical shape can be measured so as to determine its physical dimensions (particularly cross-sectional area).

In addition to being used for electrical and mechanical studies, myocardial tissue preparations can also be used for biochemical studies. In fact, much of the earlier work concerned with relating cAMP production to stimulation of cardiac beta-adrenoceptors was performed in isolated cardiac tissue (Sutherland and Rall, 1960; Tuttle *et al.*, 1976).

4.3 ISOLATED WHOLE HEARTS

The most physiologically relevant isolated cardiac tissue

Table 2.1 Ionic composition of physiological salt solutions used for cardiac tissue in comparison with rat serum and interstitial fluid

Solution	Cations				Anions						Gas (%)	
	Na	K	Ca	Mg	Cl	HCO_3	H_2PO_4	SO_4	Urea	Glucose	O_2	CO_2
Total ion concentration in plasma	152	3.7	2.7	1.06	114	26.5	1.7	0.69	7.0	5.8		
Ionized ion concentration in plasma	150	3.6	1.6	0.76	112	25.7	1.6	0.69	7.0	5.8		
Ionized ion concentration in interstitial fluid	147	3.5	1.5	0.72	115	26.3	1.7	0.73	7.0	5.8		
Tyrode	149	2.7	1.8	1.05	145	11.9	0.42	0	0	5.6	95	5
Locke	159	5.6	2.1	0	167	1.8	0	0	0	5.6	various	
Ringer	155	2.71	1.8	0	160	1.2	0	0	0	0	100	0
DeJalon	160	5.6	0.5	0	161	6.0	0	0	0	2.8	95	5
Sund	160	5.6	0.5	1.0	163	6.0	0	0	0	2.8	95	5
Krebs–Henseleit	143	5.9	2.5	0	128	24.9	1.18	1.64	0	5.6	95	5

The concentrations of anions and cations are given in mM. For Locke, Ringer and Tyrode solutions the pH varies with the content of CO_2 in the gassing mixture. Note that 2.0 mM sodium pyruvate is added to Krebs–Henseleit solutions.

preparation is the isolated perfused whole heart. This preparation has been used extensively (see review by Broadley, 1979). Levi and Mullane (1990) have reviewed the use of the isolated heart for assessing the cardiac actions of substances which are possibly released during immunological reactions; substances such as prostaglandins, leukotrienes and platelet activating factor. Ideally isolated hearts are used in heart/lung preparations where blood is used as the perfusing medium (Hilton and Eichholtz, 1925; Mendler *et al.*, 1972). Other perfusing solutions include washed platelets, prepared from platelet-enriched plasma (Purchase *et al.*, 1986), or the standard Ringer (Ringer, 1883) solutions such as Locke's (1901) or Krebs-Henseleit (1932) solutions modified in some studies by addition of EDTA (Neely *et al.*, 1967; MacConaill, 1985).

Difficulties with heart/lung preparations has led to use of isolated perfused heart preparations, the most well-known being that due to Langendorff (1895). In this preparation, the aorta is cannulated and fluid allowed to flow through the ostia into the coronary arteries in the normal antrograde manner (a common error in many publications refers to the preparation as being perfused in a "retrograde" manner; this is not so). The pressure within the aortic root ensures that the aortic valves are normally closed.

A major improvement on the Langendorff perfusion technique is the working heart preparation due to Neely *et al.* (1967) in which the heart is made to perform external work. Hearts can be cannulated such that perfusion fluid flows in and out of the atria through the ventricles allowing for perfusion over a range of ventricular pressures.

Isolated whole hearts can be used for a large number of electrical, mechanical and biochemical studies. Normally the heart is at least partially denervated by the surgery required for cardiac excision, although it is possible to make preparations in which the cardiac nerves are left relatively intact (Hukovic and Muscholl, 1962).

The function of the isolated heart is normally assessed in terms of cardiac output (working preparation) or pressure production in the left ventricle (Langendorff preparation). Many methods can be used to measure contractility in the left ventricle; usually either an apical force transducer (Gottlieb and Magnus, 1904; Beckett, 1970) is used to record tension or a balloon catheter is inserted into the left ventricle to record pressure (Gottlieb and Magnus, 1904; Curtis *et al.*, 1986). In addition, a surface electrogram (*in vitro* equivalent of the ECG) and coronary flow can readily be recorded. Electrograms can be recorded using wick, silver sphere (Curtis *et al.*, 1986), or suction electrodes. A non-contact method for recording the ECG involves placing electrodes in close approximation to the epicardial surface of a heart submerged in a bath of physiological solution (Doring and Dehnert, 1988).

Coronary flow can be measured by various means including use of flow probes, but simple timed collection of coronary effluent (perfusate) is perfectly adequate. Only flow probes give an accurate record of the different phases of flow during the cardiac cycle, and rapid flow changes due to rapid alterations in coronary vascular resistance, however.

Various improvements to the Langendorff isolated perfused heart preparation have been developed over the years. For example, use of a special perfusion apparatus facilitates exchange of drugs, etc. (Curtis *et al.*, 1986) and polycarbonate tubing reduces oxygen loss (Krzeminski *et al.*, 1991). By use of such preparations it is easy to assess directly the pharmacological actions of drugs of interest in immunopharmacology. Computerized data acquisition systems allow for the ready analysis of all aspects of cardiac function (Krzeminski *et al.*, 1991).

In addition to facilitating assessment of drug responses, isolated perfused hearts have also been used to determine cardiac functional integrity in the setting of various pathological states. Hearts which have been taken from hypertensive, diabetic or immunologically challenged animals may be perfused and their functionality measured in terms of isovolumic contraction and Frank-Starling curves (Doring and Dehnert, 1988; Fein and Scheuer 1990; Hahn and Bernauer, 1970). The isolated heart thus allows functionality to be examined in the absence of complicating factors such as variations in blood flow, nervous elements, etc. In addition, isolated hearts can be examined histologically. Cardiotoxic actions, such as those that might occur as a result of immunological damage, can be assessed in isolated hearts in terms of functional, metabolic and morphological changes as shown for doxorubicin by Pelikan *et al.* (1986) and for adriamycin by Kusuoka *et al.* (1991).

5. *Summary*

Immunological processes themselves and the biologically active materials released as a result of immunological reactions are all capable of disturbing cardiac function. Such disturbances of function need to be investigated in both clinical and experimental settings. Many of the methods for assessing cardiac function can readily be applied either clinically or experimentally although there are techniques whose complexity and cost generally confines their use to the clinic. Some methods can only be applied *in vitro* to isolated hearts, or are destructive of tissue, and therefore can only be used in the experimental situation.

In view of the fact that the ultimate function of the heart is maintenance of an adequate cardiac output the measurement of cardiac output is obviously of premier importance in assessment of cardiac function. Cardiac output can be assessed a number of different ways, under a variety of conditions, in both man and other animals. In many situations it is important to monitor cardiac function over extended periods of time and this is best accomplished by use of noninvasive techniques. Many noninvasive techniques are available, each has its own strengths and weaknesses. Thus the choice of which technique to use will depend on non-scientific factors such as cost and accessibility.

In addition to cardiac output there are a host of other cardiac functions that can be measured and that relate, more or less directly, to cardiac output. A myriad of electrophysiological methods allow the electrical activity of the heart to be assessed from the level of single ion channels to the level of the whole heart. While intracellular potential measurement techniques can only be applied in experimental settings it is still possible, in a clinical setting, to obtain analyses of electrical actions, almost down to a single cell level. The main problem in measuring electrical activity is that such activity does not readily reflect the contractile function of the myocardium, except where electrophysiological function is radically disturbed by arrhythmias.

Coronary flow ultimately determines cardiac function in intact hearts. Unfortunately the methods for measuring coronary flow are not comprehensive, for although it is easy to measure coronary flow in isolated hearts or in open-chest animals, its assessment in man or intact animals is less satisfactory.

Unfortunately, most biochemical techniques used for assessing cardiac function rely on the destructive analysis of cardiac tissue and therefore are of limited value in the clinical setting. However, measurement of cardiac-specific substances being released from damaged cardiac cells is of value. It is hoped that NMR, and perhaps PET, will allow for the easy and continuous measurements of biochemicals of interest in man and experimental animals in the near future.

In conclusion a host of techniques and procedures are available whereby very good assessments of cardiac function can be made both *in vitro* and *in vivo*. Such techniques and procedures can readily be applied to immunological studies of the heart.

6. Acknowledgements

The British Columbia and Yukon Heart and Stroke Foundation are thanked for grant support as is the British Columbia Health Care Research Foundation.

7. References

Abbott, C.P. (1964). A technique for heart transplantation in the rat. Arch. Surg. 89, 645–652.

Aizawa, Y., Kamei, K. and Shibata, A. (1980). Quantitative analysis of thallium-201 uptake and effects of heart rate. Jpn. Cir. J. 44, 863–866.

Aldrich, R.W., Corey, D.P. and Stevens, C.F. (1983). A reinterpretation of mammalian sodium channel gating based on single channel recording. Nature 306, 436–441.

Antoni, H. (1971). Electrophysiological mechanisms underlying pharmacological models of cardiac fibrillation. Naunyn Schmied. Arch. Pharmacol. 269, 177–199.

Austin, M., Wenger, T.L., Harrell, F.E., Luzzi, F.A. and Strauss, H.C. (1982). Effect of myocardium at risk on outcome after coronary occlusion and release. Am. J. Physiol. 243, H340–H345.

Avkiran, M. and Curtis, M.J. (1991). Independent dual perfusion of left and right coronary arteries in isolated rat hearts. Am. J. Physiol. 261, H2082–H2090.

Bainton, D., Jones, G.R. and Hole, D. (1978). Influenzae and ischaemic heart disease: a possible trigger for acute myocardial infarction. Int. J. Epidemiol. 7, 231–239.

Barnard, C.N. (1967). A human cardiac transplant: An interim report of two successful operations performed at Groote Schurr Hospital, Capetown. S. Afr. Med. J. 4, 1271–1274.

Bauer, R. (1989). Principles and developmental status of scintigraphy and nuclear magnetic resonance tomography. Z-Kardiol. 78, 81–96.

Beaumont, J.L. and Beaumont, V. (1978). Immunological aspects of atherosclerosis. Atherosclerosis Rev. 3, 133–146.

Beckett, P.R. (1970). The isolated perfused heart preparation: two suggested improvements. J. Pharm. Pharmacol. 22, 818–822.

Beher, V.S. (1982). In "Myocardial Infarction: Measurement and Intervention" (ed G.S. Wagner), pp. 173–197. Martinus Nijhoff Publishers, Boston.

Berman, D.S., Salel, A.F., DeNardo, G.L., Bogren, H.G. and Mason, D.T. (1975). Clinical assessment of left ventricular regional contraction patterns and ejection fraction by high-resolution gated scintigraphy. J. Nucl. Med. 16, 865–874.

Bildsoe, P. (1972). Organ transplantation in the rat. Acta Pathol. Microbial. Scand. 80, 221–227.

Blattner, R., Classen, H.G., Dehnert, H. and Doring, H.J. (1978). In "Experiments on Isolated Smooth Muscle Preparations", pp. 3–7, Hugo Sachs Elektronik KG, Freiburg.

Bodenheimer, M.M., Banka, V.S., Fooshee, C.M., Hermann, G.A. and Helfant, R.H. (1978). Quantitative radionuclide angiography in the right anterior oblique view: comparison with contrast ventriculography. Am. J. Cardiol. 41, 718–725.

Boniface, K.J., Brodie, O.J. and Walton (1953). Resistance strain gauge arches for direct measurement of heart contractile force in animals. Proc. Soc. Exp. Biol. Med. 84, 263–266.

Botting, J.H., Curtis, M.J. and Walker, M.J.A. (1985). Arrhythmias associated with myocardial ischaemia and infarction. Mol. Aspects Med. 8, 311–422.

Brady, A.J. (1991). Mechanical properties of isolated cardiac myocytes. Physiol. Rev. 71, 413–428.

Brady, A.J., Tan, S.T. and Ricchiuti, N.V. (1979). Contractile force measured in unskinned isolated adult rat heart fibres. Nature 282, 728–729.

Braunwald, E. (1971). On the difference between the heart's output and its contractile state. Circulation 43, 171–174.

Bretag, A.H. (1969). Synthetic interstitial fluid for isolated mammalian tissue. Life Sci. 8, 319–329.

Broadley, K.J. (1979). The Langendorff heart preparation – reappraisal of its role as a research and teaching model for coronary vasoactive drugs. J. Pharmacol. Meth. 2, 143–156.

Buchanan, L.V., Kabell, G., Turcotte, U.M., Brunden, M.N. and Gibson, J.K. (1992). Effects of ibutilide on spontaneous and induced ventricular arrhythmias in 24-hour canine myocardial infarction: a comparative study with sotalol and encainide. J. Cardiovasc. Pharmacol. 19, 256–263.

Buja, M.L., Tofe, A.J., Kulkarini, P.V., Mukherjee, A., Parkey, R.W., Francis M.D., Bonte, F.J. and Willerson, J.T. (1977). Sites and mechanisms of localization of technecium-99m phosphorus radiopharmaceuticals in acute myocardial infarcts and other tissues. J. Clin. Invest. 60, 724–727.

Carlsson, E. and Milne, E.N.C. (1967). Permanent implantation of endocardial tantalum screws. A new technique for functional studies of the heart in experimental animals. J. Ass. Can. Radiol. 19, 304–309.

Carmeleit, E. and Zaman, Y. (1979) . Comparative effects of lignocaine and lorcainide on conduction in the Langendorff-perfused guinea-pig heart. Cardiovasc. Res. 13, 439–449.

Carrel, A. and Guthrie, C.C. (1905). The transplantation of skin and organs. Am. Med. 10, 1101–1111.

Chance, B. (1989). What are the goals of magnetic resonance research? NMR-Biomed. 2, 179–187.

Chisholm, P.M. (1987) In "Immunology and Molecular Biology of Cardiovascular Diseases" (ed C.J.F. Spry), pp. 161–173. MTP Press Ltd, Boston.

Conway, J. and Lund-Johansen, P. (1990). Thermodilution method for measuring cardiac output. Eur. Heart J. 11 (suppl. I), 17–20.

Copelas, L.M., Briggs, M., Grossman, W. and Morgan, J.P. (1987). A method for recording isometric tension development by isolated cardiac myocytes: transducer attachment with fibrin glue. Pflugers Arch. 408, 315–317.

Craige, E. (1974). In "Noninvasive Cardiology" (ed A.M. Weissler), pp. 1–39. Grune & Stratton, New York.

Curtis, M.J. (1991). The rabbit dual coronary perfusion model: a new method for assessing the pathological relevance of individual products of the ischaemic milieu: role of potassium in arrhythmias. Cardiovasc. Res. 25, 1010–1022.

Curtis, M.J., MacLeod, B.A., Tabrizchi, R. and Walker, M.J.A. (1986). An improved perfusion apparatus for small animal hearts. J. Pharmacol. Meth. 15, 87–94.

Denman, A.M. (1981). In "Immunology of Cardiovascular Disease" (ed M.J. Lessof), pp. 1–63. Marcel Dekker, Inc., New York.

Dodge, H.T., Hay, R.E. and Sandler, H. (1962). Pressure-volume of the diastolic left-ventricle of man with heart disease. Am. Heart J. 64, 503–511.

Doring, H.J. and Dehnert, H. (1988). In "The Isolated Perfused Warm-Blooded Heart according to Langendorff", pp. 40–44. Biomesstechnik-Verlag, March, Germany.

Dow, J.W., Harding, N.G.L. and Powell, T. (1981). Isolated cardiac myocytes I. Preparation of adult myocytes and their homology with the intact tissue. Cardiovasc. Res. 15, 483–514.

Dragsten, P.R. and Webb, W.W. (1978). Mechanism of the membrane potential sensitivity of the fluorescent membrane probe merocyanine 540. Biochemistry 17, 5228–5240.

Einthoven, W. (1912). The different forms of the human electrocardiogram and their significance. Lancet 1, 853–861.

El-Sherif, N., Smith, R.A. and Evans, K. (1981). Canine ventricular arrhythmias in the late myocardial infarction period. 8. Epicardial mapping of re-entrant circuits. Circ. Res. 49, 255–265.

Engemann, R. (1985). In "Microsurgical Models in Rats for Transplantation Research" (Eds A. Thiede, R. Engemann and H. Hamelmann), pp. 19–24. Springer Verlag, Berlin.

Fagard, R. and Conway J. (1990). Measurement of cardiac output: Fick principle using catheterization. Eur. Heart J. 11 (Suppl. I), 1–5.

Fallon, J.T. (1982). In "Myocardial Infarction: Management and Intervention" (ed G.S. Wagner), pp. 373–384. Martinus Nijhoff Publishers, Boston.

Fegler, G. (1954). Measurement of cardiac output in anaesthetised animals by a thermodilution method. Quart. J. Exp. Physiol. 39, 153–164.

Fein, F.S. and Scheuer, H.(1990). In "Diabetes Mellitus, Theory and Practise" (ed H. Rifkin and D. Porte, Jr.), pp. 812–823. Elsevier Science Publishing Co., New York.

Feldman, S., Glagov, S., Wissler, R.W. and Hughes, R.H. (1976). Postmortem delineation of infarcted myocardium. Coronary perfusion with nitro blue tetrazolium. Arch. Pathol. Lab. Med. 100, 55–58.

Fletcher, E. and Brennan, C.F. (1957). Cardiac complications of Coxsackie virus infection. Lancet 1, 913–915.

Fouad-Tarazi, F.M. and McIntyre, W.J. (1990). Radionuclide methods for cardiac output determination. Eur. Heart J. 11 (Suppl. I), 33–40.

Fox, U. and Montorsi, M. (1980). A technique modification of heart-lung transplantation in rat. J. Microsurgery 1, 377–380.

Franz, M.R. (1991). Method and theory of monophasic action potential recording. Prog. Cardiovasc. Dis. 6, 347–368.

Galinanes, M. and Hearse, D.J. (1991). Metabolic, functional, and histologic characterization of the heterotopically transplanted rat heart when used as a model for the study of long-term recovery from global ischaemia. J. Heart Lung Transplant. 10, 79–91.

Geddes, L.A. (1991). "Handbook of Blood Pressure Measurements" pp. 24–33. Humana Press, New Jersey.

Gerlock, A.J., Giyanani, V.L. and Krebs, C. (1988). In "Applications of Noninvasive Vascular Techniques" (Ed E. Wickland), pp. 283–299. W.B. Saunders Co., Philadelphia.

Gero, S., Szondy, E., Fust, G., Muzei, Z., Szekeley, J. and Feher, J. (1981). In "Vascular Occlusion: Epidemiological, Pathophysiological and Therapeutic Aspects" (ed M. Tisi and J.A. Dormandy), pp. 99–106. Academic Press, New York.

Goldberg, D.M. and Winifield, D.A. (1972). Diagnostic accuracy of serum assays for myocardial infarction in a general hospital population. Br. Heart J. 34, 597–604.

Gottlieb, R. and Magnus, R. (1904). Digitalis und herzarbeit. Nach versuchen am uberlebenden warmbluterherzen. Arch. Exp. Path. Pharmakol. 51, 30–39.

Gould, R.P. and Powell, T. (1972). Intact isolated muscle cells from the adult rat heart. J. Physiol. (Lond.) 225, 16P–19P.

Grayburn, P.A., Willard, J.E., Haagen, D.R., Brickner, M.E., Alvarez, L.G. and Eichorn, E.T. (1992). Measurement of coronary flow using high-frequency intravascular ultrasound imaging and pulsed Doppler velocimetry: in vitro feasibility studies. J. Am. Soc. Echocardiogr. 5, 5–12.

Grist, N.R. and Bell, E.J. (1969). Coxsackie viruses and the heart. Am. Heart J. 77, 295–300.

Grupp, G. (1984). In "Methods in Pharmacology", Vol. 5 (ed A. Schwartz), pp. 129–139. Plenum Publishing Co., New York.

Grupp, I.L. and Grupp, G. (1984). In "Methods in Pharmacology," Vol. 5 (ed A. Schwartz), pp. 111–128. Plenum Publishing Co., New York.

Guth, B.D., Schulz, R. and Heusch, G. (1990). Evaluation of parameters for the assessment of regional myocardial contractile function during asynchronous left ventricular contraction. Bas. Res. Cardiol. 85, 550–562.

Hahn, F. and Bernauer, W. (1970). Studies on heart anaphylaxis. III. Effects of antigen and histamine on perfused guinea-pig heart. Arch. Int. Pharmacodyn. 184, 129–157.

Hajjar, D.P. (1991) Warner–Lamber/Parke–Davis Award Lecture. Viral pathogenesis of atherosclerosis. Impact of molecular mimicry and viral genes. Am. J. Pathol. 139, 1195–1211.

Hangiandreou, N.J., Folts, J.D., Peppler, W.W. and Mistretta, C.A. (1991). Coronary blood flow measurement using an angiographic first pass distribution technique: a feasibility study. Med. Phys. 18, 947–954.

Hanson, W.L. (1977). In "Pan American Health Organization, Chagas' Disease" pp. 22–34. Scientific Publication No. 347, World Health Organization, New York.

Harary, I. and Farley, B. (1960). *In vitro* studies of single isolated beating heart cells. Science 131, 1674–1675.

Harper, A.M., Lorimer, A.R. and Thomas, D.L. (1974). In "Scientific Foundations of Anaesthesia" (ed C. Scurr and S. Feldman), pp. 53–70. William Heinemann Medical Books Ltd, London.

Haselgrove, J.C., Subramanian, V.H., Leigh, J.S., Gyulai, L. and Chance B. (1983). In vivo one-dimensional imaging of phosphorus metabolites by phosphorus-31 nuclear magnetic resonance. Science 220, 1170–1173.

Haworth, R.A., Goknur, A.B., Warner, T.F. and Berkoff, H.A. (1989). Some determinants of quality and yield in the isolation of adult heart cells from rat. Cell Calcium 10, 57–62.

Heck, C.F., Shumway, S.J. and Kaye, M.P. (1989). The Registry of the International Society of Heart Transplantation. Sixth Official Report – 1989. J. Heart Transplant. 8, 271–276.

Henderson, A.H. (1988) . In "Diastolic Relaxation of the Heart" (ed W. Grossman and B.H. Lorell), pp. 73–82. Martinus Nijhoff Publishing, Boston.

Herd, J.A., Hollenberg, M., Thorburn, G.D., Kopald, H.H. and Barger, A.C. (1962). Myocardial blood flow determined with krypton-85 in unanaesthetised dogs. Am. J. Physiol. 203, 122–124.

Hilton, R. and Eichholtz, F. (1925) . The influence of chemical factors on the coronary circulation. J. Physiol. 59, 413–426.

Hudson, L. and Ribeiro-dos-Santos, R. (1981). In In "Immunology of Cardiovascular Disease" (ed M.J. Lessof), pp. 141–155. Marcel Dekker, Inc., New York.

Hukovic, S. and Muscholl, E. (1962). Noradrenalin-abgabe aus dem isolierten kaninchenherzen bei sympatischer nervenreizung und ihre pharmakologische beeinflussung. Naunyn Schmied. Arch. Pharmacol. 273, 134–153.

Ideker, R.E., Hackel, D.B. and McClees, E.C. (1982). In "Myocardial Infarction: Measurement and Intervention" (ed G.S. Wagner), pp. 347–371. Martinus Nijhoff Publishers, Boston.

Jacob, H.S. (1981). The role of activated complement and granulocytes in shock states and myocardial infarction. J. Lab. Clin. Med. 98, 645–653.

Jacobson, S.L. and Piper, H.M. (1986). Cell cultures of adult cardiomyocytes as models of the myocardium. J. Mol. Cell. Cardiol. 18, 661–678.

Jansen, B.H., Larson, B.H. and Shankar, K. (1991). Monitoring of the ballistocardiogram with the static charge sensitive bed. IEEE-Trans-Biomed. Eng. 38, 748–751.

Jara, F.M., Toledo-Pereyra, L.H., Lewis, J.W. and Magilligan, D.J. (1979). Auxiliary heart transplantation. Ann. Thoracic Surg. 29, 483–488.

Jestadt, R. and Sandritter, W. (1959). Erfahrungen Mit Der TTC (triphenyltetrazoliumchlorid) Reaktion fur die pathologisch-anatomischen diagnose des frischen herzinfarktes. Kreislaufforsch. 48, 802–809.

Jochim, K., Katz, L.N. and Mayne, W. (1935). The monophasic electrogram obtained from the mammalian heart. Am. J. Physiol. 111, 177–186.

Johns, T.N.P. and Olson, B.J. (1954). Experimental myocardial infarction. I. A method of coronary occlusion in small animals. Ann. Surg. 140, 675–682.

Johnston, K.M., MacLeod, B.A. and Walker, M.J.A. (1983). Responses to ligation of a coronary artery in conscious rats and the actions of antiarrhythmics. Can. J. Physiol. Pharmacol. 61, 1340–1353.

Ju, Y.-K., Saint, D.A. and Gage, P.W. (1992). Effects of lignocaine and quinidine on the persistent sodium current in rat ventricular myocytes. Br. J. Pharmacol. 107, 311–316.

Kadota, L.T. (1985). Theory and application of thermodilution cardiac output measurement: A review. Heart Lung 14, 605–616.

Kamino, K., Hirota, A. and Fujii, S. (1981). Localization of pacemaker activity in early embryonic heart monitoring using voltage-sensitive dye. Nature. 290, 595–597.

Kaneko, H. and Goshima, K. (1982). Selective killing of fibroblast-like cells in cultures of mouse heart cells by treatment with calcium ionophore A23187. Exp. Cell. Res. 142, 407–411.

Kaplan, M.H. (1963). Immunologic relation of streptococcal and tissue antigens. I. Properties of antigen in certain strains of group A streptococci exhibiting an immunologic cross-reaction with human heart tissue. J. Immunol. 90, 595–606.

Kass, D.A. and Beyar, R. (1991). Evaluation of contractile state of maximal ventricular power divided by the square of end-diastolic volume. Circulation 84, 1698–1708.

Kass Wenger, N. (1974). In "The Heart, Arteries and Veins" (ed J.W. Hurst, R.B. Logue, R.C. Schant and N. Kass Wenger), pp. 1326–1327. McGraw-Hill Book Co., New York.

Katz, S.M., Liebert, M., Gill, T.J., Kunz, H.W., Cramer, D.V. and Guttmann, R.D. (1983). The relative roles of MHC and non-MHC genes in heart and skin allograft survival. Transplantation 36, 96–101.

Kjekshus, J.K., Maroko, P.R. and Sobel, B.E. (1972). Distribution of myocardial injury and its relation to epicardial ST-segment changes after coronary artery occlusion in the dog. Cardiovasc. Res. 6, 490–499.

Klempnauer, J., Steiniger, B., Wonigeit, K. and Gunther, E. (1985). In "Microsurgical Models in Rats for Transplantation Research" (ed A. Thiede, R. Engemann and H. Hamelmann), pp. 121–126. Springer Verlag, Berlin.

Klingel, K., Hohenadl, C., Canu, A., Albrecht, M., Seemann, M., Mall, G. and Kandolf, R. (1992). Ongoing enterovirus-induced myocarditis is associated with persistent heart muscle infection: quantitative analysis of viral replication, tissue damage, and inflammation. Proc. Nat. Acad. Sci. USA, 89, 314–318

Klurfeld, D.M., Allison, M.J., Gerszten, E. and Dalton, H.P. (1979). Alterations of host defence paralleling cholesterol-induced atherogenesis. I. Interactions of prolonged experimental hyper-cholesterolemia and infections. J. Med. 10, 35–36.

Konertz, W., Semik, M. and Brenhard, A. (1985). In "Microsurgical Models in Rats for Transplantation Research" (ed A. Thiede, E. Deitz, R. Engemann and H. Hamelmann), pp. 25–30. Springer Verlag, Berlin.

Korecky, B. and Masika, M. (1990). Use of a heterotopic cardiac transplant for pharmacological and toxicological studies. Toxicol. Pathol. 18, 541–546.

Kostuk, W.J., Ehsani, A.A., Karliner, J.S., Ashburn, W.L., Peterson, K.L., Ross, J. and Sobel, B.E. (1973). Left-ventricular performance after myocardial infarction assessed by radioisotope angiocardiography. Circulation 47, 242–249.

Krebs, H.A. and Henseleit, K. (1932). Untersuchungen uber die harnestoffbildung im tierkorper. Hoppe-Seyler's Zeit. Physiol. Chem. 210, 33–66.

Krzeminski, T., Kurcok, A., Kapustecki, A., Kowalski, J., Slowinski, Z. and Brus, R. (1991). A new concept of the isolated perfused heart preparation with on-line computerized data evaluation. J. Pharmacol. Meth. 25, 95–110.

Kusuoka, H., Futaki, S., Koresune, Y., Kitabatake, A., Suka, H., Kamada, T. and Inoue, M. (1991). Alterations of intracellular calcium homeostasis and myocardial energetics in acute adriamycin-induced heart failure. J. Cardiovasc. Pharmacol. 18, 437–444.

Labovitz, A.J. and Williams, G.A. (1992) "Doppler Echocardiography: The Quantitative Approach." Lea & Febiger, London.

Langendorff, O. (1895). Untersuchungen am uberlebenden saugethierherzen. Pflugers Arch. ges. Physiol. 61, 291–332.

Lareau, S., Moore, R., Mainwood, G.W., Labow, R.S., Keon, W.J. and Deslauriers, R. (1991) An NMR probe to study function and metabolism in isolated human cardiac tissue. Mag. Reson. Med. 20, 312–318.

Lebowitz, E., Greene, M.W., Fairchild, R., Bradley-Moore, P.R., Atkins, H.L., Ansari, A.N., Richards, P. and Belgrave, E. (1975). Thallium-201 for medical use. Int. J. Nucl. Med. 16, 151–155.

Lee, S. (1990). Historical significance in rat organ transplantation. Microsurgery 11, 115–121.

Lee, S., Macedo, A.R., Curtis, G.P., Lee, D. and Orloff, M.J. (1982). A simplified model for heterotopic rat heart transplantation. Transplantation 33, 438–442.

Levi, R. and Mullane, K.M. (1990). In "Methods in Enzymology" (ed J.N. Abelson and M.I. Simon), pp. 610–620. Academic Press, New York.

Levy, J.V. (1971). In "Methods in Pharmacology," Vol. 1 (ed A. Schwartz), pp. 77–104. Appleton-Century Crofts, New York.

Lichtig, C., Glagov, S., Feldman, S. and Wissler, R.W. (1973). Myocardial ischaemia and coronary artery atherosclerosis: a comprehensive approach to postmortem studies. Med. Clin. North Am. 57, 79–91.

Locke, F.S. (1901). Die wirkung der metalle des blutplasmas und verschiedener zucker auf die islierte saugethierherzen. Z. Physiol. 14, 670–672.

Longmore, D.B. (1989). The principles of magnetic resonance. Br. Med. Bull. 45, 848–880.

Lund-Johansen, P. (1990). The dye dilution method for measurement of cardiac output. Eur. Heart J. 11 (Suppl. I), 6–12.

MacConaill, M. (1985). Calcium precipitation from mammalian physiological salines (Ringer solutions) and the preparation of high [Ca] media. J. Pharmacol. Meth. 14, 147–155.

Manning, A.S. and Hearse, D. (1984). Reperfusion arrhythmias: mechanisms and protection. J. Mol. Cell. Cardiol. 17, 497–518.

Marcus, M.L. Go, R.T. and Ehrhardt, J.C. (1982). In "Myocardial Infarction: Management and Intervention" (ed G.S. Wagner), pp. 325–346. Martinus Nijhoff Publishers, Boston.

Marcus, M.L., Tomanek, R.J., Ehrhardt, J.C., Kerber, R.E., Brown, D.D. and Abboud, F.M. (1976). Relationships between myocardial perfusion, myocardial necrosis, and technecium-99m pyrophosphate uptake in dogs subjected to sudden coronary occlusion. Circulation 54, 647–653.

Mark, D.B., Adams, D. and Kisslo, J. (1986). In "Basic Doppler Echocardiography" (ed J. Kisslo, D. Adams and D.B. Mark) pp. 7–23. Churchill Livingstone, New York.

Maxwell, M.P., Hearse, D.J. and Yellon, D.M. (1984). The coronary collateral circulation during regional myocardial ischaemia: a comparative study in eight species. J. Mol. Cell. Cardiol. 169 (Suppl. 2), 47.

Mehra, R., Zeiler, R.H., Gough, W. B. and El-Sherif, N. (1983). Re-entrant ventricular arrhythmias in the late myocardial infarction period. 9. Electrophysiologic-anatomic correlation of re-entrant circuits. Circulation 67, 11–24.

Meier, G.D., Ziskin, M.C., Santamore, W.B. and Bove, A.A. (1980). Kinematics of the beating heart. I.E.E.E. Trans. Biomed. Eng. 27, 319–329.

Mendler, N., Hagl, S., Sebening, F. and Theobald, K.P. (1972). Metabolite des energiestoffwechsels im parabiotisch perfundierten rattenherzen wahrend und nach kardioplegie durck ischamie, kaliumchlorid und kalium-magnusium-aspartat. Arzneimittelforschung 22, 909–914.

Miles, M.A. (1979). In "Biology of the Kinetoplastida" (ed W.H.R. Lumsden and D.A. Evans), pp. 117–196. Academic Press, London.

Miles, M.A. (1981). In "Perspectives in Trypanosomiasis Research" (ed J.R. Baker) pp. 1–11. John Wiley & Sons Ltd, New York.

Mitchell, J.H., Wildenthal, K. and Mullins, C.E. (1969). Geometrical studies of the left ventricle utilizing biplane cineflourography. Fed. Proc. 28, 1334–1343.

Mogelvang, J., Stubgaard, M., Thompsen, C. and Henrickson, O. (1982). Evaluation of right ventricular volumes measured by magnetic resonance imaging. Eur. Heart J. 9, 529–533.

Moses, R.L. and Claycomb, W.C. (1982). Disorganization and reestablishment of cardiac muscle cell ultrastructure in cultured adult rat ventricular muscle cells. J. Ultrastruct. Res. 81, 358–374.

Muller, G.H. (1990). Heart transplantation model as an immunological monitor. Microsurgery 11, 122–126.

Musters, R.J., Post, J.A. and Verkleij, A.J. (1991). The isolated neonatal rat-cardiomyocyte used in an in vitro model for "ischaemia". I. A morphological study. Biochim. Biophys. Acta. 1091, 270–277.

Nadeau, R.A. and Amir-Jahed, A.K. (1965). Selective perfusion of the AV node of the dog by cannulation of the posterior septal artery. Rev. Can. Biol. 24, 291–297.

Nag, A.C. and Zak, R. (1979). Dissociation of adult mammalian heart into single cell suspension: an ultrastructural study. J. Anat. 129, 541–559.

Neely, J.R., Liebermeister, H., Battersby, E.J. and Morgan, H.E. (1967). Effect of pressure development on oxygen consumption by isolated rat heart. Am. J. Physiol. 212, 804–814.

Novitzky, D., Boniaszczuk, J., Cooper, D.K., Isaacs, S. *et al.* (1984). Prediction of acute cardiac rejection using radionuclide techniques. S. Afr. Med. J. 65, 5–7.

Nunnally, R.L. (1986). In "NMR in Medicine: The Instrumentation and Clinical Applications" (eds S.R. Thomas and R.L. Dixon), pp. 249–268. American Institute of Physics Publishing Co., New York.

Okada, R.D., Boucher, C.A. and Pohost, G.M. (1982). In "Myocardial Infarction: Measurement and Infarction" (ed G.S. Wagner), pp. 295–324. Martinus Nijhoff Publishers, Boston.

Omvik, P. (1990). Magnetic resonance imaging and cine computerized tomography as future tools for cardiac output measurement. Eur. Heart J. 11 (Suppl. I), 141–143.

Ono, K. and Lindsay, E.S. (1969). Improved technique of heart transplantation in rats. J. Thorac. Cardiovasc. Surg. 57, 225–229.

Parker, J.O., Ledwich, J.R., West, R.O. and Case, R.B. (1969). Reversible cardiac failure during angina pectoris: haemodynamic effects of atrial pacing in coronary artery disease. Circulation 39, 745–757.

Parmley, W.W. and Sonnenblick, E.H. (1971). In "Methods in Pharmacology," Vol. 1 (ed A. Schwartz), pp. 105–123. Appleton-Century Crofts, New York.

Patterson, E. and Lucchesi, B.R. (1981). Chronic ventricular tachyarrhythmias in the conscious dog: prevention by UM-272 (dimethylpropranolol). J. Cardiovasc. Pharmacol. 3, 769–780.

Patterson, E., Eller, B.T. and Lucchesi, B.R. (1983). Effects of diltiazem upon experimental ventricular dysrhythmias. J. Pharmacol. Exp. Ther. 225, 224–233.

Pelikan, P.C.D., Weisfeldt, M.L., Jacobus, W.E., Miceli, M.V., Bulkley, B.H. and Gerstenblith, G. (1986). Acute doxorubicin cardiotoxicity: Functional metabolic and morphological alterations in the isolated perfused rat heart. J. Cardiovasc. Pharmacol. 8, 1058–1066.

Pinson, A., Padieu, P. and Harary, I. (1987). In "The Heart Cell in Culture", (ed A. Pinson), pp. 8–22. CRC Press, Boca Raton, FL.

Piper, H.M., Probst, I., Schwartz, P., Hutter, F.J. and Spieckermann, P.G. (1982). Culturing of calcium stable adult cardiac myocytes. J. Mol. Cell. Cardiol. 14, 397–412.

Pitt, A., Friesinger, G.C. and Ross, R.S. (1969). Measurement of blood flow in the right and left coronary artery beds in humans and dogs using the ^{133}Xenon technique. Cardiovasc. Res. 3, 100–106.

Pohost, G.M. and Canby, R.C. (1987). Nuclear magnetic resonance imaging: current and future prospects. Circulation 75, 88–95.

Powell, T. (1988). In "Biology of Isolated Cardiac Myocytes" (ed W.A. Clark, R.S. Decker and T.K. Borg), pp. 9–13. Elsevier, Amsterdam.

Powell, T., Sturridge, M.S., Suvarna, S.K., Terrar, D.A. and Twist, V.W. (1981). Intact individual heart cells isolated from human ventricular tissue. Br. Med. J. 283, 1013–1015.

Purchase, M., Dusting, G.J., Li, D.M.F. and Read, M.A. (1986). Physiological concentrations of epinephrine potentiate thromboxane A2 release from platelets in the isolated rat heart. Circ. Res. 58, 172–176.

Rackley, C.E. and Russell, R.O. (1972). Left ventricular function in acute myocardial infarction and its clinical significance. Circulation 45, 231–244.

Radda, G.K. and Seeley, P.J. (1979). Recent studies on cellular metabolism by nuclear magnetic resonance. Ann. Rev. Physiol. 41, 749–769.

Rao, V.K. and Lisitza, M. (1985). Accessory heart transplantation to groin in the rat: a new model for retransplantation experiments. Transplantation 40, 567–569.

Reybrouck, T. and Fagard, R. (1990). Assessment of cardiac output at rest and during exercise by a carbon dioxide rebreathing method. Eur. Heart J. 11 (Suppl. I), 21–32.

Ringer, S. (1883). A further contribution regarding the influence of the different constituents of the blood on the contractions of the heart. J. Physiol. (Lond.) 4, 29–42.

Ritchie, D.M., Kellelher, G.J., Macmillan, A., Fasolak, W., Roberts, J. and Mansuthani, S. (1979). The cat as a model for myocardial infarction. Cardiovasc. Res. 13, 199–206.

Ritchie, J.L. (1978). In "Thallium-201 Myocardial Imaging" (ed J.L. Ritchie, G.W. Hamilton and F.J. Th. Wackers), pp. 1–8. Raven Press, New York.

Roberts, R. (1982). In "Myocardial Infarction: Measurement and Intervention" (ed G.S. Wagner), pp. 107–142. Martinus Nijhoff Publishers, Boston.

Roberts, R., Gowda, K.S., Ludbrook, P.A. and Sobel, B.E. (1975). Specificity of elevated serum MB creatine phosphokinase activity in the diagnosis of acute myocardial infarction. Am. J. Cardiol. 36, 433–437.

Rogers, W.J., McDaniel, H.G., Mantle, J.A. and Smith, J. (1982). In "Myocardial Infarction: Measurement and Intervention" (ed G.S. Wagner), pp. 159–172. Martinus Nijhoff Publishers, Boston.

Ross, R.S., Ueda, K., Lichtlen, P.R. and Rees, J.R. (1964). Measurement of myocardial blood flow in animals and man by selective injection of radioactive inert gas into the coronary arteries. Circ. Res. 15, 28–41.

Ross, W.N., Salzberg, B.M., Cohen, L.B. and Davila, H.V. (1974). A large change in dye absorption during the action potential. Biophys. J. 14, 983–986.

Ross, W.N., Salzberg, B.M., Cohen, L.B., Grinvald, A., Davila, H.V., Waggoner, A.S. and Wang, C.H. (1977). Changes in absorption, fluorescence, dichroism, and birefringence in stained giant axons: optical measurement of membrane potential. J. Membrane Biol. 33, 141–183.

Russell, S.J.M. and Bell, E.J. (1970). Echoviruses and carditis. Lancet 1, 784–785.

Sackner, M.A., Hoffman, R.A., Stroh, D. and Krieger, B.P. (1991a). Thoracocardiography part 1: noninvasive measurement of changes in stroke volume; comparisons to thermodilution. Chest 99, 613–622.

Sackner, M.A., Hoffman, R.A., Krieger, B.P., Stroh, D. and Sackner, J.D. (1991b). Thoracocardiography part 2: noninvasive measurement of changes in stroke volume; comparisons to impedence cardiograph. Chest 99, 896–903.

Salama, G. and Morad, M. (1976). Merocyanine 540 as an optical probe of transmembrane electrical activity in the heart. Sciences 191, 485–487.

Salari, H. and Walker, M.J.A. (1989). Cardiac dysfunction caused by factors released from endotoxin-activated macrophages. Circ. Shock. 27, 263–272.

Samson, W.E. and Scher, A.M. (1960). Mechanism of S-T segment alteration during acute myocardial injury. Circ. Res. 8, 780–787.

Sandler, H., Meier, G.D. and Alderman, E.L. (1982). In "Ventricular Wall Motion" (ed V. Sigwart and P.H. Heintzen), pp. 1–13. Thieme-Stratton, New York.

Sandritter, W. and Jestadt, R. (1957). Triphenylterazoliumchlorid (TCC) als reduction-sindikator zur makrokopischen diagnose des frischen herzinfarktes. Z. Allg. Pathol. 97, 188–189.

Satoh, K., Narimatsu, A. and Taira, N. (1981). Effects of antiarrhythmic drugs on AV nodal and intraventricular conduction as assessed in the isolated, blood-perfused AV node preparation of the dog. J. Cardiovasc. Pharmacol. 3, 753–768.

Satoh, K., Yanagisawa, T. and Taira, N. (1979). Effects on atrioventricular conduction and blood flow of enantiomers of verapamil and of tetrodotoxin injected into the posterior and the anterior septal artery of the atrioventricular node preparation of the dog. Naunyn Schmied. Arch. Pharmacol. 308, 89–98.

Schaper, W. (1971). In "The Collateral Circulation of the Heart" (ed D.A.K. Black) pp. 29–50. North Holland Co., Amsterdam.

Schaper, W. (1979). In "Pathophysiology of Myocardial Perfusion" (ed W. Schaper), pp. 415–470. Elsevier Science Publishing Co., Amsterdam.

Schaper, W., Schaper, J. and Winkler, B. (1986). The collateral circulation of the heart. J. Mol. Cell. Cardiol. 18(Suppl. 1), 60.

Schelbert, H.R. and Mody, F.V. (1991). In "Principles and Practise of Cardiovascular Imaging" (ed G.M. Pohost and R.A. O'Rourke), pp. 29–50, Little, Brown & Co., London.

Scherlag, B.J., Helfant, R.H., Haft, J.I. and Damato, A.N. (1970). Electrophysiology underlying ventricular arrhythmias due to coronary ligation. Am. J. Physiol. 219, 1665–1671.

Scheutz, A., Kemkes, B.M., Breuer, M., Brandl, U., Englehardt, M., Kugler, C., Manz, C., Hatz, R., Gokel, J.M. and Hanner, C. (1992). Kinetics and dynamics of acute rejection after heterotopic heart transplantation. J. Heart Lung Transplant. 11, 289–299.

Schwarzfeld, T.A. and Jacobson, S.L. (1981). Isolation and development in cell culture of myocardial cells of the adult rat. J. Mol. Cell. Cardiol. 13, 563–575.

Segel, L.D. and Rendig, S.V. (1982). Isolated working rat heart perfusion with perfluorochemical emulsion Fluosol-43. Am. J. Physiol. 242, H485–H490.

Selvester, R.H. (1975). Criteria for atrial enlargement and infarct size (applicable to computer diagnostic programs). Bulletin of the US Department of Health, Education and Welfare, p. 81.

Selvester, R.H., Sanmarco, M.E., Solomon, J.C. and Wagner, G.S. (1982). In "Myocardial Infarction: Measurement and Intervention" (ed G.S. Wagner), pp. 23–50. Martinus Nijhoff Publishers, Boston.

Selye, H., Bajusz, E., Grasso, S. and Mendell, P. (1960). Simple techniques for the surgical occlusion of coronary vessels in the rat. Angiology 11, 398–407.

Starling, E.H. (1918). "The Linacre Lecture on the Law of the Heart." Longmans, Green & Co., London.

Stollerman, G.H. (1975). "Rheumatic Fever and Streptococcal Infection". Grune & Stratton Inc., New York.

Sutherland, E.W. and Rall, T.W. (1960). The relation of adenosine-3',5'-phosphate and phosphorylase to the actions of catecholamines and other hormones. Pharmacol. Rev. 12, 265–299.

Swan, H.J.C., Ganz, W., Forrester, J., Marais, H., Diamond, G. and Chonette, D. (1970). Catheterization of the heart in man with use of a flow-directed balloon-tipped catheter. New Engl. J. Med. 283, 447–451.

Swan, H.J.C., Forrester, J.S., Diamond, G., Chatterjee, K. and Parmley, W.W. (1972). Haemodynamic spectrum of myocardial infarction and cardiogenic shock: a conceptual model. Circulation 45, 1097–1110.

Szekeres, L. (1971). In "Methods in Pharmacology" (ed A. Schwartz), pp. 151–190. Appleton-Century Crofts, New York.

Szekeres, L. (1979). IV. Experimental models for the study of antiarrhythmic agents. Prog. Pharmacol. 2, 25–31.

Takayama, K., Takahashi, I., Kawano, T., Hara, M., Teruya, H., Nakayama, N., Nakatsuka, T., Furuhata, H. and Yoshimura, S. (1984). Continuous measurement of cardiac output during ergometer exercise using an ultrasonic pulsed Doppler flowmeter. J. Cardiogr. 14, 833–840.

Tarr, M., Trank, J.W., Leiffer, P. and Shepherd, N. (1979). Sarcomere length-resting tension relation in single frog atrial cardiac cells. Circ. Res. 45, 554–559.

Teague, S.M. (1986). In "Basic Doppler Echocardiography" (ed J. Kisslo, D. Adams and D.B. Mark), pp. 147–157. Churchill Livingstone, New York.

Tennant, R. and Wiggers, C.J. (1935). The effect of coronary occlusion on myocardial contractility. Am. J. Physiol. 112, 351–361.

Terasaki, F., James, T.N., Nakayama, Y., Deguchi, H., Kitaura, Y. and Kawamura, K. (1992). Ultrastructural alterations of the conduction system in mice exhibiting sinus arrest or heart block during coxsackie B3 acute myocarditis. Am. Heart J. 123, 439–452.

Terry, H.J. (1980). In "Methods in Angiology" (ed M. Verstraete), pp. 21–37. Martinus Nijhoff Publishers, Boston.

Thamlikitkul, V., Kobwanthanakun, S., Pruksachatvuthi, S. and Lertluknithi, R. (1992). Pharmacokinetics of rheumatic fever prophylaxis regimens. J. Int. Med. Res. 20, 20–26.

Theroux, P.D., Franklin, D., Ross, J. and Kemper, W.S. (1974). Regional myocardial function during acute coronary occlusion and its modification by pharmacologic agents in the dog. Circ. Res. 35, 896–908.

Tillmans, H., Ikeda, S. and Bing, R.J. (1975). In "Handbook of Experimental Pharmacology" (ed J. Schmier and O. Eichler), pp. 117–130. Springer Verlag, Berlin.

Timmermann, W. (1985). In "Microsurgical Models in Rats for Transplantation Research" (ed A. Thiede, E. Deltz, R. Engemann and H. Hamelmann), pp. 31–35. Springer Verlag, Berlin.

Tsujioka, K., Ogasawara, Y. and Kajiya, F. (1986). In "New Approaches in Cardiac Mechanics" (ed K. Kitamura, H. Abe and K. Sagawa), pp. 21–31. Japan Scientific Societies Press, Tokyo.

Tuttle, R.R., Hillman, C.C. and Toomey, R.E. (1976). Differential β adrenergic sensitivity of atrial and ventricular tissue-assessment by chronotropic, inotropic, and cyclic AMP responses to isoprenaline and dobutamine. Cardiovasc. Res. 10, 452–458.

Ty Smith, N. (1974). In "Noninvasive Cardiology" (ed A.M. Weissler), pp. 39–149. Grune & Stratton, New York.

Ueda, H., Kaihara, S., Ueda, K., Shugishita, Y., Sasaki, Y. and Iio, M. (1965). Regional myocardial blood flow measured by I-131 labelled microaaggregated albumin (I-131 AA). Jpn. Heart J. 6, 534–542.

Ueda, H., Cheung, Y.C., Masetti, P., Diaco, M., Martin, M., Kupiec-Weglinski, J.W. and Tilney, N.L. (1991). Synergy between cyclosporine and anti-IL-2 receptor monoclonal antibodies in rats. Functional studies of heart and kidney allografts. Transplantation 52, 437–442.

United Kingdom and United States Joint Report (1955). The treatment of acute rheumatic fever in children: cooperative clinical trial of ACTH, cortisone, and aspirin. Circulation 11, 343–371.

Valtonen, V.V. (1991). Infection as a risk factor for infarction and atherosclerosis. Ann. Med. 23, 539–543.

Vandenberg, C.A. and Horn, R. (1984). Inactivation viewed through single sodium channels. J. Gen. Physiol. 84, 535–564.

Vatner, S.F., Young, M. and Patrick, T.A. (1985). In "Cardiovascular Ultrasonic Flowmetry" (ed S.A. Altobelli, W.F. Voyles and E.R. Greene), pp. 81–99. Elsevier Scientific Publishing Co., New York.

Vliers, A.C.A.P., Visser, K.R. and Zulstra, W.G. (1973). Analysis of indicator distribution in the determination of cardiac output by thermal dilution. Cardiovasc. Res. 7, 125–132.

Wackers, F.J. Tn., Berger, H.J. and Zaret, B.L. (1982). In "Myocardial Infarction: Measurement and Intervention" (ed G.S. Wagner), pp. 199–233. Martinus Nijhoff Publishers, Boston.

Waggoner, A.S., Wang, C.H. and Tolles, R.L. (1977). Mechanism of potential-dependent light absorption changes of lipid bilayer membranes in the presence of cyanine and oxonol dyes. J. Membrane Biol. 33, 109–140.

Wagner, G.S. (1982). "Myocardial Infarction: Measurement and Intervention." Martinus Nijhoff Publishers, Boston.

Walker, M.J.A., Curtis, M.J., Hearse, D.J., Campbell, R.W.F., Janse, M.J., Yellon, D.M., Cobbe, S.M., Coker, S.J., Harness, J.B., Harron, D.W.G., Higgins, A.J., Julian, D.G., Lab, M.J., Manning, A.S., Northover, B.J., Parratt, J.R., Riemersma, R.A., Riva, E., Russel, D.C., Sheridan, D.J., Winslow, E. and Woodward, B. (1988). The Lambeth conventions: guidelines for the study of arrhythmias in ischaemia, infarction and reperfusion. Cardiovasc. Res. 22, 447–455.

Walker, M.J.A., MacLeod, B.A. and Curtis, M.J. (1991). Animal models of human disease: myocardial ischaemia and infarction in rats. Comp. Pathol. Bull. 23, 3–4.

Wallerson, D.C., Ganau, A., Roman, M.J. and Devereux, R.B. (1990). Measurement of cardiac output by M-mode and two-dimensional echocardiography: application to patients with hypertension. Eur. Heart J. 11(Suppl I), 67–78.

Weiss, E.S., Ahmed, S.A., Welch, M.J. *et al.* (1977). Quantification of infarct in cross sections of canine myocardium in vivo with position emission transaxial tomography and ^{11}C-Palmitate. Circulation 55, 66–73.

White, C.W., Wright, C.B., Doty, D.B., Hiratza, L.F., Eastham, C.L., Harrison, D.G. and Marcus, M.L. (1984). Does visual interpretation of the coronary arteriogram predict the physiologic importance of a coronary stenosis? New Engl. J. Med. 310, 819–824.

Wijngaard, P.L.J., Gimpel, J.A., Schuurman, H.-J., van der Meulen, A., Gmelig-Meyling, F.H. and Jambroes, G. (1990). Monitoring rejection after heart transplantation: cyto-immunological monitoring on blood cells and quantitative birefringence measurements on endomyocardial biopsy specimens. J. Clin. Pathol. 43, 137–142.

Williams, K.A., Garvin, A.A. and Taillon, L.A. (1992). Clinical nuclear imaging techniques for the diagnosis and evaluation of acute myocardial infarction. Compr. Ther. 18, 6–10.

Williams, R.C. (1981). In "Immunology of Cardiovascular Disease" (ed M.J. Lessof), pp. 105–115. Marcel Dekker, Inc., New York.

Williams, T.J. and Rampart, M. (1987). In "Immunology & Molecular Biology of Cardiovascular Diseases" (ed C.J.F. Spry), pp. 119–128. MTP Press Ltd, Boston.

Wilson, R.F. (1991) Assessment of the human coronary circulation using a Doppler catheter. Am. J. Cardiol. 67, 44D–56D.

Winbury, M.M. (1975). In "Handbook of Experimental Pharmacology" (ed J. Schmier and O. Eichler) pp. 1–69. Springer Verlag, Berlin.

Winkler, B., Sass, S., Binz, K. and Schaper, W. (1984). Myocardial blood flow and myocardial infarction in rats, guinea pigs, and rabbits. J. Mol. Cell. Cardiol. 16(Suppl 2), 48.

Winslow, E. (1984). Methods in the detection and assessment of antiarrhythmic activity. Pharmacol. Therap. 24, 401–433.

Wit, A.L. and Cranefield, P.F. (1974). Effect of verapamil on the sinoatrial and atrioventricular nodes of the rabbit and the mechanism by which it arrests reentrant atrioventricular nodal tachycardia. Circ. Res. 35, 413–425.

Wolf, A.P. and Jones, W.B. (1983). Cyclotrons in biomedical radioisotope production. Radiochim. Acta 34, 1–7.

Woodruff, J.F. (1980). Viral myocarditis: A review. Am. J. Pathol. 101, 425–484.

Wotherspoon, J.S. and Dorsch, S.F. (1986). The mechanism of prolonged graft survival following removal of the regional lymph node. Transplantation 42, 532–537.

Wyatt, D.G. (1960). In "Flow Properties of Blood and Other Biological Systems" (ed A.L. Copley and G. Stainsby), pp. 390–392. Academic Press, London.

Zeitler, E. and Gross-Vorholt R. (1980). In "Methods in Angiology" (ed M. Verstraete), pp. 302–333. Martinus Nijhoff Publishing, Boston.

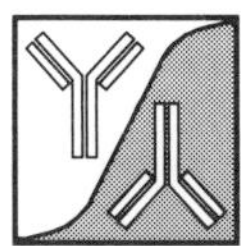

3. Immunopharmacology of the Coronary Vascular Endothelium

Allan M. Lefer

1. Normal Function and Pathophysiological Alteration of the Coronary Endothelium

The coronary circulation comprises a vital segment of the circulatory system since it is the only vascular system supplying the heart. The regulation of coronary blood flow is exceedingly important since the heart extracts virtually all of the oxygen delivered to it in the coronary arterial blood. Thus, the only way that significant quantities of additional oxygen can be delivered to the heart to meet its increased metabolic demands is by increasing coronary blood flow. This is accomplished by increasing coronary arterial vessel calibre (i.e by relaxing the vascular smooth muscle in the coronary arterial system). This is thought to occur largely by metabolic mechanisms.

Since 1980, the endothelium has been recognized as playing a vital role in the modulation vascular tone (Furchgott and Zawadski, 1980). This is achieved by metabolically producing and releasing EDRF which has been identified as NO (Palmer *et al.*, 1987; Moncada *et al.*, 1989). In addition to regulating vascular tone, NO also inhibits platelet aggregation (Furlong *et al.*, 1987) and inhibits neutrophil adhesion (McCall *et al.*, 1988; Lefer A. M. *et al.*, 1991). Thus, NO is an important endogenous anti-thrombotic and anti-inflammatory substance.

Recent investigations have also discovered a variety of adhesive molecules expressed on the surface of the vascular endothelium (Tonnesan 1989; Albelda, 1991; Berman and Calderon, 1992; Harlan *et al.*, 1992; Zimmerman *et al.*, 1992). These include a variety of polypeptides and glycoproteins which collectively play an important role in the regulation of adherence of leukocytes to the endothelium and their subsequent transmigration through the endothelial layer. These processes are important in a variety of inflammatory related circulatory disease processes including: (a) reperfusion injury following ischaemia; (b) angioplasty; and (c) atherogenesis ultimately leading to atherosclerosis.

The cardiac endothelium is a very metabolically active tissue consisting of a single cell layer lining the endocardium and the coronary vasculature. In common with most endothelial linings, the coronary endothelium

Immunopharmacology of the Heart
ISBN 0–12–200245–8

produces a variety of important substances ranging from very small molecules (like NO) having a molecular weight of 30 Da, to intermediate size molecules (including nucleosides and lipids) having molecular weights of 200–600 Da, to peptides having molecular weights of 20–40 kDa and very large surface proteins having molecular weights of 90–150 kDa. An exciting new chapter has been opened recently in the development of specific antibodies (e.g. mABs) against several of these adhesive molecules. These mABs and their fragments (F′ABs) have been very valuable research tools enabling investigation of the physiological and pathophysiological role of these adhesion molecules, as well as providing potential therapeutic application of these monoclonal antibodies in several disease states (Harlan *et al.*, 1992).

The major purpose of this chapter is to discuss the role of adherence molecules and other endothelial-derived molecules in regulating the normal physiological properties of the coronary endothelium (e.g. prevention of leukocyte adherence, inhibition of platelet aggregation, and normal modulation of vasorelaxation). Particular attention will be focused on the use of relevant antibodies or other immunomodulators which block or down-regulate adhesive molecules enabling one to elucidate their function in circulatory disease states.

2.   Endothelially Derived Adhesion and Anti-adhesion Molecules

2.1   Adhesion Molecules

There are a multiplicity of adhesion molecules, their counter receptors and their respective ligands which cooperate in the promotion of adherence of formed elements (e.g. platelets and leukocytes) in the blood to the vascular endothelium. There are a variety of adherence glycoproteins present on the neutrophil membrane (Springer, 1990; Kishimoto *et al.*, 1991; McEver, 1991a; Ruoslahti, 1991). Many of these heterodimeric proteins belong to the integrin family of adhesion molecules (e.g. the β_2 integrins). Each of these β_2 integrins contain a different alpha chain (e.g. CD11a also known as LFA-1, CD11b also known as Mac-1 or Mo-1, and CD11c also known as gp 150.95). Each of these α-chains is coupled to the same β-chain known as CD18 (Springer, 1990). The heterodimer CD11a/CD18 is present on all leuko-cytes, and is involved in leukocyte-endothelial cell adhesion. The primary endothelial cell counter receptor for CD11a/CD18 is ICAM-1 (Rothlein *et al.*, 1986). CD11b/CD18 is present largely on neutrophils and monocytes and is involved in leukocyte-endothelial adherence and phagocytosis. Fibrinogen Factor X and ICAM-1 are thought to be the endothelial counter receptor for CD11b/CD18. CD11c/CD18 is present on all granulocytes and monocytes and is involved in complement binding. Its major ligand is 2C3b (Makgoba *et al.*, 1989). The other leukocyte adherence molecule of significance is L-selectin present on lymphocyte and neutrophil membranes. These are the major adhesion molecules of leukocyte origin.

There is a great diversity of endothelial cell adhesion molecules. The major ones are listed in Table 3.1. These adhesion molecules belong primarily to the selectin (Springer and Lasky, 1991; Winkelhake, 1991) and the IgG superfamilies (Simmons *et al.*, 1988; Springer 1990). All are large molecules ranging from 95 to 140 kDa. The other endothelially derived adhesion molecule relevant to

Table 3.1   Characteristics of endothelial derived adhesion molecules

Adhesion molecule	Molecular weight (kDa)	Alternate names	Family	Time of maximal expression	Adherent cell type
E-Selectin	115	ELAM-1	Selectin (formerly LECAM)	4 h	PMNs, platelets, monocytes
P-Selectin	140	GMP-140 PADGEM, CD62	Selectin	20 min	PMNs, platelets, monocytes
ICAM-1	95	CD54	IgG superfamily	Constitutive and inducible after 12 h	PMNs, lymphocytes, monocytes
ICAM-2	—	—	IgG superfamily	Constitutive	PMNs, lymphocytes, monocytes
VCAM-1	110	INCAM-110 Athero-ELAM	IgG superfamily	8 h	Platelets, monocytes, tumor cells
PECAM	130	CD31	IgG superfamily	Constitutive	Platelets, leukocytes
ATHERO-ELAM	98–118	—	?	>18 h	Monocytes
PAF	0.523	AGEPC, PAF acether	Phospholipid	Seconds	PMNs, platelets

this discussion is PAF which is a low molecular weight phospholipid (Lorant *et al.*, 1991). The time of maximal expression of these adherence molecules varies over a wide range of time from constitutive molecules (e.g. ICAM-2, PECAM and ICAM-1, low levels) to adhesion molecules induced over seconds (PAF), minutes (e.g. P-selectin) and hours (e.g. E-selectin and VCAM-1; Phillips *et al.*, 1990; McEver 1991b; Taichman *et al.*, 1991). Thus, the adherence molecules present on endothelial cells function cooperatively in promoting adherence of cells to the endothelial lining of blood vessels. If an infection is present, leukocytes migrate to that area via a chemotactic gradient, where they adhere to the endothelium, migrate across the endothelium to the site of the infection, and become activated whereupon they release a variety of cytotoxic mediators (Entman *et al.*, 1991; Pober and Cotran, 1991). In the case of a localized site of infection, the cytotoxic mediators kill the bacteria present at the infected site. However, in myocardial ischaemia–reperfusion, neutrophils release their mediators inappropriately thereby injuring cardiac myocytes (Mullane and Smith, 1990; Entman *et al.*, 1991). These neutrophil-derived mediators include oxygen-derived free radicals (e.g. superoxide radicals, hydrogen peroxide), cytokines (e.g. TNFα, IL-1β), proteases (e.g. elastase), and lipids (e.g. platelet-activating factor). These mediators can be delivered at close range when the neutrophils are very near the target site or when they adhere to the target site. In this regard, Entman and colleagues (Entman *et al.*, 1990; Smith *et al.*, 1991a) have shown that neutrophils adhere to cardiac myocytes after the PMNs undergo transendothelial migration (see also Chapter 5) This myocyte adherence is ICAM-1-dependent, and allows for neutrophils to bind to cardiac myocyte membranes when they can release cytotoxic mediators which can injure and eventually kill the myocytes (Dreyer *et al.*, 1991; Entman *et al.*, 1991).

The adhesion molecules appear to act in concert in attracting leukocytes to the reperfused coronary endothelium and in promoting adherence, transendothelial migration and activation of the leukocytes (Smith, 1992). Thus, L-selectin (also known as LECAM-1, LAM-1, MEL-14 and the homing receptor) which is constitutively present on leukocyte membranes is important in both lymphocyte and neutrophil trafficking. L-Selectin is important in the earliest contact between the leukocyte and the endothelium, since it is apparently responsible for the "rolling" phenomenon (Kishimoto *et al.*, 1989; Smith *et al.*, 1991b). Rolling slows down the progress of the neutrophil, and allows other adhesive mechanisms to operate. In this manner, the slowly rolling neutrophil can adhere to the endothelium via the CD11/CD18 adherence glycoprotein complex which allows for a more permanent adherence. As the CD11/CD18 complex becomes activated, L-selectin is shed from the neutrophil membrane (Kishimoto *et al.*, 1989; Smith *et al.*, 1991b).

In addition to temporal events on the neutrophil,

endothelial adherence molecules also show staging of the adherence phenomenon. Thus ICAM-2 is constitutively present at high levels of activity, whereas ICAM-1 is constitutively present at low levels of activity (Berman and Calderon, 1992) but becomes upregulated slowly over a 12 h period. However, PAF can be released in seconds and interacts with P-selectin which is upregulated to the endothelial cell surface in 10–20 min (Geng *et al.*, 1990; McEver, 1991c) from Weibel–Palade bodies inside endothelial cells. This interaction between PAF and P-selectin (Lorant *et al.*, 1991; Zimmerman *et al.*, 1992) acts to tether neutrophils to endothelial cell membranes once endothelial cells are activated by either histamine or thrombin (McEver, 1991c; Zimmerman *et al.*, 1992). This process can be further amplified by activation of the CD11/CD18 complex on the neutrophil. Much later (i.e. after about 4 h), E-selectin becomes activated by cytokines (e.g. TNFα, IL-1β), and reinforces the adherence of neutrophils to endothelial cell surfaces. Later still (i.e 12–24 h), the endothelium-derived ICAM-1 becomes fully activated, also by cytokines (i.e, IL-1β, IL-6), and further reinforces the adherence process (Entman *et al.*, 1991).

Other adhesion molecules present on endothelial cells appear to be related to atherosclerosis (e.g. VCAM-1 which is induced over an 8 h or longer time course; Zimmerman *et al.*, 1992; Cybulsky and Gimbrone, 1991) and a newly discovered ATHERO-ELAM which requires at least 15 h to be induced (Sporn and Roberts, 1989), or to platelet adherence to other cell types (PECAM; Simmons *et al.*, 1990).

2.2 ANTI-ADHESION MOLECULES

The endothelium also produces a variety of endogenous anti-adherence molecules summarized in Table 3.2. In contrast to the large adhesive proteins and carbohydrate molecules, most of these antiadhesive substances are very small molecules ranging from 30 to 352 Da. TGFβ_1, at 25 kDa, is a substantial polypeptide, but is still much smaller than the large protein adhesive molecules (Gamble and Vadas, 1988; Sporn and Roberts, 1989; Lefer *et al.*, 1990). Most of the anti-adhesive molecules are largely non-protein in nature, NO being an inorganic gas, PGI$_2$ an eicosanoid, and adenosine, a nucleoside.

In contrast to the adherence proteins and glycoproteins which are largely inactive in influencing coronary vascular tone, the three small molecular anti-adhesive molecules (i.e. NO, PGI$_2$ and adenosine) are all powerful coronary vasodilators. Moreover, these three substances all exert anti-platelet (e.g. anti-thrombotic) and anti-neutrophil effects which may be of considerable significance in mediating their anti-adhesion properties (Higgs *et al.*, 1978; Palmer *et al.*, 1987; Engler and Gruber, 1991). Moreover, these three substances all exert their actions via a cycle nucleotide second messenger (e.g. cGMP in the case of NO, and CAMP in the cases of PGI$_2$ and

Table 3.2 Characteristics of endogenous endothelial anti-adhesion molecules

Anti-adhesion molecule	Molecular weight (Da)	Abbreviation	Class of substance	Second messenger	Other actions
Nitric oxide	30	NO	Gas	cGMP	Vasodilator, anti-platelet, anti-neutrophil, quenches superoxide
Prostacyclin	352	PGI_2	Eicosanoid	cAMP	Vasodilator, anti-platelet, anti-neutrophil, membrane stabilizer
Adenosine	267	Ado	Nucleoside	cAMP	Vasodilator, anti-platelet, anti-neutrophil, enhances PGI_2 formation
Transforming growth factor β_1	25 000	$TGF\beta_1$	Homodimeric peptide	—	Angiogenesis, regulation of myogenesis, opposes cytokines

adenosine). $TGF\beta_1$ is without these vasodilator or anti-platelet effects, but clearly exerts anti-PMN adherence in other ways, probably related to its anti-cytokine effect (Gamble and Vadas, 1988; Lefer *et al.*, 1990). This anti-adherence action is translated into a significant cardio-protection effect in ischaemia–reperfusion (Lefer *et al.*, 1990).

2.3 THE EDRF/NO SYSTEM

Although much is known about prostacyclin and adeno-sine formation, NO biosynthesis and action is relatively new, NO being discovered as the major EDRF in 1987 (by Palmer *et al*). It is now clear that NO is generated along with L-citrulline from the amino acid L-arginine and molecular oxygen, by an NO synthase (Hibbs *et al.*, 1990). There are at least two NO syntheses (Werner-Felmayer *et al.*, 1991). There is a constitutive NO synthase which is calcium and calmodulin dependent and is present in endothelial cells and some neurons. The other enzyme is an inducible NO synthase which is not calcium dependent but is present in macrophages and vascular smooth muscle cells, and which requires about 8 h for maximal activation. The NO produced by the endothelium is physiologically important for normal vascular function, whereas the induced NO can become excessive and can potentially lead to pathophysiological consequences if produced unchecked for long durations.

NO is produced basally by vascular endothelial cells in small quantities, but this NO release can be stimulated by endothelial-dependent vasodilators (e.g. ACh, ADP, the calcium ionophore A23187, bradykinin, etc.). Figure 3.1, illustrates how one can demonstrate basal release of NO by the vascular endothelium. The major key to determining basal NO release is to block it using a specific NO synthase competitive antagonist which replaces L-arginine in the endothelial cell but is incapable of releasing NO. Such arginine analogs are L-NMMA and

L-NAME (Moncada and Palmer, 1990). Thus, the increase in vascular tone exerted by L-NAME or L-NMMA is due to the turning off of basal NO release by unmasking basal vascular tone. This phenomen only occurs in the presence of a functionally intact endothelium. As can be seen, basal NO release is reduced following ischaemia and reperfusion of a vascular bed signifying the occurrence of endothelial dysfunction. Similarly, agonist-induced reduction in NO also parallels this decrease in basal NO release following ischaemia and reperfusion (Lefer *et al.*, 1991a). This marked decrease in NO release and its consequences is discussed below.

3. *The Coronary Endothelium in Disease Processes and Its Immunopharmacology*

The coronary endothelium plays a key role in regulating coronary blood flow via release of EDRF and adenosine (Engler and Gruber, 1991; Ueeda *et al.*, 1992), in promoting and maintaining an anti-thrombotic surface via EDRF, PGI_2 (Moncada and Vane, 1978; Moncada *et al.*, 1991), in quenching oxygen-derived free radicals via EDRF (Gryglewski *et al.*, 1986; Rubanyi and Vanhoutte, 1986), and in opposing the inflammatory promoting actions of granulocytes, particularly neutro-phils (McCall *et al.*, 1988; Kubes *et al.*, 1991; Lefer *et al.*, 1991a; Siegfried *et al.*, 1992a). This latter role is important as a "trigger mechanism" for reperfusion injury (Bulkley, 1989; Lefer *et al.*, 1991a). The normal anti-inflammatory actions of the endothelium are also important in interactions with monocytes and the endothelium, as an early step in the chronic process of atherogenesis of coronary arteries, as well as in the acute processes involved in intimal hyperplasia following

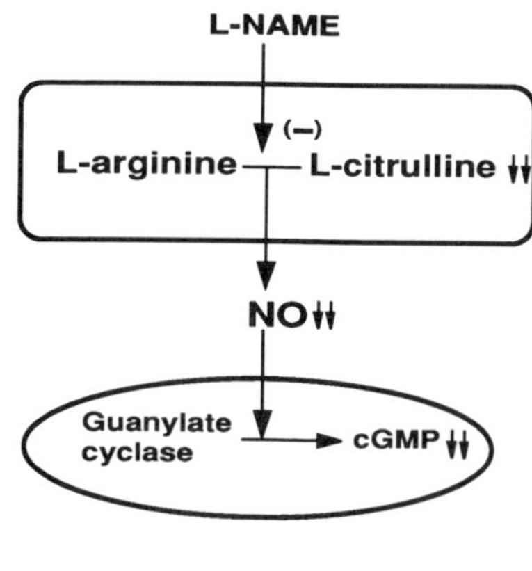

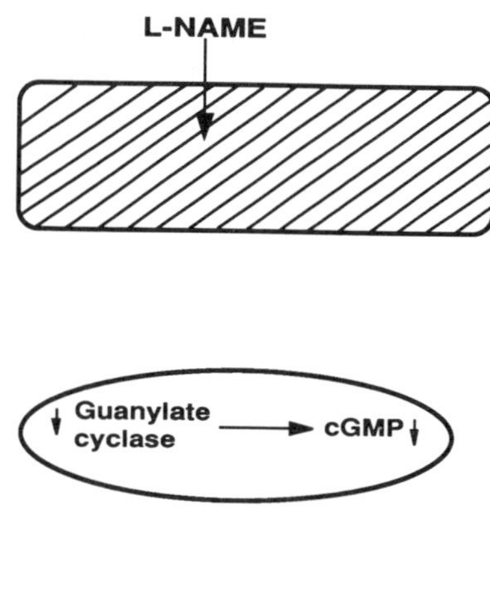

Figure 3.1 Diagram of coronary vascular endothelium showing vascular smooth muscle cells and release of NO. In Panel A, there is normal basal release of NO. The arginine analogue L-NAME is used to block the basal synthesis and release of NO. This results in a significant contraction due to the loss of the vasodilation induced by basal NO release. In Panel B, following ischaemia and 4.5 h of reperfusion, there is almost no basal release of NO, and thus L-NAME fails to produce any significant contraction.

catheter-induced coronary injury (i.e. restenosis following balloon angioplasty).

The early pathophysiology of all of these important clinical entities involving coronary artery injury, centres around the endothelium. Two critical areas of the coronary endothelium that play pivotal roles in these pathophysiological processes are: (a) the adhesive interactions between leukocytes and the endothelium (i.e. adhesive molecules of the integrin, selectin and immunoglobulin families): and (b) modulation of the endothelium-derived relaxing factor/NO (EDRF/NO) system. Thus, the following discussion will focus on the adherence molecules and endothelially derived NO as the key elements of the disease processes.

3.1 MYOCARDIAL ISCHAEMIA–REPERFUSION AND THE CORONARY ENDOTHELIUM

A reduction in coronary artery flow to critical levels due to a thrombus, vasospasm, or a physical obstruction of the coronary lumen, results in ischaemia to the myocardium supplied by that artery. If not corrected, this severe ischaemia will lead to myocardial cell injury and necrosis over time, and eventually, endothelial cell injury will ensue. This endothelial dysfunction occurs by 4.5–6 h (Viehman *et al.*, 1991). However, if the ischaemic coronary artery is reperfused so that blood flow

and oxygen is restored, there is an early manifestation of "reperfusion injury", whereby the endothelium experiences a significant dysfunction 2.5–5 min after the start of reperfusion (Tsao and Lefer, 1990; Ma *et al.*, 1993). Because many studies in which reperfusion takes place fail to differentiate between injury caused by reperfusion and injury caused by ischaemia itself, the term "ischaemia–reperfusion" has been used to describe the procedure in most of the studies to be cited in this chapter. The ability of the endothelium to release NO in response to an endothelium-dependent vasodilator (e.g. ACh) is dramatically impaired 2.5–5 min after the start of reperfusion, and is greatly aggravated compared to that of coronary arteries undergoing permanent occlusion (i.e. not reperfused) at equivalent times (Tsao and Lefer 1990; Tsao *et al.*, 1990; Viehman *et al.*, 1991). Not until after 4.5 h of permanent occlusion is there a significant endothelial dysfunction in the non-reperfused group. This endothelial reduction in NO release also applies to basally released NO (i.e. NO continuously released from the endothelium without stimulation by endothelium-dependent dilators). This has been observed in coronary artery rings given a NO synthase inhibitor, an analog of L-arginine, L-NAME (Ma *et al.*, 1993).

Recently, evidence has accumulated to suggest that NO exerts important anti-leukocyte actions which may be of great significance in preserving endothelial function in ischaemia–reperfusion. First, Johnson *et al.* (1991) reported that intravascular infusion of NO preserves myocardial integrity and retards neutrophil infiltration into ischaemic–reperfused hearts. The latter was measured by cardiac MPO activity, an index of neutrophil migration, since MPO is virtually exclusively a neutrophilic enzyme (Bradley *et al.*, 1982; Mullane *et al.*, 1985). Second, Kubes *et al.* (1991) reported that the NO synthase inhibitor L-NMMA increased leukocyte adherence in mesenteric venules measured by intravital microscopy. This *in vivo* measure of enhanced leukocyte stickiness clearly suggests a physiological role for NO as a regulatory inhibitor of PMN adherence. Thirdly, a variety of NO donors (e.g. SIN-1, C87-3754 and SPM-5185), markedly attenuate PMN adherence to cat coronary endothelium (Lefer, D. J. *et al.*, 1991; Siegfried *et al.*, 1992a,b). Moreover, equivalent control compounds containing the same organic chemical backbone but lacking the NO moiety, did not attenuate PMN adherence. Moreover, nitroprusside, another NO donor, has been shown to override the increased endothelial permeability induced by L-NAME. Fourth, infusion of L-arginine, the substrate of NO, also preserves the endothelium, limits PMN adherence to the coronary endothelium and attenuates cardiac necrosis following reperfusion of an ischaemic coronary bed (Weyrich *et al.*, 1992a). Fifth, several NO donors (e.g. SIN-1, C87-3754 and SPM-5185) when given at threshold vasodilator effects all exert a marked endothelial protective and a cardioprotective effect, significantly attenuating myocardial necrosis produced by reperfusion of an ischaemic area (Siegfried *et al.*, 1992a,b). This effect is a sustained one, since it occurs 24 h following reperfusion of the ischaemic coronary bed (Lefer *et al.*, 1993). Thus NO is an important endogenous substance produced by the endothelium which protects the coronary endothelium against reperfusion injury. The precise mechanisms of the endothelial preserving and cardioprotective actions of NO are not well understood, but may involve the down regulation of adherence molecules on either the neutrophil (i.e. the CD11/CD18 glycoprotein complex) or endothelial counter receptors (e.g. P-selectin, E-selectin, ICAM-1; Lefer, 1993).

A variety of monoclonal antibodies directed against specific adherence molecules are now available. Several of these have been tested in ischaemia–reperfusion or related shock states. The most widely tested antibodies have been those directed against the leukocyte adherence glycoprotein complex (i.e. CD11/CD18). The first reported study using a monoclonal antibody against the PMN adherence glycoproteins in myocardial ischaemia–reperfusion was reported by Simpson *et al.* (1988) who showed that a monoclonal antibody (i.e. mAB 904) to CD11b (i.e Mac-1, Mo-1) significantly reduces myocardial necrosis after 90 min of ischaemia followed by 6 h of reperfusion in dogs. This antibody was given midway through the ischaemic period (i.e. 45 min prior to reperfusion). The antibody reduced infarct size by 46% ($P < 0.01$) compared to a control non-blocking antibody.

The results of Simpson *et al.* (1988) were followed shortly by the studies of Seewaldt-Becker *et al.* (1990). These investigators employed monoclonal antibodies against either CD11a (LFA-1; i.e. mAB R3.1) or against the common β-chain, CD18 (i.e. mAB R3.3) in a rabbit model consisting of 60 min of myocardial ischaemia followed by 5 h of reperfusion. When given just prior to occlusion, mAB R3.3 reduced infarct size by 68% and mAB R3.1 reduced it by 50%. However, when these antibodies were injected intravenously just prior to reperfusion, no significant protection was observed. This is a puzzling finding, and needs to be clarified, particularly since mAB R3.3 but not mAB R3.1 inhibited adherence of human or rabbit granulocytes to HUVEC, and neither antibody decreased circulating granulocyte counts in rabbits subjected to myocardial ischaemia and reperfusion. These results suggest that blocking CD18 action inhibits neutrophil adherence and this translates into a cardioprotective effect. Recently, more potent and higher affinity monoclonal antibodies against CD18 have become available.

These newer anti-CD18 antibodies have been designated as mAB R15.7, mAB 60.3 and mAB IB4. One of these antibodies, mAB R15.7, has been tested in myocardial ischaemia–reperfusion in both cats (Ma *et al.*, 1991) and primates (Winquist *et al.*, 1992). In cats, mAB R15.7 given just prior to reperfusion dramatically reduced myocardial necrosis, blocked PMN infiltration into the reperfused myocardium and preserved coronary endothelial integrity. mAB R15.7 did not exert any significant haemodynamic effect in these anesthetized cats. In primates, mAB R15.7 also protected against myocardial necrosis without exerting haemodynamic effects. In this case, the antibody was administered prior to coronary occlusion rather than prior to reperfusion. Both studies demonstrated clear and significant cardioprotective effects with blocking doses of a CD18 antibody. mAB R15.7 has also been found to have a protective effect in rabbit spinal cord injury (Clark *et al.*, 1990) and in rabbit pulmonary injury following ischaemia–reperfusion (Welbourne *et al.*, 1991).

Other CD18 antibodies have been studied in related models of ischaemia–reperfusion or shock. Thus mAB R60.3 was found to be effective in ischaemia–reperfusion of the rabbit ear (Vedder *et al.*, 1990), haemorrhagic shock or total body ischaemia–reperfusion in the rabbit (Vedder *et al.*, 1988, 1989). Also mAB IB4 has been found to protect in feline intestinal ischaemia–reperfusion (Hernandez *et al.*, 1987); and skeletal muscle ischaemia–reperfusion (Carden *et al.*, 1990) as well as in pulmonary ischaemia–reperfusion (Horgan *et al.*, 1990). Thus, there is ample evidence to support an important role of the neutrophil adherence glycoprotein complex in a variety of ischaemia–reperfusion states both locally and systemically. Blockade of CD18 by a monoclonal antibody can be highly effective in these ischaemia–reperfusion states.

Other granulocyte adherence molecules including L-selectin (LECAM-1, LAM-1) have not been widely studied. L-Selectin is responsible for leukocyte "rolling" along the endothelium of venules prior to more permanent sticking (i.e. adherence). Although several monoclonal antibodies have been raised against L-selectin, only preliminary evidence exists that this selectin adherence molecule may play a role in myocardial ischaemia–reperfusion (Kishimoto *et al.*, 1990; Weyrich *et al.*, 1992b). The antibody used against L-selectin, DREG–200, is a murine IgG_1 monoclonal antibody (Kishimoto *et al.*, 1990). Although DREG-200 exerts a significant cardioprotective effect, the magnitude of the DREG-200 effect is significantly less than that of anti-CD18 antibodies, suggesting that L-selectin is not as critical an adhesion molecule as the CD11/CD18 complex in myocardial reperfusion injury.

In the last few years, attention has focused on the role of endothelial adhesion molecules in regulating the adherence of leukocytes to the endothelial surface of blood vessels and the significance of this phenomenon in reperfusion injury. Most of the earlier work has been done employing monoclonal antibodies to ICAM-1. In this regard, Seewaldt-Becker *et al.* (1990) reported that mAB R6.5 significantly attenuated myocardial necrosis (i.e. 51%) when given to rabbits just prior to coronary artery occlusion. This was confirmed by Winquist *et al.* (1992) using mAB R6.5 in primates subjected to 90 min of ischaemia and 5 h of occlusion. Ma *et al.* (1992) also found clear and significant protection by mAB RR1/1 in cats subjected to 90 min of ischaemia and 4.5 h of reperfusion. Not only was there cardioprotection when mAB RR1/1 was given 10 min prior to reperfusion, but there was less granulocyte infiltration and the coronary endothelium was protected against post-reperfusion endothelial dysfunction (i.e. reduced EDRF release in response to endothelium-dependent vasodilators; Ma *et al.*, 1992). Figure 3.2, illustrates the results obtained in rings of cat coronary arteries isolated from cats subjected to myocardial ischaemia given either mAB RR1/1 or a control isotope antibody. Coronary arteries isolated from these cats given an anti-ICAM-1 antibody exhibited a much greater vasorelaxation to ACh than untreated vessels, indicating that blockade of ICAM-1 preserves EDRF release from the coronary endothelium. This antibody has also been found to protect the ischaemic–reperfused cat intestine (Granger *et al.*, 1991) and the ischaemic–reperfused rabbit lung (Horgan *et al.*, 1991). Thus, ICAM-1, the major counter receptor for CD11/CD18 on the endothelium, also appears to be an important receptor for regulating granulocyte adherence to endothelial cells in ischaemia–reperfusion. Consequently, blockade of ICAM-1 effectively protects the ischaemic–reperfused tissues from injury. Of course, there are other endothelial receptors to which CD11/CD18 can bind, so that blockade of ICAM-1 does not appear to be as effective a cardioprotective regimen as

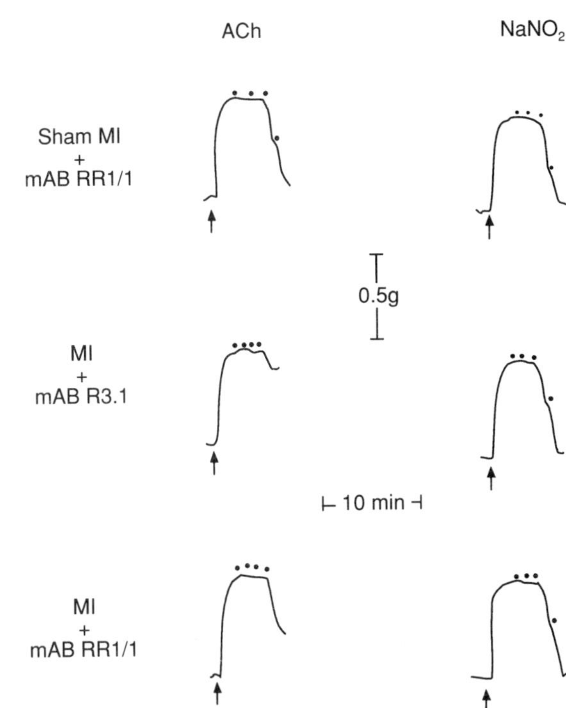

Figure 3.2 Representative recordings of isolated cat left anterior descending coronary artery rings from a sham-operated control nonischaemic cat (top panel), a cat subjected to MI/R and given a non-blocking anti-ICAM antibody (mAB R3.1; middle panel) and a cat subjected to MI/R and given an anti-ICAM-1 antibody (mAB RR1/1; bottom panel). Note the endothelial preservation produced by the mAB RR1/1 as evidenced by a significantly greater vasorelaxation to ACh compared to the non-functional antibody shown in the middle panel.

blockade of CD18. Nevertheless, ICAM-1 represents an important site for potential therapeutic intervention in a variety of regional ischaemic states.

Less is known about the role of other endothelial adhesion molecules (e.g E-selectin, P-selectin) in ischaemia–reperfusion and related states. In the case of E-selectin (i.e. ELAM-1), one published report on uncomplicated reperfusion injury (Winquist *et al.*, 1992) found that CL-2, an anti-ELAM antibody failed to reduce myocardial infarct size in cynomolgous monkeys. Similarly, there was little or no expression of E-selectin in baboons subjected to traumatic or hypovolumic shock, but E-selectin was upregulated on pulmonary vessels of baboons in septic shock (Redl *et al.*, 1991). This is not totally surprising since lipopolysaccharide (i.e. bacterial endotoxin) induces ELAM-1 over a time course of 4–6 h. Moreover, the lung appears to be more a critical site for ELAM-1 action than the heart. Anti-ELAM antibodies have been studied in acute airway inflammation with accompanying airway obstruction in monkeys (Gundel *et al.*, 1991) and in neutrophil-mediated lung

36 A.M. LEFER

injury in rats (Mulligan *et al.*, 1991). In these settings, CL-2 protected against late-phase airway obstruction in monkeys (Gundel *et al.*, 1991) and CL-3 markedly reduced vascular permeability and hemorrhagic extravasation induced by immune complex deposition in rat lungs. Both processes are neutrophil-dependent. Thus, ELAM-1 appears to play a role in delayed pulmonary responses after immunological challenge, but does not appear to be critically involved in coronary vascular dysfunction following myocardial ischaemia and reperfusion.

P-Selectin (i.e. GMP-140) is known to be induced over a period of 10–20 min (Lasky, 1991; Lorant *et al.*, 1991; McEver, 1991a,b) and too be an important adhesive force on both endothelial cells and platelets (Cybulsky and Gimbrone, 1991; McEver, 1991a). Moreover, Lorant *et al.* (1991) found that when P-selectin and PAF are coexpressed on endothelial cells, there is maximal neutrophil adherence to the endothelium. Both P-selectin and PAF act in a cooperative manner to tether PMNs to the endothelium even without activation of the CD11/CD18 adherence glycoprotein system. This P-selectin-mediated process was blocked by the anti-P-selectin monoclonal antibody G-1 (McEver, 1991c). Recently, Weyrich *et al.* (1992a), injected mAB PB1.3, a specific monoclonal antibody to P-selectin, 10 min prior to reperfusion. mAB PB1.3 markedly attenuated myocardial necrosis in cats subjected to 90 min of reperfusion and 4.5 h of reperfusion. The attenuation of myocardial injury was 55%. Moreover, mAB PB1.3 dramatically attenuated neutrophil adherence to ischaemic–reperfused cat coronary endothelium, and also preserved coronary endothelial release of EDRF following reperfusion. Figure 3.3 is a photomicrograph of a cat heart exposed to mAB PB1.3 showing that the exclusive area of binding of this antibody is to the endothelial lining of the coronary vasculature. These results suggest that P-selectin is an important endothelial adhesion molecule in myocardial ischaemia–reperfusion, comparable to that of ICAM-1.

When evaluating all the available evidence in ischaemia–reperfusion states, the endothelium appears to be a critical site for the pathophysiology of reperfusion injury. The initial step in the endothelial dysfunction occurs 2.5–5 min after reperfusion and appears to involve both a decrease in EDRF release and an enhanced release of superoxide radicals (Tsao and Lefer 1990; Tsao *et al.*, 1990). This is followed 20 min later by enhanced adherence of granulocytes to the coronary endothelium which continues to increase over the next several hours. Although many PMNs adhere to the ischaemic–reperfused coronary artery endothelium, even more PMNs (i.e. about 3–4-fold higher PMN adherence) occur on coronary venous endothelium; Lefer *et al.*, 1992). This PMN adherence leads to myocardial cell injury (i.e reperfusion injury) by releasing a variety of mediators from the PMN granules.

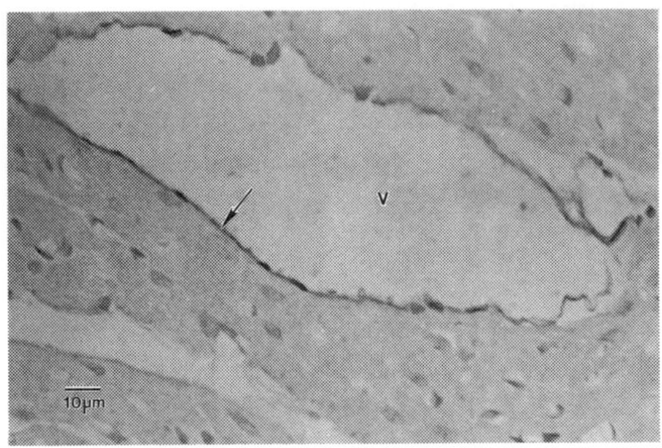

Figure 3.3 Representative photomicrograph of immunohistology showing an avidin–biotin stain of cat heart tissue exposed to mAB PB1.3, an anti-P-selectin antibody. Note the only positive staining is seen on the coronary endothelium (indicated by the arrow). This micrograph was prepared by Dr K. H. Albertine of Thomas Jefferson University.

Immunopharmacological means of preventing PMN adherence to the coronary vascular endothelium appears to be an effective means of dealing with reperfusion injury. At present, monoclonal antibodies to leukocyte adherence glycoproteins (e.g. CD18) appear to be somewhat more effective than antibodies to endothelial counter receptors (e.g. ICAM-1 and P-selectin) in attenuating infarct extension following reperfusion. This may be due to the fact that an anti-CD18 antibody blocks essentially all the neutrophil-adherence mechanisms whereas there are multiple counter receptors on the endothelial cell surface. This possibility must be tested further before its validity can be established.

3.2 ANGIOPLASTY AND RESTENOSIS

Another problem in which the coronary endothelium plays a vital role is in the vascular injury induced by balloon catheterization of the coronary arteries. This is usually an invasive intervention employed to unplug the atherosclerotic or obstructed coronary artery allowing for normal blood flow. The two major problems associated with this intervention are: (a) short-term problems occur which are similar to reperfusion injury upon establishing flow including haemorrhage, cardiac arrhythmias, deficits in cardiac contraction and vascular injury characterized by endothelial denudation, and (b) long-term effects occur which are characterized by stimulation of vascular smooth muscle growth such that the rapid growth of the blood vessel wall fills the lumen of the vessel and essentially obstructs the flow through it (i.e. restenosis) leading to a secondary narrowing or closure of the vessel after a period of 6–18 months. Unfortunately, little is known about the role of the endothelium in these pathophysiological processes. However, the absence of the endothelium certainly results in medial hyperplasia and excessive growth of the vascular smooth muscle layer of the coronary artery.

Figure 3.4 illustrates these phenomena in an endothelial denuded rat carotid artery 2 weeks following endothelial denudation compared to a control carotid artery (Osborne *et al.*, 1989a). One approach to prevent this restenosis is by the use of heparans (i.e. heparin); Osborne *et al.* (1989a) found that infusion of heparin following air-induced endothelial injury of rat carotid arteries caused endothelial dysfunction characterized by a loss of endothelium-dependent vasodilation to ACh, with little or no loss of response to an endothelium-independent vasodilation to acidified $NaNO_2$. Moreover, heparin restored the endothelium responses 2 weeks later. Thus, intimal injury leading to hyperplasia is associated with a defect in EDRF/NO, and this defect as well as the intimal thickening can be overcome by treatment with heparin. Equivalent results were obtained by Araki *et al.* (1993) who observed a large leukocyte adherence following similar injury to rat arteries using intravital microscopy.

Recently, Siegfried and Lefer (1992) reported that 6 days of treatment with an NO-donor (C87-3754) to rats subjected to carotid artery injury, produces significant attenuation of endothelial dysfunction. The low doses of C87-3754 used (i.e. 30 μg/rat/day) did not alter basic haemodynamics, but improved relaxation to 100 nM ACh from $17 \pm 6\%$ to $67 \pm 8\%$ ($P < 0.001$). Figure 3.5 illustrates these findings showing the endothelial preserving effect of C87-3754 given at 30 μg/day over 6 days to a rat subjected to air injury on the first day. Thus, in arterial restenosis, NO appears to play a significant role in preventing hyperplasia and attenuating the endothelial dysfunction.

3.3 HYPERCHOLESTEROLAEMIA AND ATHEROSCLEROSIS

There is now ample evidence that a diet rich in cholesterol (i.e. more than 0.5%) can exert profound alterations in vascular endothelial structure and function (Lefer and Sedar, 1991). The vascular endothelium is very sensitive to injury, and one of the first steps in atherogenesis is endothelial injury (Stemerman, 1981). Ross and Harker (1976) have shown that animals fed a high cholesterol diet develop an increased permeability to a variety of materials including lipids (e.g. cholesterol and cholesterol esters) in endothelial cells (Hajjar *et al.*, 1981). Normally, endothelial cell membrane cholesterol levels are maintained within very narrow limits (Inbar *et al.*, 1974, Whetton *et al.*, 1981). However, excess influx of free cholesterol into cells causes cholesterol esterification. Increases in cell membrane content of cholesterol result in altered permeability properties of the lipid-laden cell membrane (Inbar *et al.*, 1974). The increased cholesterol content of cell membranes can also alter biochemical properties of the cell (Graham and Green, 1970). Moreover, higher cholesterol levels increase concentrations of LDL, a major cholesterol transporting lipoprotein, and probably down regulate LDL receptors on endothelial cells. Elevated cholesterol content in endothelial cells can retard the production of prostacyclin (PGI_2), an important eicosanoid in cardiovascular control mechanisms (Pomerantz and Hajjar, 1989). Thus, increased endothelial cholesterol accumulation can profoundly alter the normal physiological properties of endothelial cells. Moreover, increased cholesterol-ester formation in endothelial cells appears to be a key early step in atherogenesis and the formation of atherosclerotic plaques (Pomerantz and Hajjar, 1989).

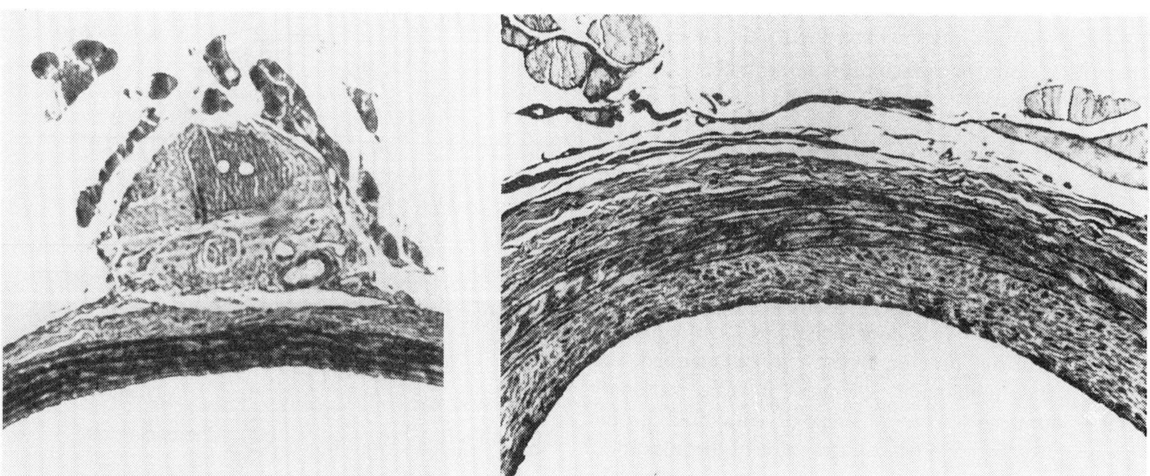

Figure 3.4 Representative photomicrograph of two carotid arteries 14 days following endothelial denudation. Note the medial hyperplasia in the endothelial derived artery (left panel) and the prevention of this by the continuous infusion of heparin (right panel).

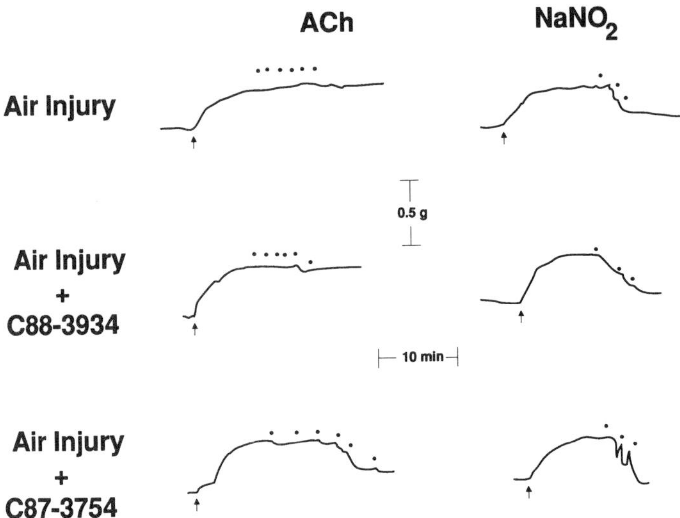

Figure 3.5 Representative recordings of rat carotid artery rings isolated from a rat subjected to air injury and given no treatment (top panel), a rat subjected to air injury and given an inactive NO donor (C88-3934; middle panel), and from a rat subjected to an injury and given the NO donor, C87-3754 (bottom panel). Note the artery isolated from the C87-3754-treated rat exhibits endothelial protection as shown from the improved vasorelaxation to ACh.

There is a link between cholesterol deposition in the endothelium and endothelial dysfunction. Thus, even in the absence of atherosclerotic plaques, endothelial dysfunction is present in conjunction with cholesterol deposition, as evidenced by a reduced vasorelaxation to ACh but not to acidified sodium nitrite (Osborne *et al.*, 1989b). This endothelial dysfunction can be prevented by chronic treatment with lovastatin (i.e. an HMG CoA reductase inhibitor; Osborne *et al.*, 1989c).

Recently, with the development of monoclonal antibodies to VCAM-1 and ICAM-1; (Oppenheimer-Marks *et al.*, 1991; Taichman *et al.*, 1991) which neutralize these adhesion molecules on the endothelial surface of rabbit arteries in the early phases of atherogenesis, prospects are emerging for the possibility of immunotherapy against the adherence of monocytes and the infiltration of PMNs and macrophages into arterial walls and the subsequent development of atherosclerosis. (Lefer and Ma, 1993).

4. *Summary and Conclusions*

This chapter is intended to provide an overview of the critical role the coronary endothelium plays in regulating normal circulatory function in this vital vascular bed, and how pathophysiological changes in the coronary endothelium can lead to a variety of acute and chronic disease states. Figure 3.6 illustrates some of the important pathophysiological changes occurring in the coronary endothelium compared to a normal coronary artery (Panel A) where administration of ACh elicits normal NO release, and a marked vasorelaxation response. Panel

B shows the results of hypercholesterolaemia with the cholesterol droplets (black dots) in the endothelial cells. ACh elicits no significant coronary vasorelaxation under such conditions. Panel C shows a marked atherosclerotic plaque under the endothelium (black layer) which frequently results in a paradoxical vasoconstriction to

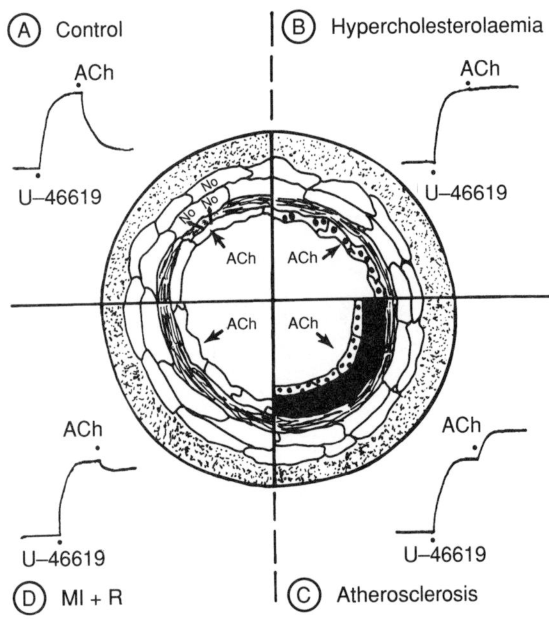

Figure 3.6 Schematic diagram of changes in endothelial structure and function in hypercholesterolaemia, atherosclerosis and myocardial ischaemia–reperfusion.

ACh. Panel D illustrates a severe endothelial dysfunction following myocardial ischaemia followed by reperfusion (MI + R). The hypercholesterolaemia response occurs after days to weeks of increased cholesterol levels and the response to atherosclerosis requires weeks to months of cholesterol feeding, but the response to reperfusion occurs 2.5–5 min after reperfusion of an ischaemic bed. Much work remains to be done to relate these findings to the adherence molecules. In recent years, the immunology of the coronary endothelium has become clarified, and the use of immunomodulators (i.e. monoclonal antibodies to adhesion molecules, and modulators of the endothelium-dependent NO system) have been shown to exert profound effects on the physiology and pathophysiology of the coronary endothelium. This infrastructure will undoubtedly provide a focus for future work and will help further unravel the secrets of the endothelium. This area will be an exciting subject of medicine as we enter the next millennium.

5. Acknowledgement

Supported in part by Research Grant No. Gm-45434 from the National Institute of General Medical Sciences of the NIH.

6. References

Albelda, S.M. (1991). Endothelial and epithelial cell adhesion molecules. Am. J. Respir. Cell Mol. Biol. 4, 195–203.

Araki, H., Muramoto, J., Nishi, K., Jougasaki, M. and Inoue, M. (1993). Heparin adheres to the damaged arterial wall and inhibits its thrombogenicity. Circulation Res., in press.

Berman, J.W. and Calderon, T.M. (1992). The role of endothelial cell adhesion molecules in the development of atherosclerosis. Cardiovasc. Pathol. 1, 17–28.

Bradley, P.P., Priebat, D.A., Christensen, R.D. and Rothstein, G. (1982). Measurement of cutaneous inflammation: estimation of neutrophil content with an enzyme marker. J. Invest. Dermatol. 78, 206.

Buckley, G.B. (1989). In: "Splanchnic Ischaemia and Multiple Organ Failure" (ed E. Marston, G.B. Bulkley, R.G. Fiddian-Green, and U.H. Haglund) pp. 191–193, Edward Arnold Publishers, London.

Carden, D.L., Smith, J.K. and Korthuis, R.J. (1990). Neutrophil-mediated microvascular dysfunction in postischaemic canine skeletal muscle. Circ. Res. 66, 1436–1444.

Clark W.M., Madden, K.P., Rothlein, R. and Zivin, J.A. (1990). Reduction of central nervous system ischaemic injury in rabbits using leukocyte adhesion antibody treatment. Stroke 22, 877–883.

Cybulsky, M.I. and Gimbrone, M.A. Jr (1991). Endothelial expression of a mononuclear leukocyte adhesion molecule during atherogenesis. Science 251, 788–791.

Dreyer, W.J., Michael, L.H., West, M.S., Smith, C.W., Rothlein, R., Rossen, R.D., Anderson, D.C. and Entman, M.L. (1991). Neutrophil accumulation in ischaemic canine myocardium. Insights into time course, distribution, and mechanism of localization during early reperfusion. Circulation 84, 400–411.

Engler, R.L. and Gruber, H.E. (1991). In "The Heart and Cardiovascular System", 2nd edn (ed H.A. Fozzard, E. Haber, R.B. Jennings, A.M. Katz and H.E. Morgan), pp. 1745–1764. Raven Press, New York.

Entman, M.L., Youker, K., Shappell, S.B., Siegel, C., Rothlein, R., Dreyer, W.J., Schmalstieg, F.C. and Smith, C.W. (1990). Neutrophil adherence to isolated adult canine myocytes. Evidence for a CD 18-dependent mechanism. J. Clin. Invest. 85, 1497–1506.

Entman, M.L., Michael, L., Rossen, R.D., Dreyer, W.J., Anderson, D.C., Taylor, A.A. and Smith, C.W. (1991). Inflammation in the course of early myocardial ischaemia. FASEB J. 5, 2529–2537.

Furchgott, R.F. and Zawadski, J.V. (1980). The obligatory role of endothelial cells on the relaxation of arterial smooth muscle by acetylcholine. Nature (Lond.) 288, 373–376.

Furlong, B., Henderson, A.H., Lewis, M.J. and Smith, J.A. (1987). Endothelium-derived relaxing factor inhibits in vitro platelet aggregation. Br. J. Pharmacol. 90, 687–692.

Gamble, J.R. and Vadas, M.A. (1988). Endothelial adhesiveness or blood neutrophils is inhibited by transforming growth factor-β. Science 242, 97–99.

Geng, J.-G., Bevilacqua, M.P., Moore, K.L., McIntyre, T.M., Prescott, S.M., Kim, J.M., Bliss, G.A., Zimmerman, G.A. and McEver, R.P. (1990). Rapid neutrophil adhesion to activated endothelium mediated by GMP-140. Nature (Lond.) 343, 757–760.

Graham, J.M. and Green, C. (1970). The properties of mitochondria enriched in vitro with cholesterol. Eur. J. Biochem. 12, 58–64.

Granger, D.N., Russell, J., Arfors, K.E., Rothlein, R. and Anderson, D.C. (1991). Role of CD11/CD18 and ICAM-1 in ischaemia–reperfusion induced leukocyte adherence and emigration in mesenteric venules. FASEB J. 5, A1753.

Gryglewski, R.J., Palmer, R.M.J. and Moncada, S. (1986). Superoxide anion is involved in the breakdown of endothelium-derived vascular relaxing factor. Nature (Lond.) 320, 454–456.

Gundel, R.H., Wegner, C.D., Torcellini, C.A., Clarke, C.C., Haynes, N., Rothlein, R., Smith, C. W. and Letts, L.G. (1991). Endothelial leukocyte adhesion molecule-1 mediates antigen-induced acute airway inflammation and late-phase airway obstruction in monkeys. J. Clin. Invest. 88, 1407–1411.

Hajjar, D.P., Falcone, D.J., Fowler, S. and Minick, C.R. (1981). Endothelium modifies the altered metabolism of the injured aortic wall. Am. J. Pathol. 102, 28–39.

Harlan, J.M., Winn, R.K., Vedder, N.B., Doerschuk, C.M. and Rice, C.L. (1992). In "Adhesions: Its Role in Inflammatory Disease" (ed J.M. Harlan and D.Y. Liu), pp. 117–150. W.H. Freeman & Co., New York.

Hernandez, L.A., Grisham, M.B., Twohig, B., Arfors, K.E., Harlan, J.M. and Granger, D.N. (1987). Role of neutrophils in ischaemia–reperfusion-induced microvascular injury. Am. J. Physiol. 253, H699–H703.

Hibbs, J.B., Jr, Taintor, R.R., Vavrin, Z., Granger, D.L., Drapier, J.-C., Amber, I.J. and Lancaster, J.R., Jr (1990). In

"NO from L-Arginine: A Bioregulatory System" (ed S. Moncada and E.A. Higgs), pp. 189–223. Elsevier Science Publishers B.V., Amsterdam.

Higgs, E.A., Moncada, S., Vane, J.R., Caen, J.P. and Michel, H. (1978). Effect of prostacyclin (PGI$_2$) on platelet adhesion to rabbit arterial subendothelium. Prostaglandins 16, 17–22.

Horgan, M.J., Wright, S.D. and Malik, A.B. (1990). Antibody against leukocyte integrin (CD18) prevents reperfusion-induced lung vascular injury. Am. J. Physiol. 259, L315–L319.

Horgan, M.J., Ge, M., Gu, J., Rothlein, R. and Malik, A.B. (1991). Role of ICAM-1 in neutrophil-mediated lung vascular injury after occlusion and reperfusion. Am. J. Physiol. 261, H1578–H1584.

Inbar, M., Shinitzky, M. and Sachs, L. (1974). Microviscosity in the surface membrane lipid layer of intact normal lymphocytes and leukemic cells. FEBS Lett. 38, 268–274.

Johnson, G., III, Tsao, P.S. and Lefer, A.M. (1991). Cardioprotective effects of authentic NO in myocardial ischaemia with reperfusion. Crit. Care Med. 19, 244–252.

Kishimoto, T.K., Jutila, M.A., Berg, E.L. and Butcher, E.C. (1989). Neutrophil Mac-1 and MEL-14 adhesion proteins inversely regulated by chemotactic factors. Science 245, 1238–1241.

Kishimoto, T.K., Jutila, M.A. and Butcher, E.C. (1990). Identification of a human peripheral lymph node homing receptor: a rapidly down-regulated adhesion molecule. Proc. Nat. Acad. Sci. USA 87, 2244–2248.

Kishimoto, T.K., Warnock, R.A., Jutila, M.A., Butcher, E.C., Lane, C., Anderson, D.C. and Smith, C.W. (1991). Antibodies against human neutrophil LECAM-1 (LAM-1/Leu-8/DREG-56 antigen) and endothelial cell ELAM-1 inhibit a common CD18-independent adhesion pathway in vitro. Blood 78, 805–811.

Kubes, P., Suzuki, M. and Granger, D.N. (1991). NO: an endogenous modulator of leukocyte adhesion. Proc. Nat. Acad. Sci. USA 88, 4651–4655.

Lasky, L.A. (1991). Lectin cell adhesion molecules (LEC-CAMs): a new family of cell adhesion proteins involved with inflammation. J. Cell. Biochem. 45, 1–8.

Lefer, A.M. and Ma, X.-I. (1993) Decreased basal nitric oxide release in hypercholesterolenia increases neutrophil adherence to rabbit coronary artery endotheliun. Arterio sclerosis, Thrombosis. 13, 771–776.

Lefer, A.M. and Sedar, A.W. (1991). Endothelial alterations in hypercholesterolaemia and atherosclerosis. Pharmacol. Res. 23, 1–12.

Lefer, A.M., Tsao, P.S., Lefer, D.J. and Ma, X.-I. (1991). Role of endothelial dysfunction in the pathogenesis of reperfusion injury after myocardial ischaemia. FASEB J. 5, 2029–2034.

Lefer, A.M., Tsao, P., Aoki, N. and Palladino, Jr., M.A. (1990). Mediation of cardioprotection by transforming growth factor-β. Science 249, 61–64.

Lefer, D.J. (1993). Unpublished observations.

Lefer, D.J., Nakanishi, K., Johnston, W.E. and Vinten-Johansen, J. (1991). Cardioprotective effects of SPM-5185, a novel NO-donor, in myocardial ischaemia-reperfusion. Circulation 84, II-2465.

Lefer, D.J., Nakanishi, K., Vinten-Johansen, J., Ma, X.-I. and Lefer, A.M. (1992). Cardiac venous endothelial dysfunction after myocardial ischaemia and reperfusion in dogs. Am. J. Physiol. 263, H850–H856.

Lefer, D.J., Ma, X.-I., Guo, J.-P. and Lefer, A.M. (1993). Unpublished observations.

Lorant, D.E., Patel, K.D., McIntyre, T.M., McEver, R.P., Prescott, S.M. and Zimmerman, G.A. (1991). Coexpression of GMP-140 and PAF by endothelium stimulated by histamine or thrombin: a juxtacrine system for adhesion and activation of neutrophils. J. Cell Biol. 115, 223–234.

Ma, X.-I., Tsao, P.S. and Lefer, A.M. (1991). Antibody to CD-18 exerts endothelial and cardiac protective effects in myocardial ischaemia and reperfusion. J. Clin. Invest. 88, 1237–1243.

Ma, X.-I., Lefer, D.J., Lefer, A.M. and Rothlein, R. (1992). Coronary endothelial and cardiac protective effects of a monoclonal antibody to intercellular adhesion molecule-1 in myocardial ischaemia and reperfusion. Circulation 86, 937–946.

Ma, X.-I., Weyrich, A.A., Lefer, D.J. and Lefer, A.M. (1993). Diminished basal Nitric Oxide release after myocardial ischaemia and reperfusion promotes neutrophil adherence to coronary endothelium. Circ. Res. 72, in press.

Makgoba, M.W., Sanders, M.E. and Shaw, S. (1989). The CD2-LFA and LFA-1-ICAM pathways: relevance to T-cell recognition. Immunol. Today 10, 417–422.

McCall, T., Whittle, B.J.R., Boughton-Smith, N.K. and Moncada, S. (1988). Inhibition of FMLP-induced aggregation of rabbit neutrophils by NO. Br. J. Pharmacol. 95, 517P.

McEver, R.P. (1991a). Leukocyte interactions mediated by selectins. Thrombosis Haemostasis 66, 80–87.

McEver, R.P. (1991b). GMP-140: a receptor for neutrophils and monocytes on activated platelets and endothelium. J. Cell Biochem. 45, 156–161.

McEver, R.P. (1991c). GMP-140, a receptor that mediates interactions of leukocytes with activated platelets and endothelium. Trends Cardiovasc. Med. 1, 152–156.

Moncada, S. and Vane, J.R. (1978). Pharmacology and endogenous roles of prostaglandin endoperoxides, thromboxane A$_2$, and prostacyclin. Pharmacol. Rev. 30, 293–338.

Moncada, S. and Palmer, R.M.J. (1990). In "NO from L-Arginine: a Bioregulatory System" (ed S. Moncada and E.A. Higgs), pp. 19–33. Elsevier Science Publishers B.V., Amsterdam.

Moncada, S., Palmer, R.M.J. and Higgs, E.A. (1989). Biosynthesis of NO from L-arginine: a pathway for the regulation of cell function and communication. Biochem. Pharmacol. 38, 1709–1715.

Moncada, S., Palmer, R.M.J. and Higgs, E.A. (1991). NO: physiology, pathophysiology, and pharmacology. Pharmacol. Rev. 43, 109–142.

Mullane, K.M. and Smith, C.W. (1990). In "Pathophysiology of Severe Ischemic Myocardial Injury" (ed H.M. Piper), pp. 239–267. Kluwer Academic Publishers, Dordrecht.

Mullane, K.M., Kraemer, R. and Smith, B. (1985). Myeloperoxidase activity as a quantitative assessment of neutrophil infiltration into ischaemic myocardium. J. Pharmacol. Meth. 4, 157.

Mulligan, M.S., Varani, J., Dame, M.K., Lane., C.L., Smith, C.W., Anderson, D.C. and Ward, P.A. (1991). Role of endothelial-leukocyte adhesion molecule 1 (ELAM-1) in neutrophil-mediated lung injury in rats. J. Clin. Invest. 88, 1396–1406.

Oppenheimer-Marks, N., Davis, L.S., Bogue, D.T., Ramberg,

J. and Lipsky, P.E. (1991). Differential utilization of ICAM-1 and VCAM-1 during the adhesion and transendothelial migration of human T lymphocytes J. Immunol. 147, 2913–2921.

Osborne, J.A., Lefer, A.M., Dryski, M. and Bjornsson, T.D. (1989a). Inhibition of EDRF release by intimal hyperplasia is prevented by heparin. Circulation 80, II-6.

Osborne, J.A., Siegman, M.J., Sedar, A.W., Mooers, S.U. and Lefer, A.M. (1989b). Lack of endothelium-dependent relaxation in coronary resistance arteries of cholesterol-fed rabbits. Am. J. Physiol. 256, C591–C597.

Osborne, J.A., Lento, P.H., Siegfried, M.R., Stahl, G.L., Fusman, B. and Lefer, A.M. (1989c). Cardiovascular effects of acute hypercholesterolaemia in rabbits: reversal with lovastatin treatment. J. Clin. Invest. 83, 465–473.

Palmer, R.M.J., Ferrige, A.G. and Moncada, S. (1987). NO release accounts for the biological activity of endothelium-derived relaxing factor. Nature (Lond.) 327, 524–526.

Phillips, M.L., Nudelman, E., Gaeta, F.C.A., Perez, M., Singhal, A.K., Hakomori, S. and Paulson, J.C. (1990). ELAM-1 mediates cell adhesion by recognition of a carbo-hydrate ligand, Sialyl-Le^x. Science 250, 1130–1132.

Pober, J.S. and Cotran, R.S. (1991). Immunologic interactions of T lymphocytes with vascular endothelium. Adv. Immunol. 50, 261–302.

Pomerantz, K.B. and Hajjar, D.P. (1989). Eicosanoids in regulation of arterial smooth muscle cell phenotype, pro-liferative capacity, and cholesterol metabolism. Arthero-sclerosis 9, 413–429.

Redl, H., Dinges, H.P., Buurman, W.A., Van der Linden, C.J., Pober, J.S., Cotran, R.S. and Schlag, G. (1991). Expression of endothelial leukocyte adhesion molecule-1 in septic but not traumatic/hypovolemic shock in the baboon. Am. J. Pathol. 139, 461–466.

Ross, R. and Harker, L. (1976). Hyperlipidemia and athero-sclerosis. Science 193, 1094–1100.

Rothlein, R., Dustin, M.L., Marlin, S.D. and Springer, T.A. (1986). A human intercellular adhesion molecule (ICAM-1) distinct from LFA-1. J. Immunol. 137, 1270–1274.

Rubanyi, G.M. and Vanhoutte, P.M. (1986). Superoxide anions and hyperoxia inactivate endothelium-derived relaxing factor. Am. J. Physiol. 250, H822–H827.

Ruoslahti, E. (1991). Integrins. J. Clin. Invest. 87, 1–5.

Seewaldt-Becker, E., Rothlein, R. and Dammgen, J.W. (1990). In "Leukocyte Adhesion Molecules: Structure, Function, and Regulation" (ed T.A. Springer, D.C. Anderson, A.S. Rosenthal and R. Rothlein), pp. 138–148. Springer-Verlag, Heidelberg.

Siegfried, M.R. and Lefer, A.M. (1992). Recovery of endothelial function in the rat by C87-3754, a sydnonimine type NO donor after carotid vascular injury. Circulation 86, I–21.

Siegfried, M.R., Erhardt, J., Rider, T, Ma, X.-I. and Lefer, A.M. (1992a). Cardioprotection and attenuation of endo-thelial dysfunction by organic NO donors in myocardial ischaemia-reperfusion. J. Pharmacol. Exp. Therap. 260, 668–675.

Siegfried, M.R., Carey, C., Ma, X.-I. and Lefer, A.M. (1992b). Beneficial effects of SPM-5185, a cysteine containing NO donor in feline myocardial ischaemia–reperfusion injury. Am. J. Physiol. 263, H771–H777.

Simmons, D., Makgoba, M.W. and Seed, B. (1988). ICAM, an adhesion ligand of LFA-1, is homologous to the neural cell adhesion molecule NCAM. Nature (Lond.) 331, 624–627.

Simmons, D.L., Walker, C., Power, C. and Pigott, R. (1990). Molecular cloning of CD31, a putative intercellular adhesion molecule closely related to carcinoembryonic antigen. J. Exp. Med. 171, 2147–2152.

Simpson, P.J., Todd, R.F., III, Fantone, J.C., Mickelson, J.K., Griffin, J.D. and Lucchesi, B.R. (1988). Reduction of experimental canine myocardial reperfusion injury by a monoclonal antibody (Anti-Mo1, Anti-CD11b) that inhibits leukocyte adhesion. J. Clin. Invest. 81, 624–629.

Smith, C.W. (1992). In "Adhesion: Its Role in Inflammatory Disease" (ed J.M. Harlan and D.Y. Liu), pp. 83–115. W.H. Freeman & Co., New York.

Smith, C.W., Entman, M.L., Lane, C.L., Beaudet, A.L., Ty, T.I., Youker, K., Hawkins, H.K. and Anderson, D.C. (1991a). Adherence of neutrophils to canine cardiac myocytes in vitro is dependent on intercellular adhesion molecule-1. J. Clin. Invest. 88, 1216–1223.

Smith, C.W., Kishimoto, T.K., Abbass, O., Hughes, B., Rothlein, R., McIntire, L.V., Butcher, E. and Anderson, D.C. (1991b). Chemotactic factors regulate lectin adhesion molecule 1 (LECAM-1)-dependent neutrophil adhesion to cytokine-stimulated endothelial cells in vitro. J. Clin. Invest. 87, 609–618.

Sporn, M.B. and Roberts, A.B. (1989). Transforming growth factor-β. Multiple actions and potential clinical applications. J. Am. Med. Assoc. 262, 938–941.

Springer, T.A. (1990). Adhesion receptors of the immune system. Nature (Lond.) 346, 425–434.

Springer, T.A. and Lasky, L.A. (1991). Sticky sugars for selectins. Nature (Lond.) 349, 196–197.

Stemerman, M.B. (1981). Effects of moderate hypercholestero-laemia on rabbit endothelium. Artherosclerosis 1, 25–32.

Taichman, D.B., Cybulsky, M.I., Djaffar, I., Longenecker, B.M., Teixido, J., Rice, G.E., Aruffo, A. and Bevilacqua, M.P. (1991). Tumor cell surface $\alpha^4\beta_1$ integrin mediates adhesion to vascular endothelium: demonstration of an interaction with the N-terminal domains of INCAM-110/VCAM-1. Cell Regulation 2, 347–355.

Tonnesen, M. (1989). Neutrophil-endothelial cell interactions: mechanisms of neutrophil adherence to vascular endothelium. J. Invest. Dermatol. 93, 53S–58S, Suppl. 2.

Tsao, P.S. and Lefer, A.M. (1990). Time course and mechanism of endothelial dysfunction in isolated ischaemic and hypoxic perfused rat hearts. Am. J. Physiol. 259, H1660–H1666.

Tsao, P.S., Aoki, N., Lefer, D.J., Johnson, G., III and Lefer, A.M. (1990). Time course of endothelial dysfunction and myocardial injury during myocardial ischaemia and reper-fusion in the cat. Circulation 82, 1402–1412.

Ueeda, M., Silvia, S.K. and Olsson, R.A. (1992). NO modulates coronary autoregulation in the guinea pig. Circulation 70, 1296–1303.

Vedder, N.B., Winn, R.K., Rice, C.L., Chi, E.Y., Arfos, K.-E. and Harlan, J.M. (1988). A monoclonal antibody to the adherence-promoting leukocyte glycoprotein, CD18, reduces organ injury and improves survival from haemorrhagic shock and resuscitation in rabbits. J. Clin. Invest. 81, 939–944.

Vedder, N.B., Fouty, B.W., Winn, R.K., Harlan, J.M. and Rice, C.L. (1989). Role of neutrophils in generalized reper-fusion injury associated with resuscitation from shock. Surgery 106, 509–516.

Vedder, N.B., Winn, R.K., Rice, C.L., Chi, E.Y., Arfors, K.-E. and Harlan, J.M. (1990). Inhibition of leukocyte adherence by anti-CD18 monoclonal antibody attenuates reperfusion injury in the rabbit ear. Proc. Nat. Acad. Sci USA 87, 2643–2646.

Viehman, G.E., Ma, X.-I., Lefer, D.J. and Lefer A.M. (1991). Time course of endothelial dysfunction and myocardial injury during coronary arterial occlusion. Am. J. Physiol. 261 (Heart Circ. Physiol. 30), H874–H881.

Welbourn, R., Goldman, G., Hill, J., Lindsay, T., Shepro, D. and Hechtman, H.B. (1991). Lung injury following hindlimb ischaemia is mediated by neutrophil CD 18 adherence receptors. FASEB J. 5, A1492.

Werner-Felmayer, G., Werner, E.R., Fuchs, D., Hausen, A., Reibnegger, G. and Wachter, H. (1991). On multiple forms of NO synthase and their occurrence in human cells. Res. Immunol. 142, 555–561.

Weyrich, A.S., Ma, X.-I. and Lefer, A.M. (1992a). The role of L-arginine in ameliorating reperfusion injury following myocardial ischaemia in the cat. Circulation 86, 279–288.

Weyrich, A.S., Ma, X.-I. and Lefer, A.M. (1992b). Protective effects of a P-selectin specific monoclonal antibody in myocardial ischaemia and reperfusion. Circulation 86, I–80.

Whetton, A.D., Gordon, L.M. and Houstaz, M.D. (1981). Elevated membrane cholesterol concentrations inhibit glucagon-stimulated adenylate cyclase. Biochem. J. 210, 437–445.

Winkelhake, J.L. (1991). Will complex carbohydrate ligands of vascular selectins be the next generation of non-steroidal anti-inflammatory drugs? Glycoconjugate J. 8, 381–386.

Winquist, R.J., Frei, P.P., Letts, L.G., Van, G.Y., Andrews, L.K., Rothlein, R., Dreyer, W.J., Smith, C.W. and Hintze, T.H. (1992). A monoclonal antibody directed against inter-cellular adhesion molecule-1 limits necrosis and improves function after myocardial ischaemia and reperfusion in anesthetized monkeys. J. Vasc. Res. 29, 227.

Zimmerman, G.A., Prescott, S.M. and McIntyre, T.M. (1992). Endothelial cell interactions with granulocytes: tethering and signaling molecules. Immunol. Today 13, 93–100.

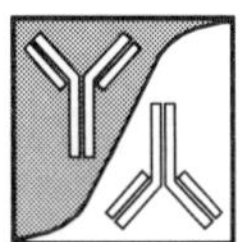

4. The Role of Macrophages and the Modification of LDL in the Pathogenesis of Atherosclerosis

Victor M. Darley-Usmar *and* David G. Hassall

1. Introduction

Myocardial infarction and stroke are common causes of death in industrialized societies and are caused by ischaemic episodes resulting from the formation of thrombi that block the arteries supplying the heart or brain. Almost without exception the underlying disease that precipitates these occlusive events is atherosclerosis. The predisposition of individuals to develop atherosclerosis has been under continuous scrutiny over the past 30 years and a number of well-defined risk factors have been determined. These studies have indicated how individuals could alter their lifestyle to minimize the risk of developing coronary heart disease. Recommendations include changes in dietary habits to reduce total fat intake, a reduction in smoking, control of obesity, an increase in exercise and control of a number of other clinical problems such as diabetes or hypertension. In most cases it is still unclear how these factors contribute to the pathogenesis of the disease at the molecular level. One possible exception is the observation that elevated plasma cholesterol is a risk factor for developing coronary heart disease. Many studies have shown that individuals with increased plasma LDL levels are at increased risk from developing coronary heart disease (Anderson *et al.*, 1987). This observation is often associated with a reduction in the plasma concentration of HDL (Johnson *et al.*, 1991).

Atherosclerotic lesions develop in medium-sized vessels such as the coronary, carotid, basilar and vertebral arteries as well as the aorta and arteries that supply the lower extremities. Spontaneous atherosclerosis is usually a slow, degenerative and progressive process, beginning in childhood, which does not manifest itself clinically until middle age or later. Accelerated lesions, with some similarities to atherosclerosis, may also develop as a consequence of therapeutic interventions such as angioplasty or saphenous-vein bypass grafting. The biological processes involved in these clinical situations appear to arise directly from physical injury to the endothelium and have recently been reviewed (Fuster *et al.*, 1992).

Here we will focus on the role of monocyte-derived macrophages and modified lipoproteins in atherogenesis.

Immunopharmacology of the Heart
ISBN 0–12–200245–8

We have indicated the direction in which current research appears to be moving in these areas and where opportunities for new therapies may arise. In the next section we will briefly outline the principal structural characteristics of the artery wall and the major changes which occur during the development of an atherosclerotic lesion.

2. *Normal Artery, Fatty Streaks and Fibrous Plaque*

Normal healthy muscular arteries are formed from concentric layers of smooth muscle cells. The smooth muscle cells are bounded by an internal and external elastic lamina and the lumen of the vessel is lined with a single layer of endothelial cells. The outside of the artery is protected by the adventitia, a tough coating, which also contains a number of important conduits for nerves, lymphatic drainage and in larger arterioles, blood vessels that supply the artery (vasa vasorum). Evidently, the artery is a complex organ and the development of atherosclerosis involves interactions between the different cell types endogenous to this tissue and others, such as monocytes and lymphocytes, which may infiltrate from the blood. The current model used by many investigators studying the biological processes involved in the initiation of an atherosclerotic lesion is "the response to injury hypothesis" proposed and developed by Ross (Ross and Glomsett, 1976). In this review we will explore the idea developed by Steinberg and co-workers (Steinberg *et al.*, 1989; Witztum and Steinberg, 1991) that toxic substances formed during the oxidation of lipids in the artery wall can "damage" cells and so initiate the development of the lesion.

Luminal atherosclerotic lesions can vary in both size and composition and have been divided into two main types, fatty streaks and fibrous plaque (Stary, 1989). Fatty streak lesions, are formed throughout life and are characterized by the formation of fat-filled "foam cells" under an intact endothelium (Gerrity, 1981a,b). The oily droplets in the cytoplasm of foam cells that gives them their characteristic appearance are formed from cholesteryl ester (Brown *et al.*, 1980a). The morphological changes that characterize the evolution of human lesions have been determined from studies of postmortem tissue and a number of stages have been described (Stary, 1990). Advanced lesions may be predominantly "fatty" or "fibrous" in nature. Interestingly, although fibrotic calcified lesions can cause severe occlusion of coronary arteries, they rarely result in infarction whereas disruption of lipid-rich plaques, which initially may cause only moderate stenosis, can precipitate the formation of a thrombus and infarction (Fuster *et al.*, 1992).

Fibrous plaque is a calcified lesion of atherosclerosis which causes narrowing of the artery, is prothrombotic,

and leads to weakening of the muscle and aneurysmal dilatation. There is now general agreement that, as the lesion develops, smooth muscle cell proliferation and the further infiltration of both monocytes and T-cells occurs (Jonasson *et al.*, 1986; Hansson *et al.*, 1988). These cells synthesize and secrete extracellular matrix molecules such as collagen and, through foam cell formation, cause an increase in tissue cholesterol and cholesteryl ester. These cellular interactions are highly complex and presumably involve a wide variety of different biological mediators during the development of early and advanced lesions.

The finding that an increase in blood cholesterol is directly correlated with the number of deaths from cardiovascular disease has had a dramatic impact on the clinical approach to the treatment of coronary heart disease. The body synthesizes approximately 60% of the cholesterol it requires, the remainder coming from catabolism and recycling of lipoprotein cholesterol and dietary intake. The amount of cholesterol entering from the diet is, therefore, a minor component and this fact probably explains the limited effect modulation of dietary cholesterol has on the concentration of plasma cholesterol (Havel, 1988). These observations have led both clinicians and the pharmaceutical industry to adopt the alternative therapeutic strategy of lowering plasma lipids. This has resulted in a number of treatments including the use of bile acid sequestrants, nicotinic acid, fibrates and HMG CoA reductase inhibitors (Illingworth, 1987; Grundy, 1988). These first generation therapies are all currently available and their effect upon a reduction in total mortality from cardiovascular disease will remain of great interest over the next decade.

The primary prevention strategies mentioned previously, coupled with pharmacological control of plasma cholesterol and modulation of the thrombotic event are the current approaches to treating coronary heart disease. These approaches are directed at modulating the major risk factors for developing coronary heart disease. New and possibly more focused therapies are, in our view, likely to arise from a deeper understanding of the biological processes giving rise to atherosclerotic lesions within the artery wall. Some mechanistic aspects of these events will be reviewed in the rest of this article.

3. *Macrophages in the Vessel Wall*

The involvement of cells belonging to the immune system (monocyte-derived macrophages and T-cells) in both early and late atherosclerotic lesions has led some to suggest that atherosclerosis is an inflammatory disease (Munro and Cotran, 1988). It has long been thought that the most important inflammatory cell in the atherosclerotic lesion is the macrophage yet it is still far from clear whether it performs a protective function which must be overwhelmed before the lesion may develop or itself actively contributes to the pathophysiology of the disease (Stary, 1987; Witztum and Steinberg, 1991). Nevertheless, its central role in mediating key biological responses in atherosclerotic lesions is not in dispute and some of the ideas relating to macrophage cell function and the disease will be discussed in this and following sections. The involvement of macrophages in atherosclerosis has been clearly established by the use of monoclonal antibodies specific for both monocytes and macrophages (Aquel *et al.*, 1984, 1985; Bowyer and Mitchinson, 1989). These studies also confirmed that the monocyte-derived macrophage was the major progenitor of foam cells in atherosclerotic lesions.

3.1 MONOCYTE ADHERENCE

Normal artery contains a resident population of macrophages which are presumed to have been derived from monocytes which migrated into the intima after first interacting with the endothelial cells. This process is likely to occur at a slow rate in healthy tissue as part of the normal function of these cells. The greatly increased numbers of macrophages in atherosclerotic tissue suggests active recruitment of monocytes has occurred during the development of the lesion. In support of this idea it has been demonstrated that the adherence of monocytes to the vascular endothelium occurs at an early stage in the development of atherosclerosis in fat-fed animals (Gerrity, 1981a,b). The factors that regulate the recruitment of these and other cells into the developing lesion are still unclear.

The first step in this process is the attachment of blood-borne monocytes to endothelium, mediated by adherence molecules (for a review, see Springer (1990) and Chapters 3 and 5 of this book). Leukocytes express adherence glycoproteins from the integrin family (CD11/CD18 or LFA-1/Mac-1/P150,95 family; Hynes, 1987; Hogg, 1989). These molecules bind complimentary ligands on the endothelial cell surface, belonging to the immunoglobulin supergene family or the selectins. Such interactions play an important role in the adherence of cells to vascular endothelium and indicate a highly selective process (Harlan, 1985; Berliner *et al.*, 1990; Lawrence and Springer, 1991). Some leukocyte integrin molecules undergo a conformational change during activation that facilitates adherence to their receptors on endothelial cells. This activation process is thought to involve intracellular signal transduction mechanisms, including the activation of protein kinase C (Vedder and Harlan, 1988; Gladwin *et al.*, 1990; Hassall *et al.*, 1991).

Recently, an adherence protein has been identified, ATHERO-ELAM, which is induced on the endothelial cell surface in the atherosclerotic lesions of cholesterol-fed animals (Gimbrone, 1991). It is structurally similar to VCAM-1, an adherence protein known to interact with its complimentary ligand VLA-4 on leucocytes (Elices *et al.*, 1990). This suggests that monocyte infiltration

into atherosclerotic areas may be orchestrated by site-specific adherence reactions leading to the targeting of monocytes to the luminal surface of the atherosclerotic lesion (Berliner *et al.*, 1990; Martin and Hassall, 1990). Differences in the time of maximal expression of different adherence molecules suggest also that mechanisms of adherence and infiltration may vary according to the duration of ischaemia.

These findings offer the possibility of inhibition of monocyte infiltration into the lesion and this could, in principle, arrest the progression of the lesion. For example, monoclonal antibodies targeted to specific adhesion molecules could be used to achieve this objective. Alternatively, the modulation of second messenger pathways within the target cells (endothelial cells or monocytes) that activate/regulate the adherence process may be possible. It is important to recognize, however, that inhibition of monocyte infiltration, while leading to the prevention of foam cell formation, may deprive this region of the artery of essential detoxification pathways for modified lipoproteins which may accumulate in atherosclerotic lesions.

3.2 FOAM CELLS AND MODIFIED LDL

We have discussed in the previous section the evidence that suggests that the foam cells found in the artery wall are derived from macrophages. Now we will consider the mechanisms which could lead to foam cell formation from these cells. Cholesterol, is transported in blood as soluble lipoproteins that are divided into three major classes; VLDL, LDL and HDL. The majority of cells have mechanisms for lipoprotein uptake, the main purpose of which is to provide cholesterol for the biosynthesis of membranes. The principal mechanism is uptake of LDL via a specific receptor for native LDL, that can be found on most cell types (Brown *et al.*, 1974; Schneider, 1989). Uptake of LDL leads to an increase in free cholesterol within the cell and the modulation of a number of biochemical processes including: (1) inhibition of cholesterol biosynthesis by a reduction in the activity of HMG-CoA reductase; (2) a down regulation of the number of native LDL receptors at the cell surface, thereby reducing the further uptake of LDL; (3) the activation of ACAT, the enzyme responsible for the formation of cholesteryl esters. Free cholesterol is liberated from the cholesteryl ester pool by the action of the enzyme NCEH and is then available for removal from the cell by HDL (Johnson *et al.*, 1991). All these processes participate in cholesterol homeostasis within the cell (Brown and Goldstein, 1983).

One of the most important biochemical changes which occurs when monocytes become macrophages is that they lose specific "native" LDL receptors and express "scavenger receptors" in the plasma membrane (Hara *et al.*, 1987; Hassall, 1992). The scavenger receptor has a broad ligand specificity for large polyanionic macromolecules including modified forms of LDL, polysaccharides and bacterial lipopolysaccharides (Brown *et al.*, 1980b; Krieger, 1992). The characteristics of LDL uptake by this receptor are quite different to those of the native LDL receptor. The principal pathways involved in the catabolism of modified lipoproteins are shown in Fig. 4.1. Uptake of modified lipoproteins results in an increase in intracellular cholesterol but this does not result in down regulation of scavenger receptors which continue to function. The increase in the free cholesterol content of the cell cannot be compensated for by inhibition of intracellular cholesterol biosynthesis. The net effect is an expansion of a specific intracellular cholesterol pool (Xu and Tabas, 1991). This is, in part, compensated for by an increase in the activity of the enzyme ACAT leading to the formation and storage of large pools of cholesteryl ester within the cytoplasm of the macrophage (Fig. 4.1).

Experimentally this can be demonstrated *in vitro* by monitoring the rate of incorporation of ^{14}C-oleate into cholesteryl esters and is illustrated in Fig. 4.2. In this series of experiments LDL was oxidized with copper and the extent of modification assessed by measurement of electrophoretic mobility relative to native LDL (REM). It was then incubated with macrophages and the incorporation of ^{14}C-oleate stimulated by oxidized LDL measured. To determine the effect of increasing modification of the LDL particle on uptake by the scavenger receptor we have plotted ^{14}C-oleate incorporation by the cells as a function of REM. It is clear that extensive loading of cells with cholesteryl esters can occur through this mechanism and the response is critically dependent on the extent of LDL modification. The uptake of oxidized LDL by macrophages may also proceed by mechanisms other than the scavenger receptor (Fig. 4.1). For example, aggregated LDL is taken up by macrophages by a scavenger receptor-independent pathway and

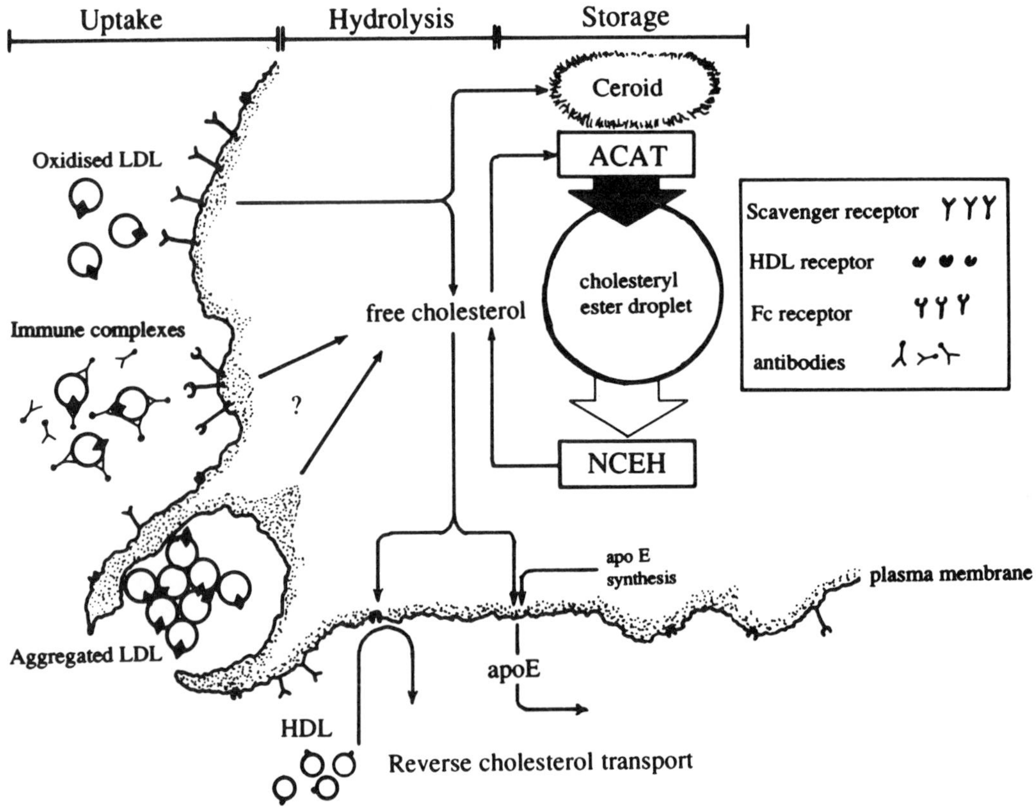

Figure 4.1 Foam cell formation in macrophages. This scheme illustrates some of the mechanisms involved in foam cell formation in the macrophage. Oxidized lipoproteins can enter the cell by scavenger receptor-dependent mechanisms, phagocytosis of aggregated particles or possibly as immune complexes. Uptake results in an increase in intracellular free cholesterol and activation of ACAT leading to the formation of cholesteryl ester stores. These are the fat droplets which give these macrophages their characteristic "foamy" appearance. The cholesteryl ester store may be depleted after conversion to free cholesterol by NCEH and transport from the cell by HDL. Other products of lipid and protein oxidation are retained within the cell as residues known as ceroid.

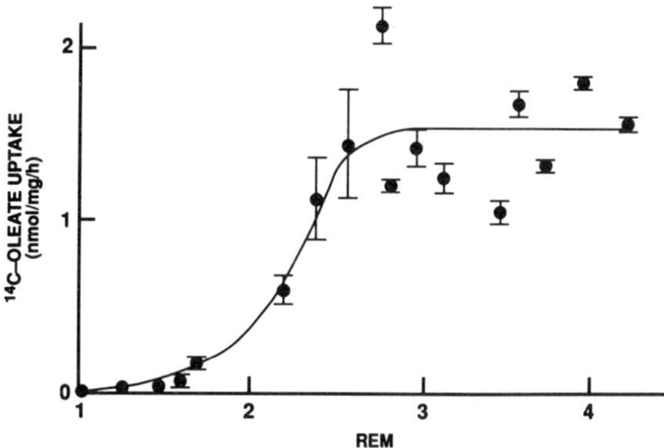

Figure 4.2 Oxidized LDL and its uptake by macrophages. The oxidation of human LDL (1 mg/ml) was achieved by the addition of 20 μM CuSO$_4$ and samples taken at various times (2–24 h) for determination of REM. These samples were then diluted to 50 μg/ml and incubated with mouse peritoneal macrophages (1 × 10^6/ml) and the ^{14}C-oleate incorporation determined. The results are shown as the mean ± S.D. (adapted from Darley-Usmar et al., 1990).

evidence has been presented which suggests that oxidized LDL may also enter the cell through other routes in addition to the scavenger receptor (Sparrow *et al.*, 1989; Tertov *et al.*, 1989; Hoff *et al.*, 1990). Once oxidized LDL has entered the cell it results in the formation of lipid decomposition products known as ceroid (Mitchinson *et al.*, 1990).

Some variant forms of plasma lipoproteins are associated with increased risk for developing atherosclerosis. One such example is LP(a), an LDL particle linked to a large glycoprotein tail with some homology to plasminogen, which is probably retained in the artery wall and subsequently modified to an atherogenic form (Beisiegel *et al.*, 1990). This lipoprotein may also be thrombogenic and so precipitate cardiac ischaemia (Mbewu and Durrington, 1990). Some forms of modified LDL, particularly oxidized LDL (see Section 4.4) may induce antibody responses to the altered particles leading to uptake of lipoproteins by the macrophage through Fc receptors (Klimov *et al.*, 1985, 1988) present on the surface of the cell (Fig. 4.1). It is clear that several different mechanisms may be involved in foam cell formation and serve to remove potentially cytotoxic particles such as oxidized LDL and extracellular debris from lesions.

Inhibition of macrophage ACAT, which should prevent intracellular cholesteryl ester deposition *in situ* may be one therapeutic approach which could inhibit foam cell formation (Sliskovic and White, 1991). This approach assumes that the HDL-dependent processes for the removal of free cholesterol from the cell can occur in the atherosclerotic lesion (Badimon *et al.*, 1989). An alternative strategy might be to reduce the size of the

cholestelyl ester pool within pre-existing foam cells. As shown in Fig. 4.1, these stores are regulated by NCEH (Aviram *et al.*, 1989; Goldberg and Khoo, 1990; Small *et al.*, 1991).

4. *Mechanisms of LDL Modification*

In the previous section we discussed the effect on macrophages of the uptake of modified LDL through the scavenger receptor pathway. In this section we will discuss the changes which must occur to LDL before recognition by this receptor is possible.

The most significant alteration in the physical properties of the LDL molecule which leads to uptake by the scavenger receptor is a change in the net negative charge of the protein, apo-B. This can be achieved by direct chemical modification of the positively charged ε-amino groups of lysine residues on the protein causing a conformational change in the LDL molecule (Brown *et al.*, 1980b; Haberland *et al.*, 1984). This process does not necessarily have to involve the lipid component of the LDL molecule. For example, both acetylation or reaction with aldehydes results in a form of LDL which is not recognized by the native LDL receptor but which can be taken up by the scavenger receptor (Brown *et al.*, 1980b). The consistent association of foam cells with atherosclerotic lesions and a plausible biochemical mechanism for their formation *in vivo* naturally led to the proposal that LDL modification may be occurring in the artery wall. A number of suggestions have been made to explain how modification might occur, including proteolysis, aggregation and oxidation (Steinbrecher *et al.*, 1989; Tertov *et al.*, 1989; Leake *et al.*, 1990). It is evident that these processes, while mechanistically diverse, are not necessarily exclusive. Currently, it is thought that the most likely mechanism, relevant to the human disease, is that of LDL oxidation (Witztum and Steinberg, 1991). The evidence that LDL oxidation can, and indeed does, form an atherogenic particle is extensive and, as discussed in the following sections, is based on the results of studies both *in vivo* and *in vitro*.

4.1 LIPID PEROXIDATION IN THE LDL PARTICLE

Oxidation of the fatty acids in an LDL particle shares many of the characteristics associated with lipid peroxidation in other biological or chemical systems. The reaction starts, as shown in Fig. 4.3, with the abstraction of a hydrogen atom from a bis-allylic carbon atom and is quickly followed by the reaction of the alkyl radical with oxygen. This first step of lipid peroxidation is known as initiation. It is easier to abstract a hydrogen atom from a fatty acid which has bis-allylic centres, like those shown in Fig. 4.3, than from a fatty acid side chain with only

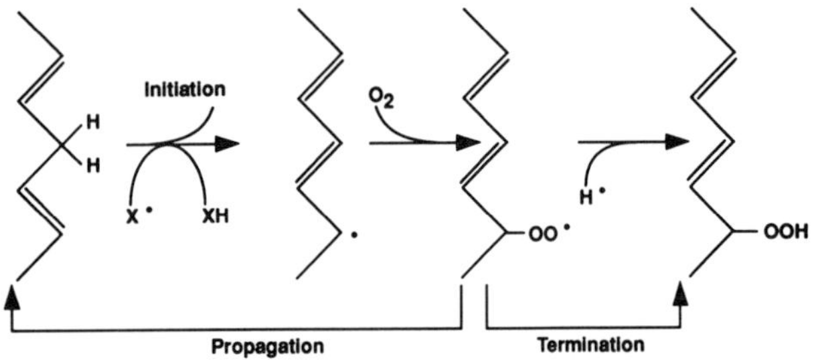

Figure 4.3 Initiation and propagation of lipid peroxidation. The chemical reactions involved in the oxidation of lipids are shown schematically in this figure. The first step of the reaction is abstraction of a hydrogen atom from an allylic carbon atom in a fatty-acid chain of the type illustrated. A number of oxidizing agents or enzymes, represented as X• in the figure, can achieve this part of the reaction. The alkyl radical then reacts with oxygen to form a peroxyl radical and rearranges to form a conjugated diene. The peroxyl radical is itself able to abstract a hydrogen atom from an allylic carbon atom and so propagate the peroxidative process as shown. Termination reactions involve the donation of a hydrogen atom to the peroxyl radical to form a peroxide which, in the absence of transition metals, is stable. In lipoproteins α-tocopherol, endogenous to the particle, can act as a hydrogen atom donor.

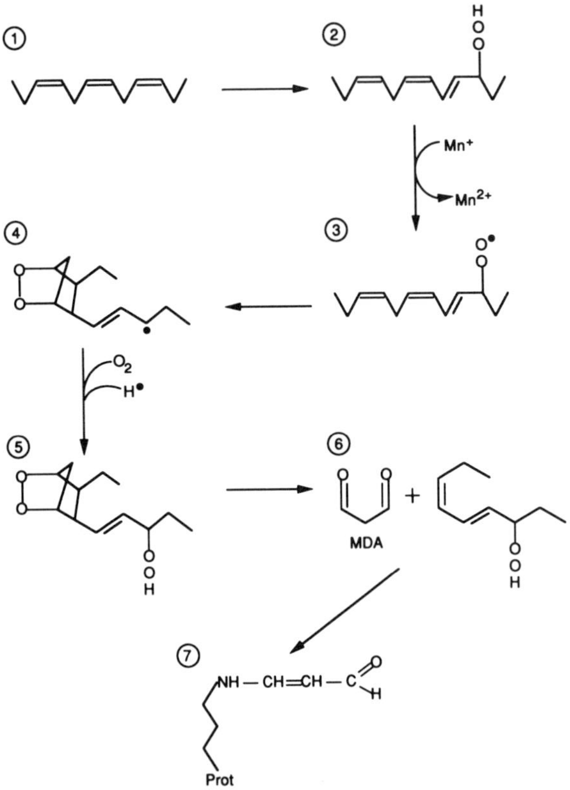

Figure 4.4 The modification of proteins by aldehydes formed during lipid peroxidation. In the following scheme adapted from Esterbauer et al. (1991) we show one of the proposed mechanisms which result in the formation of aldehydes during the lipid peroxidation reaction. In the first step a peroxide is formed from an unsaturated fatty acid as described in Fig. 4.2. In the presence of a transition metal M⁺, which could be iron or copper either bound to protein or in a low molecular weight complex, the peroxide is reduced to a peroxyl radical. In the subsequent reactions cyclization occurs with the further incorporation of oxygen. In this scheme we have shown how cleavage of the fatty acid chain results in two products one of which is the aldehyde, malondialdehyde. This compound can then react with lysine residues on the protein as shown. It has been demonstrated that oxidized LDL contains malondialdehyde conjugated to the lysine groups of the protein in a manner similar to that shown in this diagram.

one double bond. It follows therefore that fatty acid composition should influence the ability of an LDL molecule to act as a substrate for peroxidation. Recently it was shown that enrichment of LDL with oleic acid, which has only one double bond, increases its resistance to oxidation in both humans and laboratory animals (Parthasarathy *et al.*, 1990; Reaven *et al.*, 1991).

Once formed, the peroxyl radical completes a cycle of reactions in which it regenerates itself and forms peroxides (Fig. 4.3). This is referred to as the propagation stage and, although it could in principle continue autocatalytically, it is limited by a number of termination reactions which decrease the efficiency of the peroxyl radical as a mediator of propagation. Termination reactions are those which result in the loss of the radicals upon which propagation depends. A further constraint on the extent of the peroxidation reaction in LDL is the limited availability of substrates for oxidation and the presence of antioxidants endogenous to the LDL particle. This latter form of protection while effective can be overwhelmed if the LDL molecule is subjected to a consistent assault from an oxidant capable of inducing initiation.

The reactions described so far do not require the involvement of the apo-B protein neither would they necessarily result in a significant amount of protein modification. However, the peroxyl radical can attack the fatty acid to which it is attached to cause scission of the alkyl chain. One of the proposed reaction schemes for this process is shown in Fig. 4.4; the concomitant

formation of the aldehyde malondialdehyde (Esterbauer *et al.*, 1991) is also shown. Similar schemes have also been proposed for the formation of 4-hydroxynonenal (Pryor and Porter, 1990). Indeed, complex mixtures of aldehydes have been detected during the oxidation of LDL and it is clear that they are capable of reacting with lysine residues on the surface of the apo-B molecule to convert the molecule to a ligand for the scavenger receptor (Brown *et al.*, 1980b; Haberland *et al.*, 1984;

Esterbauer *et al.*, 1990). The type of product which may be formed is also shown in Fig. 4.4.

The chemical adducts formed by reaction of aldehydes with lysine residues form highly immunogenic epitopes and antibodies have been prepared that are specific for malondialdehyde and 4-hydroxynonenal conjugated LDL (Steinbrecher *et al.*, 1984; Palinski *et al.*, 1989; Yla-Hertuala *et al.*, 1989). These antibodies have proved to be powerful tools for investigating the potential role of lipid peroxidation in the pathogenesis of atherosclerosis and will be discussed in a later section.

4.2 MECHANISMS OF INITIATION AND PROPAGATION OF LIPID PEROXIDATION

Although the chemical processes which occur during lipid peroxidation are well defined, the precise mechanism(s) of initiation and propagation which may occur in the artery wall are largely unknown. Several suggestions have been made and are currently being explored. Among these is the suggestion that the enzyme 15-lipoxygenase, present in macrophages, causes the oxidation of LDL (Yla-Hertuala *et al.*, 1989). Two mechanisms are currently being explored. In the first case it is suggested that free radicals, perhaps superoxide or a peroxyl radical released from the active site of the lipoxygenase enzyme, diffuse into the LDL molecule and initiate peroxidation (McNally *et al.*, 1990). In the second mechanism, it is proposed that direct oxygenation of fatty acid side chains in the LDL molecule results in the formation of a lipid peroxide which in the absence of transition metals is stable. It is difficult to distinguish these two possibilities and data are reported in the literature which could support either (McNally *et al.*, 1990; Rankin *et al.*, 1991; O'Leary *et al.*, 1992). Nevertheless, the association of 15-lipoxygenase and its lipid products with athero-sclerotic lesions provides compelling evidence for an important role for this enzyme in the disease (Yla-Hertuala *et al.*, 1990). The enzyme 15-lipoxygenase is capable of introducing peroxides into LDL and it is possible that circulating lipoproteins contain low levels of endogenous peroxides (Rankin *et al.*, 1991). However, peroxides present in the LDL molecule are stable in the absence of transition metals and only on their addition does propagation of the peroxidation reaction occur (Thomas and Jackson, 1991; O'Leary *et al.*, 1992). Thus, if lipoxygenase is involved in promoting the oxidation and modification of LDL *in vivo* it is necessary that iron or copper are also present in the artery wall.

Many investigators have used monocultures of smooth muscle cells, endothelial cells and macrophages to model the oxidative reactions leading to LDL oxidation *in vivo* (Morel *et al.*, 1984; Parthasarathy *et al.*, 1989; McNally *et al.*, 1990). The conditions required for oxidation are rather stringent involving incubation of the LDL molecule with serum-free medium containing iron and copper and necessarily exclude interactions between different cell types. It is clear that LDL modification under these conditions involves lipid peroxidation and can be inhibited by peroxyl radical scavengers but the precise mechanisms of initiation in these systems remain unclear (Jessup *et al.*, 1991; Sparrow and Olszewski, 1992).

Other mechanisms for initiation are less well studied but are also plausible. For example, it is known that hydroxyl radicals are powerful initiators of lipid peroxidation. There are several possible ways in which hydroxyl radicals could be formed *in vivo*. The hydroxyl radical is so reactive that, in a biological environment, it is likely to react in close proximity to its site of generation. One mechanism for its formation is the iron-dependent decomposition of hydrogen peroxide in a complex series of reactions involving a reductant, hydrogen peroxide and iron itself (Halliwell and Gutteridge, 1986). This idea has been extremely successful in increasing our understanding of the role of free radicals in human pathophysiology and has resulted in a series of transition metal chelators, the "lazaroides", with potential therapeutic applications in CNS trauma (Braughler *et al.*, 1989). However, current thinking suggests that free iron would not be available in the artery wall to catalyse this reaction. An alternative mechanism for the generation of hydroxyl radical involves the inter-action of the free radicals nitric oxide and superoxide to form peroxynitrous acid (Beckman *et al.*, 1990). This compound, unlike the hydroxyl radical, is moderately stable and could diffuse from the site of formation yet decomposes to form the hydroxyl radical (Hogg *et al.*, 1992). Nitric oxide is produced continuously in the artery wall and functions as EDRF (Moncada *et al.*, 1991). The mechanism of superoxide generation and its biological function in the artery wall is less obvious but it is likely that it is being formed by endothelial cells (Nakazono *et al.*, 1991). The simultaneous generation of these two radical species results in the oxidation of LDL and provides an additional route for oxidation which depends on the interaction of cells within the artery wall (Darley-Usmar *et al.*, 1992; Jessup *et al.*, 1992).

Other investigators have suggested that the haem proteins, myoglobin or haemoglobin whose normal function is to transport oxygen, may initiate lipid peroxi-dation (Dee *et al.*, 1991). In addition it has been shown that the iron porphyrin complex, hemin, can also catalyse this reaction (Balla *et al.*, 1991). For this to occur these iron complexes must be oxidized to a low oxidation state or ferryl form by either hydrogen peroxide or possibly lipid hydroperoxides. If this mechanism were to occur *in vivo* it requires the lysis of either smooth muscle cells to release myoglobin or red blood cells to release haemo-globin, and is, therefore, likely to be particularly relevant in the later stages of coronary heart disease involving ischaemia–reperfusion.

Patients suffering from diabetes are known to be at increased risk from developing atherosclerosis and it has been proposed that this may be associated with glycation of LDL (Lopez-Virella *et al.*, 1988). Interestingly, experimental evidence has been reported that suggests that the glycation reaction is itself associated with oxidant generation leading to the oxidation of LDL (Hunt *et al.*, 1990).

4.3   THE BIOLOGICAL PROPERTIES OF OXIDIZED LDL

The biological properties of oxidized LDL have been intensively investigated and it is now recognized that "oxidized LDL" is a highly heterogeneous substance. Variation depends on the extent of oxidation which determines the relative lipid peroxide, aldehyde content and amount of modification of the apo-B protein (Esterbauer *et al.*, 1990). Aldehydes formed as a by-product of lipid peroxidation react with lysine residues on the LDL protein as shown in Fig. 4.4. As many as nine chemically distinct aldehydes have been detected in oxidized LDL (Esterbauer *et al.*, 1990).

It is not surprising, therefore, that this chemical heterogeneity is reflected in a similar degree of variation when oxidized LDL is assessed as a biological effector. Despite these difficulties it has clearly been demonstrated that oxidixed LDL is: (a) cytotoxic to endothelial cells; (b) chemotactic for monocytes; and (c) stimulates smooth muscle cell proliferation (Quinn *et al.*, 1987; Steinbrecher *et al.*, 1989). Low levels of modification result in inhibition of binding to the native LDL receptor which would presumably increase the plasma half-life of the molecule and residence time in the artery wall (Haberland *et al.*, 1984). Interestingly, LDL in which minor modification of the lipid phase has occurred, often referred to as minimally modified LDL, has a number of properties distinct from native LDL which are potentially atherogenic (Berliner *et al.*, 1990; Drake *et al.*, 1991).

The cytotoxicity of oxidized LDL is likely to be an important property associated with its ability to mediate atherogenesis. The agents present in the LDL particle which mediate this toxicity are most probably products of lipid decomposition such as aldehydes or lipid hydroperoxides (Esterbauer *et al.*, 1991). The ability of a cell to combat an oxidative insult from aldehydes is critically dependent on the glutathione content of the cell; this low molecular weight intracellular antioxidant is essential for the detoxification of lipid peroxidation products. Glutathione synthesis is induced in macrophages on exposure to oxidized LDL or 4-hydroxynonenal (Darley-Usmar *et al.*, 1991). Prevention of glutathione synthesis in both endothelial cells and macrophages results in pronounced cell death if the cells are subsequently challenged with oxidized LDL.

4.4   THE IMMUNOLOGICAL PROPERTIES OF OXIDIZED LDL

It has been shown that chemical modification of lysine residues within protein renders them highly immunogenic (Steinbrecher *et al.*, 1984; Gonen *et al.*, 1987). LDL shares this property and antibodies have been raised against malondialdehyde and hydroxynonenal conjugated LDL (Palinski *et al.*, 1989; Jurgens *et al.*, 1990). These antibodies cross-react with oxidized LDL prepared *in vitro* and also with material in human atherosclerotic lesions (Yla-Hertuala *et al.*, 1989). Furthermore it is likely that these epitopes are present on modified LDL in lesioned tissue. It is probable that the source of aldehydes forming these adducts *in vivo* is lipid peroxidation within the LDL particle in the artery wall (reviewed in Yla-Hertuala, 1991).

In support of the hypothesis that aldehyde-modified LDL may be formed *in vivo* is the observation that human sera contains auto-antibodies against homologous malondialdehyde-modified LDL (Palinski *et al.*, 1989). Indeed the titre of antibodies was found to be an independent predictor of the progression of carotid atherosclerosis (Salonen *et al.*, 1992). This finding raises the possibility that atherosclerosis may involve an autoimmune component (reviewed in Orekhov, 1991).

5.   *Inhibition of Foam Cell Formation*

The current focus of research in the field of atherosclerosis centres upon the processes which occur in the artery wall and lead to the development of an atherosclerotic lesion. In this short overview we have attempted to reflect the apparent general consensus among investigators in this field that reducing lipid accumulation in atherosclerotic lesions is likely to inhibit development of the disease. Many approaches are possible which could be reasonably expected to achieve this objective. Lowering plasma lipids is an indirect approach to this problem and is currently being explored in the clinic.

The central role macrophage-derived foam cells have in the atherosclerotic process suggests the possibility of modulating the biology of this cell to prevent excessive lipid accumulation. This could be achieved by either inhibiting the modification of LDL in the artery wall or increasing cholesterol efflux via HDL from these cells. If oxidation of LDL in the artery wall does indeed contribute to the progression of the atherosclerotic lesion then antioxidants, specifically peroxyl radical scavengers, could have anti-atherosclerotic properties. This has already been demonstrated in animal models of the disease with compounds such as probucol (Steinberg, 1986; Carew *et al.*, 1987). Trials in humans with probucol are currently in progress (Regnstrom *et al.*, 1990). Alternatively protection could be achieved in humans by enhancing the antioxidant status of LDL using endogenous antioxidants such as α-tocopherol or β-carotene. Indeed, the available epidemiological evidence suggests α-tocopherol deficiency in northern European populations may result in an increased incidence of coronary heart disease in these populations (Gey, 1986). Trials with these agents are either in progress or planned and the results are awaited with interest.

In conclusion, our current understanding of the development of atherosclerosis has increased dramatically over the past two decades. The recognition that LDL is modified in the artery wall and the key role the macrophage plays in lipid accumulation is altering our view of current therapeutic strategies. It is likely, therefore, that over the next two decades we will see a change in therapeutic direction from cholesterol lowering to targeted treatment of the disease in the artery wall.

6. *References*

Anderson, K.M., Castelli, W.P. and Levy, D. (1987). Cholesterol and mortality, 30 years follow up from the Framingham study. JAMA 257, 2176–2180.

Aquel, N., Ball, R.Y., Waldmann, H. and Mitchinson, M.J. (1984). Monocyte origin of foam cells in human atherosclerotic plaques. Atherosclerosis 53, 265–271.

Aquel, N., Ball, R.Y., Waldmann, H. and Mitchinson, M.J. (1985). Identification of macrophages and smooth muscle cells in human atherosclerosis using monoclonal antibodies. J. Pathol. 146, 197–204.

Aviram, M., Bierman, E.L. and Oram, J.F. (1989). High density lipoprotein stimulates sterol translocation between intracellular and plasma membrane pools in human monocyte-derived macrophages. J. Lipid Res. 30, 65–76.

Badimon, J.J., Badimon, L., Galvez, A., Dische, R. and Fuster, V. (1989). High density lipoprotein plasma fractions inhibit aortic fatty streaks in cholesterol fed rabbits. Lab. Invest. 60, 455–461.

Balla, G., Jacobs, H.S., Eaton, J.W., Belcher, J.D. and Vercellotti, G.M. (1991). Hemin: a physiological mediator of low-density lipoprotein oxidation and endothelial injury. Arteriosclerosis 11, 1700–1711.

Beckman, J.S., Beckman, T.W., Chen, J., Marshall, P.A. and Freeman, B.A. (1990). Apparent hydroxyl radical production by peroxynitrite: implications for endothelial injury from nitric oxide and superoxide. Proc. Natl Acad. Sci. 87, 1620–1624.

Beisiegel, U., Niendorf, A., Wolf, K., Reblin, T. and Rath, M. (1990). Lipoprotein (a) in the arterial wall. Eur. Heart J. 11 (Suppl. E), 174–183.

Berliner, J.A., Territo, M.C., Sevanian, A., Ramin, S., Kim, J.A., Bamshad, B., Esterson, M. and Fogelman, A.M. (1990). Minimally modified low density lipoprotein stimulates monocyte endothelial interactions. J. Clin. Invest. 85, 1260–1266.

Bowyer, D.E. and Mitchinson, M.J. (1989). In "Human monocytes" (ed M. Zembala and G. L. Asherson), pp. 429–438. Academic Press, London.

Braughler, J.M., Hall, E.D., Jacobs, E.J., McCall, J.M. and Means, E.D. (1989). The 21-aminosteroids: potent inhibitors of lipid peroxidation for the treatment of central nervous system trauma and ischaemia. Drugs Future 14, 143–152.

Brown, M.S. and Goldstein, J.L. (1983). Lipoprotein metabolism in the macrophage: implications for cholesterol deposition in atherosclerosis. Ann. Rev. Biochem. 52, 223–261.

Brown, M.S., Dana, S.E. and Goldstein, J.L. (1974). Regulation of 3-hydroxy-3-methylglutaryl coenzyme A reductase activity in cultured human fibroblasts: comparison of cells from a normal subject and from a patient with homozygous familial hypercholesterolaemia. J. Biol. Chem. 249, 789–796.

Brown, M.S., Ho, Y.K. and Goldstein, J.L. (1980a). The cholesteryl ester cycle in macrophage foam cells. J. Biol. Chem. 255, 9344–9352.

Brown, M.S., Basu, S.K., Falck, J.R., Ho, Y.K. and Goldstein, J.L. (1980b). The scavenger cell pathway for lipoprotein degradation: specificity of the binding site that mediates the uptake of negatively charged LDL by macrophages. J. Supramolec. Struc. 13, 67–81.

Carew, T.E., Schwenke, D.C. and Steinberg, D. (1987). Antiatherogenic effects of probucol unrelated to its hypocholesterolaemic effects. Proc. Natl. Acad. Sci. 84, 7725–7729.

Darley-Usmar, V.M., Lelchuk, R., O'Leary, V.J., Knowles, M., Rogers, M.V. and Severn, A. (1990). Odixation of low density lipoprotein and macrophage derived foam cells. Biochem. Soc. Trans. 18, 1064–1066.

Darley-Usmar, V.M., Severn, A., O'Leary, V.J. and Rogers, M. (1991). Treatment of macrophages with oxidised low density lipoprotein increases their intracellular glutathione content. Biochem. J. 278, 429–434.

Darley-Usmar, V.M., Hogg, N., O'Leary, V.J., Wilson, M.T. and Moncada, S. (1992). The simultaneous generation of superoxide and nitric oxide can initiate lipid peroxidation in human low density lipoprotein. Free Rad. Res. Comm. 17, 9–20.

Dee, G., Rice-Evans, C., Obeyesekera, S., Meraji, S., Jacobs, M. and Bruckdorfer, K.R. (1991). The modulation of ferryl myoglobin formation and its oxidative effects on low density lipoprotein by nitric oxide. FEBS Lett. 294, 38–42.

Drake, T.A., Hannani, K., Fe, H., Lavi, S. and Berliner, J.A. (1991). Minimally oxidized low density lipoprotein induces tissue factor expression in cultured human endothelial cells. Am. J. Pathol. 138, 601–607.

Elices, M.J., Osborn, L., Takada, Y., Crouse, C., Luhowskyj, S., Chi-Rosso, G. and Lobb, R.R. (1990). VCAM-1 on activated endothelium interacts with the leukocyte integrin VLA-4 at a site distinct from the VLA-4/fibronectin binding site. Cell 60, 577–583.

Esterbauer, H., Dieber-Rothender, M., Waeg, G., Streigl, G. and Jurgens, G. (1990). Biochemical, structural and functional properties of oxidised low density lipoprotein. Chem. Res. Tox. 3, 77–92.

Esterbauer, H., Schaur, R.J. and Zollner, H. (1991). Chemistry and biochemistry of 4-hydroxynonenal, malondialdehyde and related aldehydes. Free Rad. Biol. Med. 11, 81–128.

Fuster, V., Badimon, L., Badimon, J.J. and Chesbro, J.H. (1992). The pathogenesis of coronary artery disease and the acute coronary syndromes. N. Engl. J. Med. 326, 310–318.

Gerrity, R.G. (1981a). The role of the monocyte in atherogenesis. I. Transition of blood-borne monocytes into foam cells in fatty lesions. Am. J. Pathol. 103, 181–190.

Gerrity, R.G. (1981b). The role of the monocyte in atherogenesis. II. Migration of foam cells from atherosclerotic lesions. Am. J. Pathol. 103, 191–200.

Gey, G.F. (1986). On the antioxidant hypothesis with regard to arteriosclerosis. Biblthca. Nutr. Dieta 37, 53–91.

Gimbrone, M. (1991). Cell–cell interactions in the vessel wall. Ninth International Symposium on Atherosclerosis, Rosemont, IL, October 6–11 (unpublished observations).

Gladwin, A.-M., Hassall, D.G., Martin, J.F. and Booth, R.F.G. (1990). Mac-1 mediates adherence of human monocytes to endothelium via a protein kinase C dependent mechanism. Biochem. Biophys. Acta 1085, 273–298.

Goldberg, D.I. and Khoo, J.C. (1990). Stimulation of neutral cholesteryl ester hydrolase by cAMP system in P388D1 macrophages. Biochem Biophys. Acta 1042, 132–137.

Gonen, B., Fallon, J.J. and Baker, S.A. (1987). Immunogenicity of malondialdehyde-modified low density lipoproteins. Atherosclerosis 65, 265–272.

Grundy, S.M. (1988). HMG CoA reductase inhibitors for the treatment of hypercholesterolaemia. N. Engl. J. Med. 319, 24–33.

Haberland, M.E., Olch, C.L. and Fogelman, A.M. (1984). Role of lysines in mediating interaction of modified low density lipoproteins with the scavenger receptor of human monocyte macrophages. J. Biol. Chem. 259, 11305–11311.

Halliwell, B. and Gutteridge, J. (1986). Oxygen free radicals and iron in relation to biology and medicine: some problems and concepts. Arch. Biochem. Biophys. 246, 501–514.

Hansson, G.K., Jonasson, L., Lojsthed, B., Stemme, S., Kocher, O. and Gabbiani, G. (1988). Localisation of T lymphocytes and macrophages in fibrous and complicated human atherosclerotic plaque. Atherosclerosis 72, 135–141.

Hara, H., Tanishita, H., Yokoyama, S., Tajima, S. and Yamamoto, A. (1987). Induction of acetylated low density lipoprotein receptors and suppression of low density lipoprotein receptors on the cells of human monocytic leukaemia cell line (THP-1 cell). Biochem. Biophys. Res. Comm. 146, 802–808.

Harlan, J.M. (1985). Leukocyte-endothelial interactions. Blood 65, 513–525.

Hassall, D.G. (1992). Three probe flow cytometry of a human foam cell forming macrophage. Cytometry 13, 381–388.

Hassall, D.G., Bath, P.M.W., Gladwin, A.-M. and Beesley, J.E. (1991). CD11/CD18 cell surface adhesion glycoproteins: discordance of monocyte function and expression in response to stimulation. Exp. Cell Res. 196, 346–352.

Havel, R.J. (1988). Lowering cholesterol 1988: rationale, mechanisms and means. J. Clin Invest. 81, 1653–1660.

Ho, Y.K., Brown, M.S., Bilheimer, D.W. and Goldstein, J.L. (1976). Regulation of low density lipoprotein receptor activity in freshly isolated human lymphocytes. J. Clin. Invest. 58, 1465–1474.

Hoff, H.F., O'Neil, J., Pepin, J.M. and Cole, T.B. (1990). Macrophage uptake of cholesterol containing particles derived from LDL and isolated from atherosclerotic lesions. Eur. Heart J. 11 (Suppl. E), 105–115.

Hogg, N. (1989). In "Human Monocytes" (ed M. Zembala and G.L. Asherson), pp. 37–48. Academic Press, London.

Hogg, N., Darley-Usmar, V.M., Wilson, M.T. and Moncada, S. (1992). Production of hydroxyl radicals from the simultaneous generation of superoxide and nitric oxide. Biochem. J. 281, 419–424.

Hunt, J.V., Smith, C.C.T. and Wolf, S.P. (1990). Antioxidative glycosylation and possible involvement of peroxides and free radicals in LDL modification by glucose. Diabetes 39, 1420–1424.

Hynes, R.O. (1987). Integrins: a family of cell surface receptors. Cell 48, 549–554.

Illingworth, R.D. (1987). Lipid lowering drugs. Drugs 33, 259–279.

Jessup, W., Darley-Usmar, V.M., O'Leary, V.J. and Bedwell, S. (1991). 5-Lipoxygenase is not essential in macrophage mediated oxidation of low density lipoprotein. Biochem. J. 278, 163–169.

Jessup, W., Mohr, D., Gieseg, S.P., Dean, R.T. and Stocker, R. (1992). The participation of nitric oxide in cell free and its restriction of macrophage mediated oxidation of low density lipoprotein. Biochem Biophys. Acta 1180, 73–82.

Johnson, W.J., Mahlberg, F.H., Rothblat, G.H. and Phillips, M.C. (1991). Cholesterol transport between cells and high density lipoproteins. Biochem. Biophys. Acta 1085, 273–298.

Jonasson, L., Holm, J., Skalli, O., Gabbiani, G. and Hansson, G.K. (1986). Regional accumulations of T-cells, macrophages and smooth muscle cells in the human atherosclerotic plaque. Arteriosclerosis 6, 131–136.

Jurgens, G., Ashy, A. and Esterbauer, H. (1990). Detection of new epitopes formed upon oxidation of low density lipoprotein, lipoprotein (a) and very low density lipoprotein. Biochem. J. 265, 605–608.

Klimov, A.N., Denisenko, A.D., Popov, A.K., Nagornev, V.A., Pleskov, V.M., Vinogradov, A.G., Denisenko, T.V., Magracheva, E. Ya., Kheifes, G.M. and Kuznetzov, A.S. (1985). Lipoprotein–antibody immune complexes their catabolism and role in foam cell formation. Atherosclerosis 58, 1–15.

Klimov, A.N., Denisenko, A.D., Vinogradov, A.G., Nagornev, V.A., Pivovarova, Y.I., Sitikova, O.D. and Pleskov, U.M. (1988). Accumulation of cholesteryl esters in macrophages incubated with human lipoprotein–antibody autoimmune complex. Atherosclerosis 74, 41–46.

Kreiger, M. (1992). Molecular flypaper and atherosclerosis: structure of the macrophage scavenger receptor. TIBS 17, 141–146.

Lawrence, M.B. and Springer, T.A. (1991). Leukocytes roll on a selectin at physiological flow rates: distinction from and prerequisite for adhesion through integrins. Cell 65, 859–873.

Leake, D.S., Rankin, S.M. and Collard, M.J. (1990). Macrophage proteases can modify low density lipoproteins to increase their uptake by macrophages. FEBS Lett. 269, 209–212.

Lopez-Virella, M.F., Klein, R.L., Lyons, T.T., Stevenson, H.C. and Witztum, J.L. (1988). Glycosylation of low-density lipoprotein enhances cholesteryl ester synthesis in human monocyte-derived macrophages. Diabetes 37, 550–557.

Martin, J.F. and Hassall, D.G. (1990). In "The Endothelium, An Introduction to Current Research" (ed J.B. Warren), pp. 95–105. Wiley-Liss New York.

Mbewu, A.D. and Durrington, P.N. (1990). Lipoprotein (a): structure, properties and possible involvement in thrombogenesis and atherogenesis. Atherosclerosis 85, 1–14.

McNally, A.K., Chisholm, G.M., Morel, D.W. and Cathcart, M. (1990). Activated human monocytes oxidise low density lipoprotein by a lipoxygenase dependent pathway. J. Immunol. 145, 254–259.

Mitchinson, M.J., Ball, R.V., Carpenter, K.L.H. and Enright, J.H. (1990). Macrophages and ceroid in human atherosclerosis. Eur. Heart J. 11 (Suppl. E), 116–121.

Moncada, S., Palmer, R. and Higgs, E.A. (1991). Nitric oxide: physiology, pathophysiology and pharmacology. Pharmacol. Rev. 43, 109–142.

Morel, D.W., DiCorleto, P.E., and Chisholm, G.M. (1984). Endothelial and smooth muscle cells alter low density lipoprotein in vitro by free radical oxidation. Arteriosclerosis 4, 357–364.

Munro, M.J. and Cotran, R.S. (1988). The pathogenesis of atherosclerosis: atherogenesis and inflammation. Lab. Invest. 58, 249–261.

Nakazono, K., Watanabe, N., Matsuno, K., Sasaki, J., Sato, T. and Inoue, M. (1991). Does superoxide underlie the pathogenesis of hypertension? Proc. Natl Acad. Sci. 86, 10045–10048.

O'Leary, V.J., Darley-Usmar, V.M., Russell, L.J. and Stone, D.S. (1992). Pro-oxidant effects of lipoxygenase derived peroxides on the copper initiated oxidation of low density lipoprotein. Biochem. J. 282, 631–634.

Orekhov, A.N. (1991). Lipoprotein immune complexes and their role in atherogenesis. Curr. Opin. Lipidol. 2, 329–333.

Palinski, W., Rosenfeld, M.E., Yla-Hertuala, S., Gurtner, G.C., Socher, S.S., Butler, S.W., Parthasarathy, S., Carew, T.E., Steinberg, D. and Witztum, J.L. (1989). Low density lipoprotein undergoes oxidative modification in-vivo. Proc. Natl Acad. Sci. 86, 1372–1376.

Parthasarathy, S., Wieland, E. and Steinberg, D. (1989). A role for endothelial cell lipoxygenase in the oxidative modification of low density lipoprotein. Proc. Natl Acad. Sci. 86, 1046–1050.

Parthasarathy, S., Khoo, J.C., Miller, E., Barnett, J., Witztum, J.L. and Steinberg, D. (1990). Low density lipoprotein rich in oleic acid is protective against oxidative modification: implications for dietary prevention of atherosclerosis. Proc. Natl Acad. Sci. 87, 3894–3898.

Pryor, W.A. and Porter, N.A. (1990). Suggested mechanisms for the production of 4-hydroxynonenal from auto-oxidation of polyunsaturated fatty acids. Free Rad. Biol. Med. 8, 541–543.

Quinn, M.T., Parthasarathy, S., Fong, L.G. and Steinberg, D. (1987). Oxidatively modified low-density lipoproteins: a potential role in recruitment and retention of monocyte/macrophages during atherogenesis. Proc. Natl Acad. Sci. USA 84, 2995–2998.

Rankin, S., Parthasarathy, S. and Steinberg, D. (1991). Evidence for a dominant role of lipoxygenases in the oxidation of LDL by mouse peritoneal macrophages. J. Lipid Res. 32, 449–456.

Reaven, P., Parthasarathy, S., Grasse, B.J., Miller, E., Alonazan, F., Mattson, F.H., Khoo, J.C., Steinberg, D. and Witztum, J.L. (1991). Feasibility of using an oleate rich diet to reduce the susceptibility of LDL oxidative modification in humans. Am. J. Clin. Nutr. 54, 701–706.

Regnstrom, J., Walldius, G., Carlson, L.A. and Nilsson, J. (1990). Effect of Probucol treatment on the susceptibility of low density lipoprotein isolated from hypercholesterolaemic

patients to become oxidatively modified in-vitro. Atherosclerosis 82, 43–51.

Ross, R. and Glomset, J.A. (1976). I: The pathogenesis of atherosclerosis. N. Engl. J. Med. 295, 369–377.

Salonen, J.K., Yla-Hertuala, S., Yamamoto, R., Butler, S., Korpela, H., Salonen, R., Nyyssonen, K., Palinski, W. and Witztum, J.L. (1992). Antibody against oxidised LDL and progression of carotid atherosclerosis. Lancet 339, 883–887.

Schneider, W.J. (1989). The low density lipoprotein receptor. Biochem. Biophys. Acta 988, 303–317.

Sliskovic, D.R. and White, A.D. (1991). Therapeutic potential of ACAT inhibitors as lipid lowering and anti-atherosclerotic agents. TIPS 12, 194–199.

Small, C.A., Rogers, M.P., Goodacre, J.A. and Yeaman, S.J. (1991). Phosphorylation and activation of hormone sensitive lipase in isolated macrophages. FEBS Lett. 279, 323–326.

Sparrow, C.P. and Olszewski, J. (1992). Cellular oxidative modification of low density lipoprotein does not require lipoxygenases. Proc. Natl Acad. Sci. USA 89, 128–131.

Sparrow, C.P., Parthasarathy, S. and Steinberg, D. (1989). A macrophage receptor that recognizes oxidised low density lipoprotein but not acetylated low density lipoprotein. J. Biol. Chem. 264, 2599–2604.

Springer, T.A. (1990). The sensation and regulation of interactions with the extracellular environment: the cell biology of lymphocyte adhesion receptors. Ann. Rev. Cell Biol. 6, 359–402.

Stary, H.C. (1987). Macrophages, macrophage foam cells and eccentric intimal thickening in the coronary arteries of young children. Atherosclerosis 64, 91–108.

Stary, H.C. (1989). Evolution and progression of atherosclerotic lesions in coronary arteries of children and young adults. Arteriosclerosis 9 (Suppl. I), 19–32.

Stary, H.C. (1990). The sequence of cell and matrix changes in atherosclerotic lesions of coronary arteries in the first forty years of life. Eur. Heart J. 11 (Suppl. E), 3–19.

Steinberg, D. (1986). Studies on the mechanism of action of Probucol. Am. J. Cardiol. 57, 16H–21H.

Steinberg, D., Parthasarathy, S., Carew, T.F., Khoo, J.C. and Witztum, J.L. (1989). Beyond cholesterol: modification of low-density lipoprotein that increases its atherogenicity. N. Engl. J. Med. 320, 915–924.

Steinbrecher, K.P., Fisher, M., Witztum, J.L. and Curtiss, L.K. (1984). Immunogenicity of homologous low density lipoprotein after methylation, ethylation, acetylation or carbamylation: generation of antibodies specific for derivatised lysine. J. Lipid Res. 25, 1109–1116.

Steinbrecher, U.P., Lougheed, M., Kwan, W.-C. and Dirks, M. (1989). Recognition of oxidised low density lipoprotein by the scavenger receptor of macrophages results from the derivatisation of apo lipoprotein B. J. Biol. Chem. 264, 15216–15233.

Tertov, V.V., Sobenin, I.A., Gabbasov, Z.A., Popov, E.G. and Orekhov, A.N. (1989). Lipoprotein aggregation is an essential condition of intracellular lipid accumulation caused by modified low density lipoproteins. Biochem. Biophys. Res. Comm. 163, 489–494.

Thomas, C.E. and Jackson, R.L. (1991). Lipid hydroperoxide involvement in copper dependent and independent oxidation of low density lipoproteins. J. Pharmacol. Exp. Ther. 256, 1182–1188.

Vedder, N.B. and Harlan, J.M. (1988). Increased surface expression of CD11b/CD18 (Mac-1) is not required for stimulated neutrophil adherence to cultured endothelium. J. Clin. Invest. 81, 676–682.

Witztum, J.L. and Steinberg, D. (1991). Role of oxidised low-density lipoprotein in atherogenesis. J. Clin. Invest. 88, 1785–1792.

Xu, X.-X. and Tabas, I. (1991). Lipoproteins activate acyl-coenzyme A: cholesterol acyl transferase in macrophages only after cellular pools are expanded to a critical threshold level. J. Biol. Chem. 266, 17040–17048.

Yla-Hertuala, S. (1991). Macrophages and oxidized low density lipoproteins in the pathogenesis of atherosclerosis. Ann. Med. 23, 561–567.

Yla-Hertuala, S. Palinski, W., Rosenfeld, M.E., Parthasarathy, S., Carew, T.E., Butler, S., Witztum, J.L. and Steinberg, D. (1989). Evidence for the presence of oxidatively modified low density lipoprotein in atherosclerotic lesions of rabbit and man. J. Clin. Invest. 84, 1086–1099.

Yla-Hertuala, S., Rosenfeld, M.E., Parthasarathy, S., Glass, C.K., Sigal, E., Witztum, J.L. and Steinberg, D. (1990). Localisation of 15-lipoxygenase mRNA and protein with epitopes of oxidised low density lipoprotein in macrophage rich areas of atherosclerotic lesions. Proc. Natl Acad. Sci. USA 87, 6959–6963.

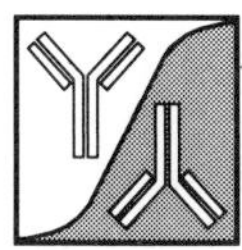

5. The Role of Leukocytes in Ischaemic Heart Disease

Mark L. Entman, Gilbert L. Kukielka, Christie M. Ballantyne and C. Wayne Smith

1. Introduction: Reaction to Injury in Ischaemia and Reperfusion

The purpose of this chapter within a volume dealing with the immunopharmacology of the heart is to discuss the potential mechanism by which inflammatory injury may complicate ischaemic heart disease. As was the case in Chapter 3, "ischaemia–reperfusion" has been used as a term to define events attributable to ischaemia or reperfusion but where differentiation between these conditions

Immunopharmacology of the Heart
ISBN 0–12–200245–8

as primary pathogenic triggers cannot be made with certainty. It is to be emphasized that no one seriously proposes that the initial injury associated with ischaemic heart disease is inflammatory in nature; rather we are describing the mechanism of a reaction to injury and will examine evidence that this secondary reaction might extend and complicate cardiac injury associated with ischaemia and reperfusion. The mechanisms of primary injury in terms of atherosclerosis and endothelial biology are discussed in detail in Chapters 3 and 4.

The concept of a reaction to injury extending the injury is classic in pathology but its application to this disease entity is relatively recent (Hillis and Braunwald, 1977). The association of inflammation with the observation of myocardial infarction is an old one and its description has mainly focused on reaction seen from 12 h to 5 days after the initial event (Mallory *et al.*, 1939). The major reason for this was that most of the information came from autopsy specimens and the relative sensitivity of the methods used at autopsy did not allow detection of neutrophil or other leukocyte infiltration at earlier time periods, although neutrophil infiltration actually can be seen histologically within the first 2 h (Mullane *et al.*, 1984) in experimental conditions. Early description, therefore, concentrated on the role of inflammation in the healing phase of myocardial infarction and, despite models in other systems, paid little attention to the possibility that the inflammatory reaction might function in a deleterious way to complicate the myocardial infarction process. In the last two decades, more attention has been paid to this latter possibility and a large literature has arisen utilizing multiple techniques to access the effects of inflammation and inflammatory injury in ischaemic models. The initial studies with anti-inflammatory drugs, which will be briefly described in this chapter, resulted in promising results and clinical trials were actually undertaken utilizing high-dose steroids to treat myocardial infarction. The failure of these clinical trials (which resulted in increased incidence of ventricular aneurysm and rupture of the ventricular wall) can be attributed to the fact that leukocytes are important in the healing phase of myocardial infarction (Roberts *et al.*, 1976). This also emphasized the need for a greater understanding of the secondary inflammatory reaction and its role in both injury and healing so that more rational and specific interventions can be developed and evaluated.

The compelling need for strategies to expose the mechanisms of inflammatory injury and carefully characterize them has come from the fact that, in the past 10 years, the development of thrombolytic therapy has allowed the reinstitution of coronary blood flow to the ischaemic zone of many patients with myocardial infarction during a period of time where such reperfusion is capable of reducing the amount of myocardial damage. This clinical advance has served as an impetus to address secondary causes of injury (since ischaemia itself has been interdicted) that might only occur during (or become significant because of) reperfusion. There is evidence that reperfusion potentiates the inflammatory reaction to injury (perhaps by increasing the number of neutrophils circulating through the ischaemic area) and substantial evidence suggests that neutrophils participate in a major way in reperfusion injury to the ischaemic heart.

The primary focus of this chapter will be the cellular and molecular mediation of the secondary inflammatory response occurring in reperfusion injury. We will begin with a brief general description of reperfusion injury as a concept (reiterating and extending some of the observations made by Ferrari in Chapter 1) and describe some of the pathophysiological consequences of this injury. The remainder of the chapter will be aimed at the cellular and molecular mediation of inflammatory injury during reperfusion.

2.   *Ischaemia–Reperfusion Injury and Its Aetiology*

2.1   PATHOPHYSIOLOGICAL PARADIGM

In both clinical and experimental models, the initial insult resulting in injury is, in all cases ischaemia. Occlusion of the coronary vessel reduces to critical levels the blood flow to the portion of the myocardium subserved, markedly impairing its energy metabolism leading to cell death. The cells in the epicentre of an occluded vascular bed are irreversibly injured by 20 min and will die regardless of any further changes. Because reduction in blood flow resulting from the occlusion of any one coronary vessel creates a distribution of degrees of ischaemia, other cells (e.g. at the perimeter and on the epicardial surface) are not as rapidly damaged and the presence of live undamaged cells adjoining dead cells and scars is commonly seen pathologically. In the absence of either spontaneous or induced reperfusion, therefore, the predominant mechanism of injury is ischaemia itself although, in relatively well-perfused areas, other sources of extension of the injury may occur.

The concept of reperfusion injury is an intriguing one and has much to recommend it; it is important to understand, however, that at the time of this writing the evidence for an important component of ultimate cell death and injury resulting from reperfusion is controversial. Although many of the abnormalities observed during reperfusion with respect to endothelial function, macromolecular efflux from the myocardial cell, leukocyte infiltration and pathological appearance of myocardial cells are surely associated with special circumstances initiated by reperfusion, evidence that this actually extends the amount of cells that ultimately will be destroyed relies on relatively few and relatively recent studies. The bulk of existing data relates to the observation that certain agents such as antioxidants (Jolly *et al.*, 1984; Mullane, 1988) and adenosine (Olafsson *et al.*, 1987) have been shown to reduce infarct size when administered only during the reperfusion period. With that reservation in mind, we wish to summarize some of the more prominent abnormalities associated with ischaemia and reperfusion from a functional standpoint in preparation for a discussion of the potential role of inflammation in these abnormalities.

There is general agreement that reperfusion of ischaemic myocardium before the onset of irreversible myocardial damage is not associated with any permanent reperfusion-associated injury (Braunwald and Kloner, 1982). Occlusions of 5–20 min duration demonstrate significant functional abnormalities upon the onset of reperfusion such as ventricular arrhythmias, contractile dysfunction for period of 24–48 h (stunned myocardium), and vasomotor dysfunction (Hearse and Bolli, 1991) (and see Chapter 2 of this book). Evidence suggests, however, that these functional abnormalities are not attended by any lethal injury to the myocardium and the ischaemic segment will ultimately recover. Because it is clear that neutrophils are not associated with the abnormalities seen during this time (Hearse and Bolli, 1991), this entity will not be discussed further in this chapter. However, the absence of any neutrophil response under these circumstances emphasizes that neutrophil-induced injury is only seen secondary to lethal injury initiated by a previous noxious stimulus (e.g. ischaemia). This relative specificity is in contrast to the role of oxygen free radicals (from other sources) which appear to influence both the short-term dysfunction (Hearse and Bolli, 1991) and, in some studies, the longer term manifestation of permanent injury (Jolly *et al.*, 1984).

The functional abnormalities seen during reperfusion consequent to lethal myocardial injury can conveniently be grouped into three general categories: myocardial dysfunction, endothelial-related vasomotor dysfunction, and increased microvascular permeability with attendant flow abnormalities. In addition, it has clearly been demonstrated that neutrophil accumulation into previously ischaemic areas is markedly augmented during early reperfusion with the highest rates seen in the first hour of reperfusion (Dreyer *et al.*, 1991b). This has led to attempts to evaluate the role of the neutrophil in these observed pathophysiological dysfunctions as a method of studying their role in reperfusion-associated lethal injury to cardiac myocytes.

There have been two general approaches to this problem: (1) use of anti-neutrophll therapy to mitigate both the functional and pathological changes associated with reperfusion; and (2) cellular and molecular investigations into the mechanism of neutrophil localization in previously ischaemic areas and the mechanism of neutrophil-induced cellular injury. Both of these subjects will be discussed in this chapter but, because of the nature of the present book, we will focus on the latter.

2.2 THE EFFECT OF ANTI-NEUTROPHIL THERAPY ON ISCHAEMIA–REPERFUSION INJURY

As stated above, although irreversible myocardial cell injury begins at 20–30 min after occlusion of a coronary vessel, the extent of myocardial damage induced by this occlusion continues to increase for up to 6 h after the onset of ischaemia. This has led to the recommendation for thrombolytic therapy in patients who present with myocardial infarctions less than 6 h old (GISSI Study Group, 1986). In fact, the facility for reperfusion of the myocardium has created considerably more interest in the possibility of reducing the actual total extent of injury induced by a coronary occlusion since the ischaemia factor is effectively removed. While the association of complement-derived components in the areas of ischaemic injury was described by Hill and Ward (1977) in the early 1970s (see also the following chapter by Rossen), the first major body of evidence suggesting a role for the neutrophil in secondary ischaemic injury came in the latter part of that decade when enormous resources were used to develop strategies for minimization of myocardial infarct size. Among the strategies examined were some which were once thought to interfere with a putative inflammatory mechanism occurring after myocardium injury. Strategies aimed at reducing the generation of chemotactic factors such as complement depletion by cobra venom toxin (Maroko *et al.*, 1978), administration of lipoxygenase inhibitors (Shappell *et al.*, 1990a), and leukotriene B_4 receptor antagonists (Taylor *et al.*, 1989) successfully reduced infarct size in some models. Experiments utilizing prostacyclin analogues (Simpson *et al.*, 1987) and adenosine (Olafsson *et al.*, 1987) to alter neutrophil function likewise were successful in some experimental models. Use of strategies, which reduced neutrophil number such as anti-neutrophil antibodies (Romson *et al.*, 1983), neutrophil-depleting chemicals (Mullane *et al.*, 1984), or neutrophil filters (Engler *et al.*, 1986a) were all successful in reducing ischaemia-related injury in some models. Finally, agents aimed at altering the effects of neutrophil localization such as cyclooxygenase inhibitors (Romson *et al.*, 1982) and free radical scavengers (Jolly *et al.*, 1984; Lucchesi and Mullane 1986) were shown to reduce infarct size or sensitivity to ischaemia in some models of experimental ischaemia. This body of data certainly pointed to a potential role of inflammation in ischaemic damage. However, the conclusions were complicated by the following considerations.

In the first place, there was no universal success with any of these strategies, and success depended, to a great degree, on the model utilized. It is important to point out, however, that when positive results were found with anti-neutrophil strategies, this occurred in many experiments even when reperfusion was not instituted (Mullane and Smith, 1990). This has led to a series of important (and incompletely resolved) questions. The question of how neutrophils in any significant number would get into the ischaemic myocardial area could (with some certainty) be addressed by the concept of areas of intermediate ischaemia that are potentially salvageable within the involved region and around its edges. One could easily visualize a strategy to reduce a "border zone" that remains at risk.

On the whole, however, the advent of thrombolytic therapy to restore coronary blood flow and alleviate myocardial damage certainly enhances the potential for mediation of secondary myocardial injury by the neutrophil. It is clear that neutrophil localization in ischaemic or previously ischaemic myocardial tissue is markedly accelerated by reperfusion and is greatest at the onset of reperfusion (Engler *et al.*, 1986b), is associated with the areas of greatest myocardial injury and lowest coronary blood flow (Rossen *et al.*, 1985; Dryer *et al.*, 1991b), and can be modulated by strategies involving the administration of therapy only upon the onset of reperfusion (Jolly *et al.*, 1984; Olafsson *et al.*, 1987; Mullane, 1988; Mullane and Smith 1990).

It is important to point out even after the early experiments the information gained in the above studies was sufficiently compelling that Hillis and Braunwald (1977) concluded that all patients with acute infarction, regardless of their haemodynamic state, may benefit from an anti-inflammatory agent. A trial of methylprednisolone for acute myocardial infarction resulted in extremely deleterious results in patients with an increased incidence of ventricular aneurysm and cardiac rupture (Roberts *et al.*, 1976). Thus, anti-inflammatory therapy also appeared to interfere with myocardial healing. This led to a second experimental approach to define the role of inflammation in myocardial injury that seeks to characterize the cellular and molecular events associated with this inflammatory reaction with an aim toward developing more site-specific interventions which conceivably could reduce secondary inflammatory injury with less risk of deleterious effects.

3. *Factors Contributing to Neutrophil Localization in Ischaemic and Reperfused Hearts*

3.1 TRAPPING OF LEUKOCYTES IN THE MICROVESSELS

Because classic studies of inflammation have emphasized the role of neutrophil accumulation in capillaries and post-capillary venules, a similar examination has been made in experimental myocardial infarction. In the experiments of Schmid-Schonbein and Engler (1987),

the earliest inflammatory changes in the ischaemia–reperfusion process occur primarily in small vessels and capillaries where neutrophils appear to be trapped. In many cases, it appears that neutrophils may actually obstruct the capillary. The accumulation of neutrophils in microvessels probably occurs as a function of a number of definable pathological processes consequent to chemotactic stimulation in the area of ischaemia and reperfusion. In the first place, it is known that chemotactic stimulation of neutrophils induces shape changes in neutrophils (they become less deformable). Additionally, these neutrophils also undergo homotypic aggregation which may further contribute to obstruction. Activated neutrophils also release a variety of vasoactive products which promote vasoconstriction such as TXA_2 (Michael *et al.*, 1990) and a variety of lipoxygenase product such as LTB_4 (Mullane *et al.*, 1988) all of which induce vasoconstriction and also activate platelet adherence and aggregation which can contribute to the obstructive process. Thus, it appears possible that neutrophils may alter microvascular circulation and, perhaps, damage the microvascular endothelium during ischaemia–reperfusion injury (Engler *et al.*, 1984, 1986b; Hernandez *et al.*, 1987; Carden *et al.*, 1990). It is also possible that the complex cellular events described above may involve a more specific factor which alters microvascular permeability, a variable which has consistently been shown to be increased during ischaemia of relatively short duration (Svendsen *et al.*, 1991).

Perhaps the most dramatic and pathologically significant microvessel abnormality, however, is the "no reflow" phenomenon associated with ischaemia–reperfusion injury, a phenomenon linked to neutrophil localization (Ambrosio *et al.*, 1989). This phenomenon is not seen after short periods of ischaemia and is generally associated with ischaemic periods of greater than 3 h which, upon reperfusion, are followed by reactive hyperaemia in the ischaemic bed. However, over the subsequent hours, several areas which are initially reperfused become ischaemic again as a result of microvascular obstruction so that there is a recurrence of severe ischaemia (Ambrosio *et al.*, 1989). This phenomenon, when it occurs, appears to be essentially an irreversible injury.

Thus, accumulation of neutrophils in the microvessels is associated with a number of pathophysiologically important changes that occur after relatively short periods of ischaemia and reperfusion but which can result in an irreversible state if prolonged. Although changes in cell shape (or deformability) and vasoconstriction may be important mechanisms for microvascular neutrophil accumulation, a large preponderance of evidence suggests that specific interactions between adhesion molecules is the most critical factor in control of leukocyte-induced pathophysiological changes in ischaemia and reperfusion.

3.2 INTERCELLULAR ADHESION AND INFLAMMATORY INJURY – CD18 AND THE LEUKOCYTE β_2 INTEGRINS

Recent evidence suggests that intercellular adhesion is a critical step in the accumulation of leukocytes at sites of inflammation (Smith, 1992). An overview of endothelial biology, given in Chapter 3 of this book, should be consulted before further reading.

In a variety of cardiovascular-related inflammatory models mABs made to the CD18 subunit of the leukocyte integrin CD11/CD18 glycoprotein complex have markedly reduced the pathophysiological consequences of cardiac ischaemia. Anti-CD18 antibodies have been demonstrated to reduce myocardial infarct size in a rapid model of 1 h ischaemia plus 5 h of reperfusion when administered systemically before the coronary occlusion (Seewaldt-Becker *et al.*, 1989). In another study in rabbits, a different anti-CD18 mAB (60.3) was applied to radiolabelled rabbit neutrophils before their introduction into a rabbit undergoing a 30 min occlusion and 3 h reperfusion protocol; the accumulation of neutrophils was markedly reduced (Williams *et al.*, 1988). Similar studies have been done in canine models of cardiac ischaemia–reperfusion in our laboratory demonstrating that the anti-CD18 mAB R15.7 reduces neutrophil accumulation in the previous ischaemic area (Dreyer *et al.*, 1991b) and anti-CD18 antibodies have been shown to reduce myocardial infarct size as well (Ma *et al.*, 1991). In Chapter 8 results of studies on the effects of anti-adhesion molecule antibodies on infarction are shown in more detail.

In addition to myocardial injury, evidence has been provided that adhesion-dependent injury to the endothelium in coronary vessels exists as part of the pathophysiology of ischaemia–reperfusion injury while the mechanism of this dysfunction is likely complex and involves other factors such as reactive oxygen radical generation (Suzuki *et al.*, 1989; Tsao and Lefer, 1990). However, depression of endothelial-derived relaxation in postischaemic vessels has been mitigated by anti-inflammatory strategies as well. Evidence suggests that large vessel injury is associated with neutrophil migration into the subendothelial layer of the vascular wall and can be obviated by mABs to CD18 (Ma *et al.*, 1991).

Finally, Hernandez *et al.* (1987) have studied ischaemia and reperfusion in the small intestine and demonstrated a marked reduction in the increased microvascular permeability occurring with ischaemia and reperfusion that could be modulated by either anti-neutrophil anti-serum or by the mAB 60.3. Korthuis and co-workers have demonstrated similar phenomena after skeletal muscle ischaemia and reperfusion has been blocked by IB4, another anti-CD18 mAB (Carden *et al.*, 1990).

These studies, and many more which have followed, suggest a critical role for the leukocyte "β_2(CD18)" integrins in neutrophil-induced inflammatory injury. The leukocyte integrins are made up of alpha (CD11) and beta (CD18) subunits which form a heterodimer. While there are three known alpha subunits (CD11a, CD11b, CD11c), there is a common CD18 subunit. The function of CD18 integrins is most dramatically demonstrated by examining neutrophil adhesion to endothelial cells in patients with a genetic defect resulting in defective or absent CD18; neutrophils from these patients show a substantial reduction in adhesion to endothelial monolayers and, perhaps more importantly, a virtual absence of neutrophil transmigration across the endothelial monolayer (Smith *et al.*, 1988, 1989). Similar findings can be demonstrated utilizing anti-CD18 mABs (Harlan *et al.*, 1985; Smith *et al.*, 1988, 1989; Furie *et al.*, 1991). The role of CD18-containing integrins in mediating neutrophil transmigration will be discussed further below.

Finally, recent experiments from our laboratory have demonstrated a CD18-dependent neutrophil adhesion to cardiac myocytes from adult dogs (Entman *et al.*, 1990) that modulates a neutrophil-specific injury to the myocardial cell; both the adhesion and the toxic effects of the neutrophil on the myocyte are completely blocked by anti-CD18 (Entman *et al.*, 1992a). Thus, the discussion above outlines two potential roles for CD18 in the inflammatory injury associated with ischaemia–reperfusion: (1) CD18 integrins are obligatory for transmigration across the post-capillary venule and tissue localization; and (2) CD18 integrins modulate tissue injury associated with neutrophil–cell adhesion.

3.3 THE ALPHA SUBUNIT OF LEUKOCYTE INTEGRINS AND THEIR SPECIFICITY FOR CD18-DEPENDENT FUNCTIONS

Although CD18 is the common beta subunit of three heterodimers found on neutrophil surfaces (also known as leukocyte integrins or beta$_2$ integrins) the species of CD11 imparts significant specificity to the functions these integrins play in the interaction of neutrophils with other cells for extracellular surfaces. The controls and functional characteristics of these integrins is an area of active research at this time and a complete review of this subject is not feasible within the context of the subject of this chapter; the reader is referred to a recent chapter in another volume of this series (Shappell and Smith, 1992) which deals with this in greater detail. The discussion herein will deal primarily with the function of LFA-1 (CD11a/CD18) and Mac-1 (CD11b/CD18). The third leukocyte integrin P150,95 (CD11c/CD18) has not been extensively investigated.

LFA-1 and Mac-1 have distinctly different characteristics from both a functional and cell biological standpoint. These characteristics influence their role in the pathophysiology of the inflammatory reaction to ischaemia–reperfusion injury. LFA-1 is found on the surface of the neutrophil and there are no intracellular storage pools; chemotactic stimulation does not alter cell surface levels of this heterodimer (Todd *et al.*, 1984; Petty *et al.*, 1987; Dougherty *et al.*, 1988; Kishimoto *et al.*, 1989b). In contrast, Mac-1 is found in low concentrations on the surface of the unstimulated neutrophil and is stored in large quantities in secondary granules (Bainton *et al.*, 1987). Upon chemotactic stimulation, there is a rapid (within minutes) translocation of Mac-1 to the cell surface.

3.3.1 Alpha Subunit Specificity and Intercellular Adhesion

The role of the leukocyte integrins in cellular adhesion was first appreciated in the observations of Schmalstieg *et al.* (1986), who found that chemotactic migration of CD18-deficient neutrophils across a protein-coated plastic surface was markedly reduced (to less than 20%) compared to normal neutrophils. This was followed by a demonstration that monoclonal antibodies against CD18 markedly diminish this migration and that the specific heterodimer mediating this adhesion-dependent migration is Mac-1 since mABs against CD11b were markedly inhibitory while those of CD11a had little effect (Schmalstieg *et al.*, 1986). Evidence suggests that stimuli such as chemotactic factors or phorbol esters exert their effect both by promoting rapid translocation to the cell surface from the intracellular stores in the secondary granules (Berger *et al.*, 1984) and also by increasing the adhesive function of constitutive surface Mac-1. Similar conclusions have been reached with regard to the increased adhesion to endothelial cells *in vitro* following chemotactic stimulation of neutrophils, which are likewise greatly inhibited by antibodies to CD11b or CD18 (Smith *et al.*, 1989).

It has been proposed that chemotactic stimulation of neutrophils also influences the distribution of Mac-1 adhesion sites on the surface of the neutrophil and plays an important role in the mechanism of cell motility. Francis *et al.* (1989) proposed a model whereby sequential expression of cell surface Mac-1 contributes to the locomotion of adhesive neutrophils. Utilizing fluorescent-labelled mABs, these investigators demonstrated that Mac-1 is up regulated during locomotion as granules fuse with a plasma membrane near the lamellipodia. As the cell migrates, the Mac-1 is swept to the uropod and accumulates in the retraction fibres. Recent experiments from our laboratory utilizing latex beads coated with albumin of different sizes to label sequential generations of Mac-1 on the surface of the neutrophil have confirmed the role of Mac-1 adhesion with chemotactic motility and further defined this

mechanism (Hughes *et al.*, 1992b). The data suggest, however, that the newly surfaced Mac-1 does not immediately participate in adhesion (Hughes *et al.*, 1992a); there is strong evidence that the constitutive Mac-1 undergoes a qualitative change following stimulation of the neutrophil which promotes participation in adhesion as the cell proceeds along a leukotactic gradient (Hughes *et al.*, 1992a). Dransfield and Hogg have produced a monoclonal antibody to a conformational state of Mac-1 which depends on magnesium and temperature (Dransfield and Hogg, 1989; Dransfield *et al.*, 1992) which may represent this "activated" state. Thus, Mac-1-dependent adhesion of neutrophils results from both qualitative and quantitative changes in surface Mac-1 that may be critical for neutrophil migration induced by chemotactic gradient.

Finally, it should be noted that homotypic aggregation of neutrophils caused by a high concentration of chemotactic factors appears to be inhibited by anti-CD11b but not anti-CD11a (Anderson *et al.*, 1986). The ligand to which Mac-1 binds during homotypic aggregation is not known. It is important to recognize that Mac-1 is relatively promiscuous and bind to a variety of proteins including iC3b and ICAM-1 (Smith *et al.*, 1989; Diamond *et al.*, 1990, 1991) and various miscellaneous cellular proteins such as keyhole limpet haemocyanin (Shappell *et al.*, 1990b) and fibrinogen (Wright *et al.*, 1989). Mac-1 adhesion at low concentrations of chemotactic agent is apparently associated in a specific way with NADP oxidase activation of neutrophils (Nathan, 1987; Entman *et al.*, 1990; Shappell *et al.*, 1990b). The possible physiological significance of this will be discussed later in this chapter.

In contrast, LFA-1 does not bind to artificial surfaces and does not appear to be quantitatively changed by chemotactic stimuli. However, LFA-1 does appear to be an important ligand for adhesion to endothelial cells and plays an important role in neutrophil transmigration across endothelial barriers (Smith *et al.*, 1989; Furie *et al.*, 1991). Cytokine stimulation of endothelial cells results in a striking increase in their adhesiveness for unstimulated neutrophils (Pober and Cotran, 1990). In contrast to the effects of chemotactic factors described above, this increase is slow to appear (3–4 h) and requires protein synthesis. Also, in contrast to the effects of chemotactic factors, this adhesion is principally inhibited by mABs against LFA-1 which, as stated above, appears to be constitutively expressed on the surface of the neutrophil and not increased by chemotactic stimulation. Subsequent experiments have demonstrated that cytokine stimulation exerts its effect on adhesion by inducing the synthesis of ICAM-1, which is a counter ligand for LFA-1 (Rothlein *et al.*, 1986; Marlin and Springer, 1987). In contrast to Mac-1, LFA-1 adhesion is specific for both ICAM-1 and ICAM-2 (the function of the latter being unknown at this time). Although ICAM-1 is expressed to some degree in unstimulated endothelial cells, its expression is markedly up regulated by cytokines. Monoclonal antibodies against ICAM-1 quantitatively inhibit binding of unstimulated neutrophils to up regulated endothelial cells to the same degree as anti-LFA-1 (Smith *et al.*, 1989). Finally, as was the case for Mac-1, there is now substantial evidence for a qualitative change in LFA-1 induced by chemotactic agents or phorbol esters which is necessary for its interaction with endothelial ICAM-1 (Dustin *et al.*, 1987); this adhesion appears to be transient and reversible (peaking within 15 min and returning to baseline within 30 min) (Dustin *et al.*, 1987; Dustin and Springer, 1989). This increased affinity of LFA-1 is associated with the appearance of an Mg^{2+}-dependent epitope on LFA-1, which is presently only in activated neutrophils (Dransfield *et al.*, 1992). Although chemotactic stimulation appears to induce the qualitative change in LFA-1 (Lo *et al.*, 1989; Smith *et al.*, 1989), it is also likely that a similar change in LFA-1 occurs in experiments where unstimulated neutrophils interact with cytokine-stimulated endothelial cells since the unstimulated neutrophils do not appear to bind to the constitutively expressed ICAM-1 on the unstimulated endothelial cells.

It is now clear that endothelial cells produce chemotactic factors in response to cytokine stimulation (Smith, 1992). One of these factors, IL-8 (Gimbrone *et al.*, 1989), and its role as a chemotactic factor has been well defined (Baggiolini *et al.*, 1989). IL-8 is produced by endothelial cells *in vitro* by cytokine stimulation over a time course similar to that describing the alterations in neutrophil adhesiveness for endothelial cells. Although it also induces an increase expression of Mac-1 (Huber *et al.*, 1991) it is an excellent candidate for the qualitative modulator of LFA-1 which accounts for the predominantly LFA-1-dependent binding of unstimulated neutrophil to cytokine-stimulated endothelial cells. Recent evidence suggests that the IL-8-induced activation may also involve cellular PAF as a co-factor (Kuijpers *et al.*, 1992).

3.3.2 Alpha Subunit Specificity and Neutrophil Migration

Examination of transendothelial migration *in vitro* and in tissue suggests that interactions of both LFA-1 and Mac-1 with ICAM-1 play important roles in this process further confirming the critical role for intercellular adhesion in the transmigration of neutrophils (Smith *et al.*, 1989; Furie *et al.*, 1991). As would be anticipated, therefore, this event can be induced *in vitro* either by establishing a concentration gradient of a chemotactic factor across an endothelial cell monolayer (Harlan *et al.*, 1985; Furie *et al.*, 1987, 1991) or by stimulation of endothelial cells by cytokines (Smith *et al.*, 1988; Furie and McHugh, 1989; Moser *et al.*, 1989; Huber *et al.*, 1991). In each case, anti-CD18 mABs almost completely inhibit migration, and CD18-deficient neutrophils exhibit very little migration (Harlan *et al.*, 1985; Smith

et al., 1988, 1989). These latter findings provide a pathophysiological explanation for the absence of extravascular neutrophils in the inflamed tissues of patients with CD18 deficiency (Anderson and Springer, 1987; Todd and Freyer, 1988). Monoclonal antibodies against both CD11a and CD11b can each partially inhibit transendothelial migration and show an additive inhibitory effect equivalent to that seen with anti-CD18 or ICAM-1 (Smith *et al.*, 1989). The presence of a chemotactic gradient increases the relative importance of Mac-1, and transendothelial migration of non-stimulated neutrophils across activated endothelial layers appears to be mediated primarily by LFA-1.

The characteristics of the migration process have been studied extensively with respect to the cellular distribution of Mac-1; equivalent studies with LFA-1 examining its distribution during migration have not been performed. The observation of a short-lived activated state of LFA-1 in neutrophils and lymphocytes, which peaking within 15 min and returning to baseline at 30 min (Dustin and Springer, 1989; Lo *et al.*, 1989) suggests the possibility that an activation–deactivation sequence may be important in allowing the ligand to adhere and detach during the process of transmigration; however, this also has not been characterized for either integrin.

In summary, although experimental paradigms can be designed to separate the actions of LFA-1 and Mac-1 on transendothelial migration, under the pathophysiological conditions ordinarily accompanying acute inflammation both chemotactic gradients and cytokine stimulation are present and it is likely that both of these integrins play an important part in this process; work from our laboratory suggests that this process may be cooperative rather than merely additive (Smith *et al.*, 1989).

3.3.3 Leukocyte Integrins and Adhesion-dependent Cytotoxicity–Mac-1 Adhesion to Parenchymal (Myocyte) ICAM-1

Recent work from our laboratory examined potential mechanisms of neutrophil adhesion to isolated adult cardiac myocytes *in vitro* (Entman *et al.*, 1990). Intercellular adhesion occurred only if the myocytes were stimulated with cytokines that induce expression of ICAM-1 (Smith *et al.*, 1991a), and when the neutrophils were stimulated to enhance CD18-dependent adhesion. Stimulation of the myocytes with cytokine such as IL-1, TNF-α, and IL-6 induced the expression of ICAM-1 (Smith *et al.*, 1991a; Youker *et al.*, 1992) and stimulation of neutrophils with zymosan activated serum (a C5a source) or PAF-promoted CD18-dependent adhesion that was blockable by both anti-CD18 (R15.7) and anti-ICAM-1 (CL18/6) mABs (Entman *et al.*, 1990; Smith *et al.*, 1991a). As will be discussed further in this chapter, cardiac lymph collected during the early reperfusion period after experimental myocardial ischaemia (Michael *et al.*, 1979) appears to contain a neutrophil-activating factor (C5a) (Dreyer *et al.*, 1991a) and potent cytokine activity capable of stimulating ICAM-1 transcription and expression (Youker *et al.*, 1992). We thus considered the hypothesis that neutrophils might induce direct cytotoxicity on cardiac myocytes under circumstances of ischaemia-reperfusion injury. We found that adhering neutrophils were apparently cytotoxic to cardiac myocytes within a few minutes so we examined the role of neutrophil–myocyte adhesion and various adhesion ligands in this process.

The binding of neutrophils to activated cardiac myocytes was found to be specific for the Mac-1–ICAM-1 interaction and was completely blocked by antibodies to ICAM-1, CD11b and CD18, but was unaffected by antibodies to CD11a which are capable of blocking neutrophil binding to an endothelial cell monolayer (Entman *et al.*, 1992a; Youker *et al.*, 1992). In separate experiments either neutrophils or cardiac myocytes were loaded with DCFH and the oxidation of this marker to DCF was monitored. Utilizing zymosan-activated serum (1%) to activate neutrophils and IL-1, TNF-α or IL-6 to stimulate ICAM-1 formation on myocytes, we were able to show that fluoresence of either cell was dependent on neutrophil–myocyte adhesion and was likewise blocked by antibodies to ICAM-1 (CL18/6), CD11b (MY904), and CD18 (R15.7). Neutrophils appeared to fluoresce immediately upon adhesion suggesting a rapid adhesion-dependent activation of the NADP oxidase of the neutrophils. In contrast, fluorescence of the cardiac myocytes appeared after several minutes and was rapidly followed by irreversible myocyte contracture. The iron chelator, desferroxamine, or the hydroxyl radical scavenger, dimethylthiourea, did not inhibit neutrophil adhesion but completely prevented either fluorescence or contracture in the cardiac myocyte. In contrast, extracellular oxygen radical scavengers such as superoxide dismutase and catalase or the extracellular ion chelator, starch-immobilized desferroxamine, did not reduce fluorescence, adhesion or the cytotoxicity. Under the experimental conditions utilized, no superoxide production could be detected in the extracellular medium during the neutrophil–myocyte adhesion under these conditions (Entman *et al.*, 1992a). We have interpreted these data as suggesting a compartmented transfer of reactive oxygen from neutrophil to myocyte dependent upon this adhesion reaction. The specific role for Mac-1 and ICAM-1 in this process suggests that neutrophil-induced cytotoxicity is specific to events produced by chemotactic activation. The possibility that other neutrophil-derived products are transferred at the neutrophil–myocyte interface via a similar compartmented mechanism has not been carefully investigated. Obviously neutrophils secrete a variety of enzymes which might similarly be toxic (Mullane and Smith, 1990). However, contracture or cell death was never seen in myocytes that had not previously fluoresced and was completely inhibited by the intracellular reactive oxygen scavenger systems despite the fact that many neutrophils adhere to the surface of these myocytes (Entman *et al.*, 1992a). The experiments were followed only for 2–3 h and the possibility of a latter toxicity mediated by other neutrophil components can not be ruled out.

The association of Mac-1 adhesion to NADP oxidase activation of neutrophils has been previously demonstrated in studies describing H_2O_2 production after neutrophil adhesion to protein-coated plastic (Nathan, 1987; Shappell *et al.*, 1990b) or cardiac myocytes (Entman *et al.*, 1990); this recent report identifies for the first time that ICAM-1 is a specific cellular ligand for Mac-1 that is necessary for the adhesion-dependent respiratory burst when neutrophils adhere to cardiac myocytes. It also suggests that neutrophil toxicity of parenchymal cells may not be something that can happen randomly but requires a highly specific set of circumstances. Not only must neutrophils localize in the extracellular fluids in response to a chemotactic gradient but also synthesis of ICAM-1 must be initiated by cytokine in order for this reaction to occur. Thus, it appears that the occurrence of chemotactic factor and the expression of ICAM-1 must relate temporally so that both activated neutrophils and ICAM-1-containing parenchymal cells co-exist.

In summary, adhesion mechanisms involving LFA-1 and Mac-1 and their interaction with ICAM-1 may mediate both the rapid activation and transmigration of neutrophils and also neutrophil-derived parenchymal cell toxicity. It would appear that the expression of ICAM-1 is obligatory for direct neutrophil-induced parenchymal injury. These mechanisms may play important roles in the overall process of cardiac injury.

3.4   THE SELECTIN FAMILY OF ADHESION LIGANDS – POTENTIAL ROLE IN ISCHAEMIA

The majority of the studies described above were done in systems where there was no flow. More recent studies have shown, however, that neutrophil adherence mechanisms are sensitive to shear stress. Lawrence *et al.* (1990) have shown *in vitro* that CD11/CD18–ICAM-1 adhesion occurs very infrequently at a wall shear stress of 0.5 dynes/cm^2 which is well below the calculated shear stress in the post-capillary venule, and Arfors *et al.* (1987) demonstrated that anti-CD18 mAB 60.3 did not inhibit neutrophil rolling on post-capillary venules in rabbits. This has added a complexity to the studies of neutrophil endothelial adhesion and has emphasized the importance of another class of adhesion molecules, the selectins (Bevilacqua *et al.*, 1991). This family of transmembrane glycoproteins each has an N-terminal lectin-like domain followed by an epidermal growth factor-like domain, and multiple repeats of a complement-binding-like domain in the extracellular space with a transmembrane region and a short cytoplasmic domain. Two of the known selectins are found on endothelial surfaces and one is found on the neutrophil; one of the endothelial selectins also is found on the surface of activated platelets. Evidence described below would suggest that selectins are critical molecules in allowing neutrophil adhesion under conditions of venular wall shear stress during physiologic blood flow.

3.4.1 P-Selectin

P-selectin (GMP-140, CO62) is found on the surface of activated platelets (site of their original description) and is also localized in the Weibel–Palade bodies of endothelial cells (McEver et al., 1989). It is rapidly shifted to the cell surface in response to such stimuli as histamine, thrombin, bradykinin, and oxygen free radicals (Geng et al., 1990; Pate et al., 1991). It is transiently retained on the cell surface (30 min to 1 h) and during this period can sustain neutrophil adhesion (Geng et al., 1990). Because P-selectin is stored intra-cellularly and can be rapidly mobilized to the endothelial (and platelet) surface, it is a prime candidate for early neutrophil adhesion. For similar reasons, P-selectin has also been suggested as a possible mediator of events occurring more acutely in ischaemic heart disease, such as cardiac arrhythmias (see Chapter 9 of this book). The ligand on the neutrophil that binds to P-selectin has not yet been fully characterized. All of the selectins contain a lectin domain and bind SLe^x carbohydrates but evidence suggests that the ligand for P-selectin and for the other endothelial selectin (E-selectin) have over-lapping specificity but are structurally distinct (Larsen et al., 1992). A specific protein component containing a sialyl LE^x-like moiety on the neutrophil appears to contribute to its specificity (Larsen et al., 1992).

3.4.2 E-Selectin

E-selectin (endothelial leukocyte adhesion molecule-1 or ELAM-1) is synthesized and expressed on the surface of endothelial cells over a 2–4 h time period after stimu-lation with LPS as well as a variety of cytokines such as IL-1 and TNF-α (Bevilacqua et al., 1989). Evidence would suggest that its expression results from new protein synthesis, and it is frequently co-expressed with ICAM-1 on endothelial cells since several of the same stimuli induce both types of adhesion molecule. When cytokine-stimulated HUVEC are exposed in vitro to unstimulated neutrophils, antibodies to ELAM-1 inhibit about 50% of the adhesion and antibodies to ICAM-1 inhibit about 40% (Luscinskas et al., 1989). As stated above, the neutrophil ligand to which the E-selectin binds has not been completely characterized, although it may be closely related to that for the P-selectin (Larsen et al., 1992). Recent evidence suggests that the third member of the selectin family, L-selectin, which is found on neutrophils is possibly one of the major ligands

binding to E-selectin (Kishimoto et al., 1991; Picker et al., 1991).

3.4.3 L-Selectin

L-Selectin (LECAM-1, leukocyte adhesion molecule-1, MEL-14 antigen, the peripheral lymph node homing receptor) is a leukocyte antigen first recognized as a lymphocyte homing receptor, and then shown to be important in murine models of peritoneal inflammation (Jutila et al., 1989). An mAB to this molecule (MEL-14) markedly inhibits neutrophil accumulation in this model. In contrast to the other selectins, L-selectin is con-stitutively expressed on the surface of neutrophils. Stimulation with either leukotactic factors, TNF-α, or GM-CSF results in a rapid disappearance of the antigen from the cell surface apparently resulting from proteolytic cleavage (Jutila et al., 1989; Kishimoto et al., 1989a). This downregulation occurs at the same time that surface Mac-1 is up regulated. It has been proposed that this ligand is important in neutrophil margination. Abbassi et al. (1991) have shown that mABs against L-selectin inhibit adhesion of canine neutrophils to cytokine-stimulated endothelial cell monolayers under conditions of flow that allow shear stress beyond 0.5 dynes/cm^2. Thus, L-selectin may allow neutrophil adhesion to venules after which the CD11/CD18 and ICAM-1 mechanism mediate transmigration. The proteolytic shedding of L-selectin may be an important detachment mechanism which allows the neutrophil to transmigrate (Kishimoto et al., 1989a; Smith et al., 1991b).

The role of selectins in myocardial reperfusion injury has yet to be determined and, other than under conditions of "no reflow" (i.e. physical trapping) that appears later in ischaemia (long after the major neutrophil influx), their activity represents the only currently defined mechanism by which shear stress can be over-come and neutrophil margination and transmigration can proceed. Some evidence for their involvement in neutrophil margination in in vitro systems will also be described below.

4. Chemotactic Factors – Signals that Initiate Inflammation in Ischaemia–Reperfusion Injury

4.1 COMPLEMENT ACTIVATION

Because our colleague, R.D. Rossen, has written a separate chapter on the role of complement in the initiation of inflammation associated with myocardial ischaemia–reperfusion (Chapter 6), the present discussion will only emphasize the points pertinent to the process of acute inflammation following coronary reperfusion. The role of complement in this process was first suggested by Hill and Ward (1971) using a rat model of myocardial ischaemia in which activation was detected after 3 h of ischaemia. Pinckard and colleagues (1973, 1975) attempted to approach the mechanism of complement activation in studies in which subcellular fractions rich in mitochondria from heart muscle were found to activate both classic and alternative complement pathways. Other mechanisms that have been suggested for complement activation are activation of lysosomal proteases (Hill and Ward, 1971) and factors associated with the release of reactive oxygen species (Shingu and Nobunaga, 1984; Vogt *et al.*, 1987). Rossen *et al.* (1988) have reported elevations of C1q binding molecules in the circulation of patients who have had acute myocardial infarction; these molecules were not found in patients with acute coronary syndromes in which there were no elevations of cardiac enzymes. The elevations seen in patients during the first few days after myocardial infarction were temporally associated with transient depressions of C3 and C4. Anti-cardiac antibodies were not found in these patients. The presence of C1q binding proteins in the serum of patients undergoing acute myocardial infarction was compatible with the reports of Pinckard *et al.* (1973, 1975) suggesting that this entity is associated with activation of the classic complement pathway in both experimental and clinical myocardial infarction.

In experimental models of myocardial ischaemia in the dog, we have demonstrated the localization of C1q in the ischaemic zone following a 45 min coronary occlusion and have correlated C1q localization with neutrophil accumulation in the same zone (Rossen *et al.*, 1985). Interestingly, small elevations of C1q localization could be demonstrated with occlusions as brief as 15 min. Michael *et al.* (1985) and Hendrickx *et al.* (1985) have previously demonstrated the presence of macromolecules originating in ischaemic myocardium that have egressed after coronary occlusions of 10–15 min duration in dogs and in baboons (Hendrickx *et al.*, 1985; Rossen *et al.*, 1985). In the former study (Rossen *et al.*, 1985) the timing of egress of macromolecules could be more sensitively measured because of the employment of an awake unanaesthetized canine model with a chronically cannulated cardiac lymph duct (Michael *et al.*, 1979). Using this method, Michael *et al.* were able to demonstrate macromolecular efflux peaking within the first hour of reperfusion regardless of the length of time of occlusion. The use of this latter model also allowed the study of inflammation in ischaemia–reperfusion injury under circumstances where there was no acute surgical trauma to obscure the mechanisms. Rossen *et al.* (1988) were thus able to demonstrate that C1q binding proteins appeared in cardiac lymph after coronary occlusions and that these proteins were present and demonstrable during the first 4 h of reperfusion. The macromolecules binding C1q ranged in size from 23 to 67 kDa, were of mitochondrial origin and were fully capable of activating the classic complement cascade (Rossen *et al.*, 1988). Subsequent studies from the same laboratory by Dreyer

et al. (1989), demonstrated that post-ischaemic cardiac lymph was strongly chemotactic with activity present within the first hour of reperfusion. Post-ischaemic cardiac lymph was able to stimulate neutrophils and induce shape change, Mac-1 surface expression, chemotaxis and increased adhesion to endothelial monolayer in neutrophils. Moreover, neutrophils found in the cardiac lymph during reperfusion demonstrated up regulation of surface Mac-1 (Dreyer *et al.*, 1989). Chemotactic activity was found within the first 4 h of reperfusion correlating very well with the time course of C1q binding protein appearance in the cardiac lymph (Dreyer *et al.*, 1989). Recent studies have demonstrated that this chemotactic activity seen early during reperfusion in the dog model is almost entirely neutralized by anti-C5a anti-sera (Dreyer *et al.*, 1991a). Finally, again in the dog model the rate of neutrophil localization in the ischaemic and reperfused myocardium is highest in the first hour and dissipates within the first 4 h (Dreyer *et al.*, 1991b).

Thus, it would appear that myocardial injury during ischaemia results in the egress of macromolecules that are capable of initiating the early inflammatory reaction ("reaction to injury") seen during reperfusion (Rossen *et al.*, 1988; Dreyer *et al.*, 1989). It is important to point out, however, that the data regarding time course of this reaction have come from experiments performed primarily in the dog because of the availability of the chronic cardiac lymph duct cannulation method. In contrast to man, the dog does not utilize a 5-lipoxygenase system for production of leukotrienes as a prominent feature in its inflammatory responses. The role of lipid-derived chemotactic factors will require study in models of ischaemia and reperfusion to be done with other species.

4.2 LIPID-DERIVED AUTACOIDS

In many species, activated neutrophils release LTB$_4$ which is a potent chemotactic agent. In the rabbit, pig and human neutrophils it is produced by activation of an enzymatic cascade that begins with PLA$_2$ and arachidonic acid entering into a pathway which involves 5-lipoxygenase. This latter enzyme is translocated to membrane components of the cell where it is enzymatically active in the production of the chemotactic mediators (Rowzer *et al.*, 1985). There are a variety of lipoxygenase enzymes found in neutrophils and their relative quantity varies according to species. Thus, 12-hydroxy isomers (12-HETE) are the major product in dog neutrophils (Mullane *et al.*, 1987) and 15-hydroxy isomers (15-HETE) are present at higher arachidonic acid concentrations in rabbit neutrophils. Because of the variation in the synthesis of lipoxygenase derivatives and the sensitivity of neutrophils to these products among species, one might expect that studies of myocardial salvage postischaemia of various inhibitors might give different results according to the species studied. In addition, a specific LTB$_4$ receptor antagonist, LY223982, was effective in inhibiting LTB₄-induced neutrophil shape change and up regulation of neutrophil Mac-1 adhesion glycoprotein complexes *in vitro* and was capable of reducing infarct size in a rabbit model of ischaemia (Taylor *et al.*, 1989). However, a similar LTB₄ antagonist, LY255283, had no effect on a dog model of ischaemia–reperfusion which is consistent with the observations regarding the relative absence of 5-lipoxygenase products in dog and the insensitivity of dog neutrophils to LTB₄ (Hahn *et al.*, 1990). In the rabbit, where cardioprotective effects were observed, the cardioprotective effect was associated with a reduction in tissue myeloperoxidase activity suggesting that the inhibition of LTB₄ action had reduced chemotactic activity in the infarcted area (Taylor *et al.*, 1989). Thus, in some species, lipoxygenase derivatives function as chemotactic agents that activate neutrophils and induce their migration to ischaemic tissue; whether or not this occurs early in ischaemia or only occurs after neutrophils have been localized to the ischaemic–reperfused area (and thus produce LTB₄) has not yet been determined.

There are other lipid-derived autacoids which may be pertinent in ischaemia although very little direct evidence is present at this writing. The peptide leukotrienes, LTC₄ and LTD₄ are produced by endothelial and vascular smooth muscle cells from LTA₄ that is secreted by leukocytes after cell adhesion (Feinmark and Cannon, 1986, 1987). Since these leukotrienes are potent vasoconstrictors they may play a role in the injury or potentiate it in some way.

Another lipid-derived molecule which is known to have chemotactic activity is PAF. The enzymatic pathway leading to its formation is induced by thrombin (Zimmerman *et al.*, 1990) in endothelial cells and recent evidence suggests that PAF species are produced non-enzymatically through the action of reactive oxygen species (see below). PAF is a potent chemotactic agent which induces increased surface expression of Mac-1 and promotes neutrophil adhesion and other Mac-1-dependent processes. As its name implies, it is also a potent activator of platelet function. The role of PAF in myocardial ischaemia and reperfusion has not been carefully studied but some interesting possibilities will be discussed in the next section where reactive oxygen species are considered as chemotactic agents. Thus far, a single study in rats suggests that PAF antagonists appear to reduce myocardial infarction after a 6 h permanent occlusion (Stahl *et al.*, 1988). PAF and other lipid-derived autacoids have been proposed as chemical mediators of sudden cardiac death by virtue of adverse effects on cardiac rhythm. This subject is examined in Chapter 9 of this book.

4.3 REACTIVE OXYGEN SPECIES AS CHEMOTACTIC ACTIVATORS

Generation and secretion of reactive oxygen species has been suggested as a major mechanism by which neutrophil-induced injury to other cells might occur. However, it is important to point out that the relationship between reactive oxygen species and neutrophil function may be more complex; there is significant evidence that reactive oxygen species may be important in neutrophil localization. The origin of this concept rests with the work of McCord and colleagues (Petrone *et al.*, 1980) who presented data suggesting that increased reactive oxygen species resulted in the formation of a "neutrophil-activating factor" present in serum. Although this factor was not identified, the possibility that an enzymatic or nonenzymatical formation of a lipid-derived autacoid must certainly be considered. Obviously, a reactive oxygen species will potentiate lipoxygenase and cycloxygenase reactions and non-enzymatic oxidation of unsaturated lipids is known to be a result of oxidative stress. In recent years, the work of Granger and his colleagues (Granger, 1988; Inauen *et al.*, 1990) has provided strong evidence for an involvement of reactive oxygen species in chemotaxis; they have demonstrated that the presence of free radical scavengers or inhibitors of free radical formation during ischaemia and reperfusion in the gut markedly reduces neutrophil infiltration (Suzuki *et al.*, 1989). Similar studies have not been done in myocardium. In addition to the activation of complement described above, two general mechanisms by which oxidative stress might induce neutrophil adhesion have been described which may be pertinent in reperfusion injury: (1) P-selectin surface expression (Patel *et al.*, 1991); and (2) the production of PAF analogues of oxidatively fragmenting phosphatidylcholine (Smiley *et al.*, 1991). These two mechanisms probably relate to the modulation of the two adhesion mechanisms described above with the former relating to the postulated early shear stress-resistant selectin-mediated adhesion and the latter relating to up regulation of Mac-1 and, perhaps, activation of LFA-1 (Smith, 1992).

4.3.1 Evidence for P-Selectin Up Regulation

Patel *et al.* (1991) demonstrated increased neutrophil adhesion in primary cultures of endothelial cells exposed to either water-soluble or lipid-soluble peroxides or induced to generate oxygen-derived free radicals by the oxidant, menadione. This formation of adhesive surface molecules occurred after a short exposure of the endothelial cells to the oxidant and clearly preceded any injury to the cells. The increased adhesion was subsequently demonstrated to be blocked by antibody to P-selectin suggesting that exposure of the endothelial cells to oxidative stress induced a prolonged expression of P-selectin on the cell surface which resulted in the increase of neutrophil adhesion. The increased adhesion was demonstrable on the endothelial cells for up to 4 h and responded to relatively a low hydrogen peroxide concentration (Patel *et al.*, 1991). These latter two properties of oxidative stress on P-selectin induced adhesion correlate with those described below relating to PAF production.

4.3.2 PAF Production and Oxidative Stress

Classic PAF (1-O-alkyl-2-acetyl-sn-glycero-3-phospho-choline) is a plasmalogenic phosphatidylcholine derivative which activates neutrophils via a receptor which specifically reacts with the sn-2 residue. There are a variety of specific antagonists to this receptor (discussed again in relation to ventricular arrhythmias, in Chapter 9). McIntyre and co-workers (Smiley *et al.*, 1991) have reported that oxidative stress is capable of generating short and intermediate length sn-2 derivatives from 1-palmitoyl-2-arachidonoyl-sn-glycero-3-phosphocholine. These derivatives appear to react with the PAF receptor to activate both platelets and neutrophils and this represents a mechanism by which oxidative stress might induce chemotactic stimulation. Gasic *et al.* (1991) demonstrated that hydrogen peroxide exposure to endothelial cell cultures or intact vessels induces a rapid and short-lived adhesion of neutrophils to endothelial cells that is blockable by anti-CD18 antibodies. When similar experiments were attempted with cardiac myocytes, it was found that neutrophil adhesion could only be stimulated if ICAM-1 was induced on the myocytes (Entman *et al.*, 1992b). However, when ICAM-1 was present on myocytes, exposure to hydrogen peroxide (in the absence of neutrophils) for 2 min induced adhesion of unstimulated neutrophils. This represented the first experiment in which unstimulated neutrophils were found to adhere to cardiac myocytes. Previous experiments had all required neutrophil activation (Entman *et al.*, 1990). In contrast to the latter conditions, the adhesion induced by hydrogen peroxide was short lived, peaking within 15 min and disappeared by 30 min and, thus, closely resembled the phenomenon described by Gasic *et al.* (1991) with endothelial cells. In each case (endothelial or myocyte adhesion) the adhesive stimulus was completely inhibited by the presence of a specific PAF-receptor antagonist WEB2086. However, the addition of exogenous PAF failed to reproduce the rapid activation and deactivation of adhesion seen with hydrogen peroxide. Subsequent experiments demonstrated that this early adhesive burst was blocked in each of these tissues by the anti-LFA-1 (anti-CD11a) monoclonal antibody, R7.1. This, again, contrasted with previous observations of experimental neutrophil myocyte adhesion all of which appeared to be Mac-1 specific (Entman *et al.*, 1992a). The rapid activation and deactivation of adhesion under these circumstances is reminiscent of the kinetics of adhesion observed in experiments delineating LFA-1 activation (see above) (Dustin *et al.*, 1987; Dustin and Springer, 1989) and suggests that oxidative stress may induce a PAF-receptor-dependent activation of LFA-1. The mechanism by which this might occur has not been delineated but recent evidence indicates that PAF and IL-8 may act

together at the endothelial surface to activate attached neutrophils (Kuijpers *et al.*, 1992).

5. *Potential Mechanisms of Neutrophil-induced Parenchymal Injury in Ischaemia–Reperfusion Injury*

The major focus of the previous sections has been the mechanism by which neutrophils are attracted to and activated in an area of injury. With respect to ischaemia–reperfusion injury, indirect evidence was cited that agents which altered neutrophil function or the number of neutrophils attracted to an area might actually prevent extension of the initial ischaemic injury.

In addition to the potential role of neutrophil-induced microvascular obstruction in extending tissue injury, there is also substantial evidence from intervention studies that neutrophils may also directly injure parenchymal cells through the release of specific products. Under conditions found *in vivo*, reactive oxygen species and proteolytic enzymes are secreted by adherent neutrophils (Suzuki *et al.*, 1989; Entman *et al.*, 1990; Inauen *et al.*, 1990; Shappell *et al.*, 1990b). Since the majority of information regarding specific neutrophil adhesion involves attachments to endothelial cells, we considered the possibility that neutrophil-derived products might be secreted consequent to neutrophil–endothelium adhesion and subsequently diffuse out of capillaries to damage the surrounding cells and extra-cellular structures. This hypothesis has little to recommend it since neutrophils obviously emigrate into close contact with parenchymal cells, and since there would be considerable dilution of the secreted products.

A more likely hypothesis is that ligand-specific adherence of neutrophil to parenchymal cells (e.g. cardiac myocytes) or endothelial cells is an important part of the secondary inflammatory injury occurring in ischaemia and reperfusion. There have been very few data *in vivo*, but recent *in vitro* evidence from our laboratories suggests that ligand-specific adhesion of neutrophils to myocardial cells is necessary for a direct toxic effect associated with neutrophil activation. Our experiments demonstrated that compartmented transfer neutrophil-derived toxic effects *in vitro* on cardiac myocytes mediated by oxygen radical species requires chemotactic activation of neutrophils with PAF or zymosan-activated canine serum (as a source of C5a) and also requires cytokine stimulation of ICAM-1 synthesis by the cardiac myocyte (Entman *et al.*, 1992a). The toxic effect could be blocked by mABs to ICAM-1, CD18 or CD11b. The resultant compart-mented transfer of reactive oxygen species appears to be critical to the rapid myocyte injury thus induced and we have suggested that this mechanism may be important in the secondary injury occurring during reperfusion (Entman *et al.*, 1992a).

The possible relevance of these observations made *in vitro* to the processes occurring during reperfusion-induced injury is strengthened by a series of studies we have done analysing cardiac lymph collected during reperfusion. Our model of the awake unanaesthetized dog with a chronically cannulated cardiac lymph duct has enabled us to examine inflammatory mediators under circumstances where the effects of surgical trauma are no longer a significant factor (Rossen *et al.*, 1988; Dreyer *et al.*, 1989; Youker *et al.*, 1992). The use of this model to define the time course and mechanism of complement activation during ischaemia and reperfusion is described in Chapter 6. In a more recent set of studies, we demonstrated that high dilutions of cardiac lymph collected during reperfusion were capable of stimulating ICAM-1 expression on the surface of cardiac myocytes *in vitro*. Preischaemic lymph contained no such activity. This cytokine activity was highest within the first hour of reperfusion but persisted up to 72 h. Post-reperfusion lymph contained cytokine activity which could stimulate the expression of ICAM-1 on both canine endothelial cells and canine myocytes. An important observation was the fact that the cytokine activity of the post-ischaemic cardiac lymph with respect to cardiac myocytes was completely inhibited by antibodies to human recombinant IL-6 at all time periods (Youker *et al.*, 1992). In contrast, the presence of anti-IL-6 antibodies did not inhibit the ability of cardiac lymph to induce ICAM-1 expression in endothelial cells thus suggesting a more complex cytokine content of extracellular fluid during reperfusion. Myocytes stimulated with postischaemic lymph are also susceptible to injury by activated neutrophils (Entman *et al.*, 1992a).

In further investigations of the pertinence of this mechanism *in vivo*, myocardial biopsies taken during reperfusion have been examined for the presence of mRNA for ICAM-1. In recent experiments following 1 h of ischaemia, we have demonstrated that ICAM-1 mRNA is present in ischaemic areas (less then 40% normal flow) within 1 h of reperfusion and continues to rise in concentration for 24 h (Anderson *et al.*, 1991). During the first 3 h of of reperfusion, there is no detectable mRNA for ICAM-1 in control areas suggesting that this is a highly selective process. At 24 h, however, ICAM-1 mRNA is found to some degree in all myocardial samples (although still highest in the ischaemic area) suggesting that circulating cytokines (e.g. IL-6) are now affecting normal areas as well as ischaemic areas (Anderson *et al.*, 1991). Thus, during the early reperfusion, there appears to be both an activation of neutrophils and induction of ICAM-1 mRNA occurring in the involved region that can be demonstrated *in vivo*. The presence of these factors suggests the possibility that, during reperfusion, a neutrophil-derived cytotoxic effect on cardiac myocytes may occur *in vivo* similar to that observed *in vitro*.

These observations and those in the discussion preceding them have led us to formulate the following hypothetical scheme to describe the cell biology of secondary inflammatory injury associated with ischaemia and reperfusion of the heart.

6. Cell Biology of Inflammatory Injury in Ischaemia–Reperfusion: A Working Hypothesis

We wish to propose the following working hypothesis which describes the events that mediate inflammatory reaction to ischaemia–reperfusion injury. While reperfusion injury is unquestionably a more complex phenomenon this construct deals specifically with the inflammatory component and makes the assumption that neutrophils are the principal determinant of this injury. Our hypothesis for the critical steps in this process is as follows:

6.1 CHEMOTAXIS

We hypothesize that the initial chemotactic event occurs when the injured myocardial cell releases macromolecules (C1q binding proteins) which activate the classical complement pathway as described. Thus, extracellular release of complement-activating macromolecules initiate production of C5a. *In vitro*, postischaemic cardiac lymph promotes chemotaxis, neutrophil-shape change, increased surface expression of Mac-1, and promotes neutrophil adhesion to endothelial cell and cardiac myocytes. Mac-1-dependent adhesion potentiates the release of oxygen free radicals. In the laboratory animal there is good evidence that macromolecular efflux from ischaemic tissue occurs after only 15 min of ischaemia and is extremely prominent in coronary occlusions of 45 min or more (Rossen *et al.*, 1985). Rossen *et al.* have suggested the primary neutrophil-activating molecules are of mitochondrial origins (Rossen *et al.*, 1988; Kagiyama *et al.*, 1989) and many of these contain cardiolipin which specifically binds C1q. Kagiyama *et al.* (1989) demonstrated that these molecules are capable of activating a classical complement pathway and result in the production of C5a. It is likely that specific lipoxygenase products such as LTB$_4$ are also important leukotactic agents in other species but they are not prominent in the canine. Whether or not these latter agents are secreted in response to the initial stimulus or are secreted as a result of the chemotactic stimulation by complement of leukocytes which then secrete LTB$_4$ as a secondary leukotactic mechanism is unclear at this time. Endothelial cells also secrete peptido-leukotrienes but the mechanism by which this would be induced in ischaemia is not known (see Chapter 9 for further discussion of peptido-leukotrienes). Regardless of the mechanism, however, evidence would suggest that leukotactic activity exists within the cardiac extracellular fluid within 15 min after a coronary occlusion and that the leukotactic activity, itself, is not dependent on reperfusion, although it appears that transmigration of neutrophils out of the intravascular space is markedly augmented by reperfusion.

6.2 NEUTROPHIL ADHESION TO ENDOTHELIUM

Activated neutrophils must first adhere to the endothelium to exert their effects on microvessels, to augment secretory functions of toxic products or vasoactive products and to emigrate into the extravascular space. Both the secretory and migratory functions require adhesion through CD11a/CD18 (LFA-1) and/or CD11b/CD18 (Mac-1). Although both are constitutively expressed to some degree, evidence cited above suggests that an additional activation step is required for both LFA-1-mediated transmigration as well as for Mac-1 adhesion and motility. It is important to point out, however, that it is not necessary to augment or activate the endothelial cell in order to produce transmigration. It is well known that neutrophils can follow a leukotactic gradient across unstimulated endothelial cells as long as the leukotactic stimulus activates the neutrophil (Harlan *et al.*, 1985; Furie *et al.*, 1987). It is apparently not necessarily required that ICAM-1 be up regulated on the endothelium during the early neutrophil transmigration associated with reperfusion; constitutively ICAM-1 may be sufficient.

It is also important to point out, however, that the above scheme, while absolutely necessary for neutrophil transmigration and neutrophil-derived toxicity does not appear to be sufficient to account for the entire adhesion transmigration process. As described previously, integrin-ICAM-1 adhesion is quite sensitive to shear stress and is unstable at shear levels above 0.5 dynes/cm^2. Calculated shear stresses in the post-capillary venule are several times that level and it appears unlikely that these shear stress levels can be sufficiently reduced by the relatively minor amount of microvascular plugging seen in early ischaemia and reperfusion. Thus, it would appear that another class of adhesion molecules is necessary to provide initial margination of neutrophils and allow transmigration mediated via ICAM-1. Neutrophils contain L-selectin which can be shown to be a very important ligand in margination in the studies of Abbassi and colleagues (Abbassi *et al.*, 1991; Smith *et al.*, 1991b) from our laboratory. However, the endothelial counter ligand identified for L-selectin is, at least in part, the E-selectin which must be synthesized on the endothelial cell in order to be expressed, i.e. it is not stored (Picker *et al.*, 1991). Since E-selectin is not stored and requires cytokine induction of several hours to be expressed on endothelial surfaces, its involvement in early margination is questionable (Bevilacqua *et al.*, 1989). This is also a reason for disregarding E-selectin as a mediator of other pathological events in acute myocardial ischaemia, as discussed in Chapter 9. While it is possible that an additional endothelial ligand for L-selectin exists, none has been identified and it is clear that the effects of L-selectin on endothelial cells require cytokine stimulation of the endothelial cell in order for margination to occur (Smith *et al.*, 1991b).

Therefore, we would postulate that the early margination event occurs through the surface expression of P-selectin on endothelial cells. P-Selectin is stored in the Weibel–Palade bodies and is rapidly translocated to the endothelial surface in response to thrombin and/or oxidative stress reactions both of which would be prominent during a reperfusion event associated with a myocardial infarction. There is apparently a specific neutrophil ligand for P-selectin which has not yet been characterized completely but differs from that for L-selectin in that it does not appear to be shed in response to leukotactic stimuli; however, it is also possible that P-selectin may interact with L-selectin as well (Picker *et al.*, 1991). The precise mechanism by which these selectins are involved in the overall adhesion transmigration process is an area of active investigation at this time.

6.3 CYTOKINE STIMULATION OF ADHESION MOLECULE SYNTHESIS

An important area for consideration in which there is relatively little progress relates to the stimuli that ultimately induce the synthesis of some of the adhesion molecules. While ICAM-1 is constitutively expressed on the endothelium and additional synthesis may not be required, the induction of ICAM-1 on parenchymal cells may be necessary in order for neutrophil-induced injury to occur (Entman *et al.*, 1992a). E-Selectin is not constitutively expressed and appears only after endothelial cells are stimulated by cytokines to induce its *de novo* synthesis (Bevilacqua *et al.*, 1989). P-Selectin is expressed in response to a variety of extracellular stimuli; however, our recent data suggest that there is an increase in mRNA for P-selectin in ischaemic areas as a function of timing of ischaemia and reperfusion (Manning *et al.*, 1992).

The data cited above suggest that IL-6 is a critical cytokine in the induction of the expression of ICAM-1 upon myocardial cells; endothelial stimulation by postischaemic cardiac lymph appears to result in response to other cytokines. The involvement of other cytokines is further supported by the observations that IL-6 secretion by cells is induced by other "primary" cytokines such as TNF-α and IL-1. The primary signal for cytokine production and the cell of origin of individual cytokines released locally in the ischaemic region has not been elucidated but is an area of significant interest. Mononuclear cells and endothelial cells are known to release cytokines capable of stimulating endothelial cells *in vitro*. In addition, recent studies have demonstrated that cytokines sufficient to activate expression of endothelial ICAM-1 and E-selectin can be supplied by activated platelets and degranulated tissue mast cells (Matis *et al.*, 1990). The signal to induce these cells to release cytokines is also unknown, although mononuclear cells are certainly activated by complement.

It is possible that the species of cytokines and growth factors secreted by, for example, the endothelial cell and the mononuclear cell, may change as a function of the duration of reperfusion. For example, during the early reperfusion phase (0–30 min), it is known that C5a levels (which stimulate monocytes) are very high in the involved region but are rapidly dissipated after 4 h of reperfusion (Dreyer *et al.*, 1989, 1991a); likewise, thrombin (which stimulates endothelium) may accumulate during occlusion and be cleared during reperfusion. Thus, some of the known primary signals for endothelial cells and mononuclear cells may change during reperfusion and one might speculate that the cytokines and growth factors secreted might also qualitatively or quantitatively change under such circumstances. It would be attractive to postulate that such a transition initiated a chronic healing phase of the inflammatory cascade. These mechanisms remain to be clarified.

6.4 CYTOTOXICITY OF NEUTROPHILS

Finally, once neutrophils have bound to the endothelium and transmigrated across the post-capillary venules they can migrate into the extracellular space. Our work with isolated cardiac myocytes and isolated neutrophils suggest that parenchymal cells expressing ICAM-1 on their surface are highly susceptible to neutrophil adhesion and neutrophil-induced injury. From the standpoint of acute injury, it would appear that reactive oxygen species are most prominent in this injury. The role of neutrophil-derived proteolytic and lipolytic enzymes cannot be easily defined utilizing studies of this type.

As emphasized in this chapter, the noxious effects of neutrophils are specifically dependent upon adhesion molecule interactions. Evidence from our laboratory (Entman *et al.*, 1990; Shappell *et al.*, 1990b) would suggest that Mac-1 is critical to the activation of NADP oxidase-related tissue injury. Our recent paper demonstrates that this is specifically a Mac-1–ICAM-1 interaction (Entman *et al.*, 1992a) and that the activation of NADP oxidase involves diglyceride-dependent activation of protein kinase C (Shappell *et al.*, 1990b). This latter finding, in addition to its possible relevance to neutrophil-associated parenchymal injury, reinforces the potential modulatory role of adhesion molecule

expression and interaction in control of specific enzymatic responses in inflammatory cells in the heart.

7. *Therapeutic Intervention into the Reaction to Injury: A Cell Biological Approach*

The insights into the cellular and molecular mechanisms that mediate the acute inflammatory reaction to injury associated with ischaemia and reperfusion has targeted new approaches to specific therapeutic interventions designed to reduce this process. These more specific and sophisticated approaches are particularly topical at this time because: (1) the past history of clinical intervention into myocardial infarct size limitation utilizing more general anti-inflammatory strategies has been catastrophic; and (2) a very high percentage of patients are now undergoing proteolytic therapy with successful reperfusion of the ischaemic myocardium during a period of time in which myocardial salvage is highly practical. Several approaches have been analysed in ischaemia models and several others are suggested that relate directly to a biological construct similar to that described above.

7.1 ANTI-COMPLEMENT APPROACHES

Early attempts at anti-complement therapy using complement depletion strategies involving cobra venom factor are obviously not practical clinically. In addition to depleting complement, this approach results in extraordinarily high levels of C5a for extended periods of time. Recombinant DNA technology has allowed production of a soluble form of CR1 (Weisman *et al.*, 1990) which is capable of binding active complement and preventing complement-induced inflammatory reactions. This molecule was successful in a rat model in markedly reducing myocardial infarction size. Such an approach does appear to have practical application to human disease, although it depends on the ability of CR1 to get into extracellular fluid of ischaemic areas sufficiently rapidly to bind up activated complement immediately upon reperfusion and depends on the scope to administer enough CR1 to bind the large majority of C5a generated proteolytically in the extracellular fluid. In addition, recently there have been a variety of studies attempting to develop inhibitors for the recently cloned and characterized C5a receptor. This approach would allow the inhibitor to find its receptor before reperfusion (neutrophils in the blood stream) but has the disadvantage of having to inhibit a high percentage of the receptors on individual neutrophils in order to block complement-derived chemotaxis.

7.2 Anti-adhesion Approaches

As previously described, monoclonal antibodies to CD18 have been used by a variety of laboratories to reduce ischaemia–reperfusion injury. Simpson and colleagues utilized monoclonal antibody to human CD11b; this appears to minimize infarct size in dogs after ischaemia and reperfusion despite the fact that it did not inhibit adherence of dog neutrophils to dog endothelial cells or cardiac myocytes (Simpson *et al.*, 1988, 1990). Interestingly, we found that this antibody also inhibited the production of reactive oxygen species by adherent dog neutrophils despite the fact that it did not interfere with neutrophil adhesion reactions. This suggests that the integrin system may also have a secondary controlling effect (in addition to adhesion) on reactive oxygen production (Shappell *et al.*, 1990b; Entman *et al.*, 1992a). In subsequent papers, Simpson and colleagues were unable to demonstrate that neutrophil infiltration was reduced by antibody 904 (Simpson *et al.*, 1990); in accordance with our observation this antibody did not inhibit adhesion *in vitro*. Antibody 904 may be acting primarily by reducing the oxidative reaction of these neutrophils.

In addition, a chimeric model of IgG and soluble L-selectin reduces neutrophil accumulation in a mouse peritonitis model (Watson *et al.*, 1991) indicating that soluble forms of single types of adhesion molecules may have therapeutic effects.

In the future it is likely that site-specific approaches to both neutrophil adhesion and neutrophil activation will be carefully studied and considered. It is possible that these approaches will be utilized in combination. While monoclonal antibodies, bioengineered proteins and peptide ligand analogue antagonists may be the initial candidates for intervention, the future will certainly bring the potential for use of site-specific small inorganic molecules.

8. *Summary*

We have attempted to present a classic cell biological scheme as a framework upon which one can examine the specific molecular processes that initiate reaction to injury occurring subsequent to reperfusion in ischaemia–reperfusion settings. The target of these investigations is the analysis of the mechanism by which an inflammatory cascade, as a reaction to injury, extends a primary injury. Understanding of the specific controlling steps in such a cascade as a function of time and degree of injury is critical to the rational design of specific site-directed interventions. The ultimate aim of such investigations is to identify specific molecular targets and devise practical methods of intervening in the biology of the process. A greater understanding of the relative importance of the specific cell biological steps during ischaemia and reperfusion is critical in order to attain a practical therapeutic approach aimed at the specific biological targets.

9. *References*

Abbassi, O., Lane, C.L., Krater, S.S., Kishimoto, T.R., Anderson, D.C., McIntyre, L.V. and Smith, C.W. (1991). Canine neutrophil margination mediated by lectin adhesion molecule-1 (LECAM-1) *in vitro*. J. Immunol. 147, 2107–2115.

Ambrosio, G., Weisman, H.F., Mannisi, J.A. and Becker, L.C. (1989). Progressive impairment of regional myocardial perfusion after initial restoration of post ischemic blood flow. Circulation 80, 1846–1861.

Anderson, D.C. and Springer, T.A. (1987). Leukocyte adhesion deficiency: an inherited defect in the Mac-1, LFA-1 and p150,95 glycoproteins. Ann. Rev. Med. 38, 175–194.

Anderson, D.C., Miller, L.J., Schmalstieg, F.C., Rothlein, R. and Springer, T.A. (1986). Contributions of the Mac-1 glycoprotein family to adherence-dependent granulocyte functions: structure–function assessments employing subunit-specific monoclonal antibodies. J. Immunol. 137, 15–27.

Anderson, D.C., Smith, C.W., Michael, L.H. and Entman, M.L. (1991). Stimulation of ICAM-1 mRNA in ischemic and reperfused canine myocardium. Circulation 84, II85 (Abstract).

Arfors, R.E., Lundberg, C., Lindbom, L., Lundberg, K., Beatty, P.G. and Harlan, J.M. (1987). A monoclonal antibody to the membrane glycoprotein complex CD18 inhibits polymorphonuclear leukocyte accumulation and plasma leakage *in vivo*. Blood 69, 338–340.

Baggiolini, M., Walz, A. and Kunkel, S.L. (1989). Neutrophil-activating peptide-1/interleukin 8, a novel cytokine that activates neutrophils. J. Clin. Invest. 84, 1045–1049.

Bainton, D.F., Miller, L.J., Kishimoto, T.R. and Springer, T.A. (1987). Leukocyte adhesion receptors are stored in peroxidase-negative granules of human neutrophils. J. Exp. Med. 166, 1641–1653.

Berger, M., O'Shea, J.J., Crow, A.S., Folks, T.M., Chused, T.M., Brown, E.J. and Frank, M.M. (1984). Human neutrophils increase expression of C3bi as well as C3b receptors upon activation. J. Clin. Invest. 74, 1566–1571.

Bevilacqua, M.P., Stengelin, S., Gimbrone, M.A., Jr and Seed, B. (1989). Endothelial leukocyte adhesion molecule 1: an inducible receptor for neutrophils related to complement regulatory proteins and lectins. Science 243, 1160–1165.

Bevilacqua, M.P., Butcher, E., Furie, B., Gallatin, M., Gimbrone, M.A., Harlan, J.M., Kishimoto, T.K., Lasky, L.A., McEver, R.P., Paulson, J.C., Rosen, S.D., Seed, B., Siegelman, M., Springer, T.A., Stoolman, L.M., Tedder, T.F., Varki, A., Wagner, D.D., Weisman, I.L. and Zimmerman, G.A. (1991). Selectins: a family of adhesion receptors. Cell 67, 233.

Braunwald, E. and Kloner, R.A. (1982). The stunned myocardium: prolonged post-ischemic ventricular dysfunction. Circulation 66, 1146–1149.

Carden, D.L., Smith, J.K. and Korthuis, R.J. (1990). Neutrophil-mediated microvascular dysfunction in postischemic canine skeletal muscle. Role of granulocyte adherence. Circulation Res. 66, 1436–1444.

Diamond, M.S., Staunton, D.E., deFougerolles, A.R., Stacker, S.A., Garcia-Aguilar, J., Hibbs, M.L. and Springer, T.A. (1990). ICAM-1 (CD54): A counter-receptor for Mac-1 (CD11b/CD18). J. Cell Biol. 111, 3129–3139.

Diamond, M.S., Staunton, D.E., Marlin, S.D. and Springer, T.A. (1991). Binding of the integrin Mac-1 (CD11b/CD18) to the third immunoglobulin-like domain of ICAM-1 (CD54) and its regulation by glycosylation. Cell 65, 961–971.

Dougherty, G.J., Murdoch, S. and Hogg, N. (1988). The function of human intercellular adhesion molecule-1 (ICAM-1) in the generation of an immune response. Eur. J. Immunol. 18, 35–39.

Dransfield, I. and Hogg, N. (1989). Regulated expression of Mg^{2+} binding epitope on leukocyte integrin alpha subunits. EMBO J. 8(12), 3759–3765.

Dransfield, I., Cabanas, C., Craig, A. and Hogg, N. (1992). Divalent cation regulation of the function of the leukocyte integrin LFA-1. J. Cell Biol. 116, 219–226.

Dreyer, W.J., Smith, C.W., Michael, L.H., Rossen, R.D., Hughes, B.J., Entman, M.L. and Anderson, D.C. (1989). Canine neutrophil activation by cardiac lymph obtained during reperfusion of ischemic myocardium. Circulation Res. 65, 1751–1762.

Dreyer, W.J., Michael, L.H., Rossen, R.D., Nguyen, T., Anderson, D.C., Smith, C.W. and Entman, M.L. (1991a). Evidence for C5a in post-ischemic canine cardiac lymph. Clin. Res. 39, 271A (Abstract).

Dreyer, W.J., Michael, L.H., West, M.S., Smith, C.W., Rothlein, R., Rossen, R.D., Anderson, D.C. and Entman, M.L. (1991b). Neutrophil accumulation in ischaemic canine myocardium: insights into the time course, distribution, and mechanism of localization during early reperfusion. Circulation 84, 400–411.

Dustin, M.L. and Springer, T.A. (1989). T-Cell receptor cross-linking transiently stimulates adhesiveness through LFA-1. Nature (Lond.) 341, 619–624.

Dustin, M.L., Sanders, M.E., Shaw, S. and Springer, T.A. (1987). Purified lymphocyte function-associated antigen-3 (LFA-3) binds to CD2 and mediates T-lymphocyte adhesion. J. Exp. Med. 165, 677–692.

Engler, R., Dahlgren, M., Schmid-Schonbein, G.W. and Dobbs, A. (1984). Leukocyte depletion prevents progressive flow impairment to ischemic myocardium. Circulation 70, II-228.

Engler, R.L., Dahlgren, M.D., Morris, D.D., Peterson, M.A. and Schmid-Schonbein, G.W. (1986a). Role of leukocytes in response to acute myocardial ischemia and reflow in dogs. Am. J. Physiol. 251, H314–H323.

Engler, R.L., Dahlgren, M.D., Peterson, M.A., Dobbs, A. and Schmid-Schonbein, G.W. (1986b). Accumulation of polymorphonuclear leukocytes during 3 h experimental myocardial ischemia. Am. J. Physiol. 251, H93–H100.

Entman, M.L., Youker, K., Shappell, S.B., Siegel, C., Rothlein, R., Dreyer, W.J., Schmalstieg, F.C. and Smith, C.W. (1990). Neutrophil adherence to isolated adult canine myocytes: evidence for a CD18-dependent mechanism. J. Clin. Invest. 85, 1497–1506.

Entman, M.L., Youker, K., Shoji, T., Kukielka, G., Shappell, S.B., Taylor, A.A. and Smith, C.W. (1992a). Neutrophil induced oxidative injury of cardiac myocytes: a compart-mented system requiring CD11b/CD18-ICAM-1 adherence. J. Clin. Invest. 90, 1335–1345.

Entman, M.L., Youker, K., Taylor, A.A. and Smith, C.W. (1992b). Activation of neutrophil LFA-1 by H_2O_2-induced endogenous platelet-activating factor allows adhesion to cardiac myocytes ICAM-1. FASEB J. 6, A1690 (Abstract).

Feinmark, S.J. and Cannon, P.J. (1986). Endothelial cell leukotriene C4 synthesis results from intercellular transfer of leukotriene A4 synthesized by polymorphonuclear leuko-cytes. J. Biol. Chem. 261, 16466–16472.

Feinmark, S.J. and Cannon, P.J. (1987). Vascular smooth muscle cell leukotriene C4 synthesis: requirement for trans-cellular leukotriene A4 metabolism. Biochim. Biophys. Acta 922, 125–135.

Francis, J.W., Todd, R.F., Boxer, L.A. and Petty, H.R. (1989). Sequential expression of cell surface C3bi receptors during neutrophil locomotion. J. Cell. Physiol. 140, 519–523.

Furie, M.B. and McHugh, D.D. (1989). Migration of neutrophils across endothelial monolayers is stimulated by treatment of the monolayers with interleukin-1 or tumor necrosis factor-alpha. J. Immunol. 143, 3309–3317.

Furie, M.B., Naprstek, B.L. and Silverstein, S.C. (1987). Migration of neutrophils across monolayers of cultured microvascular endothelial cells. J. Cell Sci. 88, 161–175.

Furie, M.B., Tancinco, M.C.A. and Smith, C.W. (1991). Monoclonal antibodies to leukocyte integrins CD11a/CD18 and CD11b/CD18 or intercellular adhesion molecule-1 (ICAM-1) inhibit chemoattractant-stimulated neutrophil transendothelial migration in vitro. Blood 78, 2089–2097.

Gasic, A.C., McGuire, G.M., Krater, S.S., Farhood, A.I., Goldstein, M.A., Smith, C.W., Entman, M.L. and Taylor, A.A. (1991). Hydrogen peroxide pretreatment of perfused canine vessels induces ICAM-1 and CD18-dependent neutrophil adherence. Circulation 84, 2154–2166.

Geng, J.G., Bevilacqua, M.P., Moore, R.L., McIntyre, T.M., Prescott, S.M., Kim, J.M., Bliss, G.A., Zimmerman, G.A. and McEver, R.P. (1990). Rapid neutrophil adhesion to activated endothelium mediated by GMP-140. Nature 343, 757–760.

Gimbrone, Jr., M.A., Obin, M.S., Brock, A.F., Luis, E.A., Hass, P.E., Hebert, C.A., Yip, Y.K., Leung, D.W., Lowe, D.G., Kohr, W.J., Darbonne, W.C., Bechtol, K.B. and Baker, J.B. (1989). Endothelial interleukin-8: a novel inhibitor of leukocyte-endothelial interactions. Science 246, 1601–1603.

GISSI Study Group (1986). Effectiveness of intravenous thrombolytic treatment in acute myocardial infarction. Lancet 1, 397–402.

Granger, D.N. (1988). Role of xanthine oxidase and granulo-cytes in ischaemia-reperfusion injury. Am. J. Physiol. 255, H1269–H1275.

Hahn, R.A., MacDonald, B.R., Simpson, P.J., Potts, B.D. and Parli, C.J. (1990). Antagonism of leukotriene B4 receptors does not limit canine myocardial infarct size. J. Pharmacol. Exp. Ther. 253, 58–66.

Harlan, J.M., Killen, P.D., Senecal, F.M., Schwartz, B.R., Yee, E.K., Taylor, R.F., Beatty, P.G., Price, T.H. and Ochs, H.D. (1985). The role of neutrophil membrane glycoprotein GP-150 in neutrophil adherent to endothelium in vitro. Blood 66, 167–178.

Hearse, D.J. and Bolli, R. (1991). Reperfusion-induced injury: manifestations, mechanisms and clinical relevance. Trends Cardiovasc. Med. 1, 233–240.

Hendrickx, G.R., Amano, J., Kenna, T., Fallon, J.T., Patrick, T.A., Manders, W.T., Rogers, G.G., Rosendorff, C. and Vatner, S.F. (1985). Creatine kinase release not associated with myocardial necrosis after short periods of coronary artery occlusion in conscious baboons. J. Am. Coll. Cardiol. 6, 1299–1303.

Hernandez, L.A., Grisham, M.B., Twohig, B., Arfors, K.E., Harlan, J.M. and Granger, D.N. (1987). Role of neutrophils in ischemia–reperfusion-induced microvascular injury. Am. J. Physiol. 238, H699–H703.

Hill, J.H. and Ward, P.A. (1971). The phlogistic role of C3 leukotactic fragment in myocardial infarcts of rats. J. Exp. Med. 133, 885–900.

Hillis, L.D. and Braunwald, E. (1977). Myocardial ischemia. N. Engl. J. Med. 296, 1093–1096.

Huber, A.R., Kunkel, S.L., Todd, III, R.F. and Weiss, S.J. (1991). Regulation of transendothelial neutrophil migration by endogenous interleukin-8. Science 254, 99–105.

Hughes, B.J., Hollers, J.C., Crockett-Torabi, E. and Smith, C.W. (1992a). Recruitment of CD11b/CD18 to the neutrophil surface and adherence-dependent cell locomotion. J. Clin. Invest. 90, 1687–1696.

Hughes, B.J., Hollers, J.C. and Smith, C.W. (1992b). In "Structure and Function of Molecules Involved in Leukocyte Adhesion II" (ed C.W. Smith, R. Rothlein, T.R. Kishimoto and P.E. Lipsky). Springer-Verlag, New York pp. 45–47.

Inauen, W., Granger, D.N., Meininger, C.J., Schelling, M.E., Granger, H.J. and Kvietys, P.R. (1990). Anoixa/reoxygenation-induced, neutrophil-mediated endothelial cell injury: role of elastase. Am. J. Physiol. 259, H925–H931.

Jolly, S.R., Kane, W.J., Bailie, M.B., Abrams, G.D. and Lucchesi, B.R. (1984). Canine myocardial reperfusion injury; its reduction by the combined administration of superoxide dismutage and catalase. Circulation Res. 54, 277–285.

Jutila, M.A.t, L., Berg, E.L. and Butcher, E.C. (1989). Function and regulation of the neutrophil MEL-14 antigen in vivo: comparison with LFA-1 and MAC-1. J. Immunol. 143, 3318–3324.

Kagiyama, A., Savage, H.E., Michael, L.H., Hanson, G., Entman, M.L. and Rossen, R.D. (1989). Molecular basis of complement activation in ischemic myocardium: identification of specific molecules of mitochondrial origin that bind C1q and fix complement. Circulation Res. 64, 604–615.

Kishimoto, T.K., Jutila, M.A., Berg, E.L. and Butcher, E.C. (1989a). Neutrophil Mac-1 and MEL-14 adhesion proteins inversely regulated by chemotactic factors. Science 245, 1238–1241.

Kishimoto, T.K., Larson, R.S., Corbi, A.L., Dustin, M.L., Staunton, D.E. and Springer, T.A. (1989b). The leukocyte integrins. Adv. Immunol. 46, 149–182.

Kishimoto, T.K., Warnock, R.A., Jutila, M.A., Butcher, E.C., Lane, C.L., Anderson, D.C. and Smith, C.W. (1991). Antibodies against human neutrophil LECAM-1 (LAM-1/Leu-8/DREG-56 antigen) and endothelial cell ELAM-1 inhibit a common CD18-independent adhesion pathway *in vitro*. Blood 78, 805–811.

Kuijpers, T.W., Hakkert, B.C., Hart, M.H.L. and Roos, D. (1992). Neutrophil migration across monolayers of cytokine-prestimulated endothelial cells: a role for platelet activating factor and IL-8. J. Cell Biol. 117, 565–572.

Larsen, G.R., Sako, D., Ahern, T.J., Shaffer, M., Erban, J., Sajer, S.A., Gibson, R.M., Wagner, D.D., Furie, B.C. and Furie, B. (1992). P-Selectin and E-selectin, distinct but overlapping leukocyte ligand specificities. J. Biol. Chem. 267, 11104–11110.

Lawrence, M.B., Smith, C.W., Eskin, S.G. and McIntire, L.V. (1990). Effect of venous shear stress on CD18-mediated neutrophil adhesion to cultured endothelium. Blood 75, 227–237.

Lo, S.K., Detmer, P.A., Levin, S.M. and Wright, S.D. (1989). Transient adhesion of neutrophils to endothelium. J. Exp. Med. 169, 1779–1793.

Lucchesi, B.R. and Mullane, K.M. (1986). Leukocytes and ischemia induced myocardial injury. Ann. Rev. Pharm. Tox. 26, 201–224.

Luscinskas, F.W., Brock, A.F., Arnaout, M.A. and Gimbrone, M.A., Jr (1989). Endothelial–leukocyte adhesion molecule-1-dependent and elukocyte (CD11/CD18)-dependent mechanisms contribute to polymorphonuclear leukocyte adhesion to cytokine-activated human vascular endothelium. J. Immunol. 142, 2257–2263.

Ma, X.-L., Tao, P.S. and Lefer, A.M. (1991). Antibody to CD18 exerts endothelial and cardiac protective effects in myocardial ischemia and reperfusion. J. Clin. Invest. 88, 1237–1243.

Mallory, G.K., White, P.D. and Salcedo-Salgar, J. (1939). The speed of healing of myocardial infarction. A study of the pathologic anatomy in seventy-two cases. Am. Heart J. 18, 647–671.

Manning, A.M., Kukielka, G.L., Dore, M., Hawkins, H.K., Sanders, W.E., Michael, L.H., Entman, M.L., Smith, C.W. and Anderson, D.C. (1992). Regulation of GMP-140 mRNA in a canine model of inflammation. FASEB J. 6, A1060 (Abstract).

Marlin, S.D. and Springer, T.A. (1987). Purified intercellular adhesion molecule-1 (ICAM-1) is a ligand for lymphocyte function-associated antigen 1 (LFA-1). Cell 51, 813–819.

Maroko, P.R., Carpenter, C.D., Chiariello, M., Fishbein, M.C., Radvany, P., Knostman, J.D. and Hale, S.L. (1978). Reduction by cobra venom factor of myocardial necrosis after coronary artery occlusion. J. Clin. Invest. 61, 661–670.

Matis, W.L., Lavker, R.M. and Murphy, G.F. (1990). Substance p induces the expression of an endothelial-leukocyte adhesion molecule by microvascular endothelium. J. Invest. Dermatol. 94, 492–495.

McEver, R.P., Beckstead, J.H., Moore, K.L., Marshall-Carlson, L. and Bainton, D.F. (1989). GMP-140, a platelet alpha-granule membrane protein, is also synthesized by vascular endothelial cells and is localized in Weibel–Palade bodies. J. Clin. Invest. 84, 92–99.

Michael, L.H., Lewis, R.M., Brandon, T.A. and Entman, M.L. (1979). Cardiac lymph from conscious dogs. Am. J. Physiol. 237, H311–H317.

Michael, L.H., Hunt, J.R., Weilbaecher, D., Perryman, M.B., Roberts, R., Lewis, R.M. and Entman, M.L. (1985). Creatine kinase and phosphorylase in cardiac lymph: coronary occlusion and reperfusion. Am. J. Physiol. 248, H350–H359.

Michael, L.H., Zhang, Z., Hartley, C.J., Bolli, R., Taylor, A.A. and Entman, M.L. (1990). Thromboxane B2 in cardiac lymph: effect of superoxide dismutase and catalase during myocardial ischemia and reperfusion. Circulation Res. 66, 1040–1044.

Moser, R., Schleiffenbaum, B., Groscurth, P. and Fehr, J. (1989). Interleukin 1 and tumor necrosis factor stimulate human vascular endothelial cells to promote transendothelial neutrophil passage. J. Clin. Invest. 83, 444–455.

Mullane, K.M. (1988). In "Human Inflammatory Disease. Clinical Immunology" (ed G. Marone, L.M. Lichtenstein, M. Condorell and A.S. Fauci), pp. 143–159. Dekker, Philadelphia, PA.

Mullane, K.M. and Smith, C.W. (1990). In "Pathophysiology of Severe Ischaemic Myocardial Injury" (ed H.M. Piper), pp. 239–267. Kluwer Academic Publishers, Dordrecht.

Mullane, K.M., Read, N., Salmon, J.A. and Moncada, S. (1984). Role of leukocytes in acute myocardial infarction in anesthetized dogs. Relationship to myocardial salvage by anti-inflammatory drugs. J. Pharmacol. Exp. Ther. 228, 510–522.

Mullane, K.M., Salmon, J.A. and Kraemer, R. (1987). Leukocyte-derived metabolites of arachidonic acid in ischemia-induced myocardial injury. Fed. Proc. 46, 2422–2438.

Mullane, K.M., Westlin, W. and Kraemer, R. (1988). In "Biology of the Leukotrienes", pp. 103–121. New York Academy of Science, New York.

Nathan, C.F. (1987). Neutrophil activation on biological surfaces: massive secretion of hydrogen peroxide in response to products of macrophages and lymphocytes. J. Clin. Invest. 80, 1550–1560.

Olafsson, B., Forman, M.B., Puett, D.W., Pou, A., Cater, C.V. and Friessinger, G.C. (1987). Reduction of reperfusion injury in the canine preparation by intracoronary perfluorochemical. Circulation 76, 1135–1145.

Patel, K.D., Zimmerman, G.A., Prescott, S.M., McEver, R.P. and McIntyre, T.M. (1991). Oxygen radicals induce human endothelial cells to express GMP-140 and bind neutrophils. J. Cell Biol. 112, 749–759.

Petrone, W.F., English, D.K., Wong, K. and McCord, J.M. (1980). Free radicals and inflammation: superoxide-dependent activation of a neutrophil chemotactic factor in plasma. Proc. Natl Acad. Sci. USA 77, 1159–1163.

Petty, H.R., Francis, J.W., Todd, R.F., III, Petrequin, P.R. and Boxer, L.A. (1987). Neutrophil C3bi receptors: formation of membrane clusters during cell triggering requires intracellular granules. J. Cell. Physiol. 133, 235–242.

Picker, L.J., Warnock, R.A., Burnt, A.R., Doergchuk, C.M., Berg, E.L. and Butcher, E.C. (1991). The neutrophil selectin LECAM-1 presents carbohydrate ligands to the vascular selectins ELAM-1 and GMP-140. Cell 66, 921–933.

Pinckard, R.N., Olson, M.S., Kelley, R.E., Detter, D.H., Palmer, J.D., O'Rourke, R.A. and Goldfein, S. (1973). Antibody-independent activation of human C1 after interaction with heart subcellular membranes. J. Immunol. 110, 1376–1382.

Pinckard, R.N., Olson, M.S., Giclas, P.C., Terry, R., Boyer, J.T. and O'Rourke, R.A. (1975). Consumption of classical complement components by heart subcellular membranes *in vitro* and in patients after acute myocardial infarction. J. Clin. Invest. 56, 740–750.

Pober, J.S. and Cotran, R.S. (1990). The role of endothelial cells in inflammation. Transplantation 50, 537–544.

Roberts, R., DeMello, V. and Sobel, B.E. (1976). Deleterious effects of methylprednisolone in patients with myocardial infarction. Circulation 53(Suppl. I), 204–206.

Romson, J.L., Hook, B.G., Rigot, V.H., Schork, M.A., Swanson, D.P. and Lucchesi, B.R. (1982). The effect of ibuprofen on accumulation of [111]Indium-labeled platelets and leukocytes in experimental myocardial infarction. Circulation 66, 1002–1011.

Romson, J.L., Hook, B.G., Kunkel, S.L., Abrams, G.D., Schork, M.A. and Lucchesi, B.R. (1983). Reduction of the extent of ischemic myocardial injury by neutrophil depletion in the dog. Circulation 67, 1016–1023.

Rossen, R.D., Swain, J.L., Michael, L.H., Weakley, S., Giannini, E. and Entman, M.L. (1985). Selective accumulation of the first component of complement and leukocytes in ischemic canine heart muscle: a possible initiator of an extra myocardial mechanism of ischemic injury. Circulation Res. 57, 119–130.

Rossen, R.D., Michael, L.H., Kagiyama, A., Savage, H.E., Hanson, G., Reisbery, J.N., Moake, J.N., Kim, S.H., Weakly, S., Giannini, E. and Entman, M.L. (1988). Mechanism of complement activation following coronary artery occlusion: evidence that myocardial ischemia causes release of constituents of myocardial subcellular origin which complex with the first component of complement. Circulation Res. 62, 572–584.

Rothlein, R., Dustin, M.L., Marlin, S.D. and Springer, T.A. (1986). A human intercellular adhesion molecule (ICAM-1) distinct from LFA-1. J. Immunol. 137, 1270–1274.

Rowzer, C.A. and Kargman, S.A. (1985). Translocation of 5-lipoxygenase to the membrane in human leukocytes challenged with ionophore A23187. J. Biol. Chem. 263, 10980–10988.

Schmalstieg, F.C., Rudloff, H.E., Hillman, G.R. and Anderson, D.C. (1986). Two dimensional and three dimensional movement of human polymorphonuclear leukocytes: two fundamentally different mechanisms of location. J. Leuk. Biol. 40, 677–691.

Schmid-Schonbein, G.W. and Engler, R.L. (1987). Granulocytes as active participants in acute myocardial ischemia and infarction. Am. J. Cardiovasc. Pathol. 1, 15–30.

Seewaldt-Becker, E., Rothlein, R. and Dammgen, J.W. (1989). In "Leukocyte Adhesion Molecules: Structure, Function, and Regulation" (ed T.A. Springer, D.C. Anderson, A.S. Rosenthal and R. Rothlein), pp. 138–148. Springer-Verlag, New York.

Shappell, S. and Smith, C.W. (1992). In "Adhesion Molecules" (ed C.D. Wegner). Academic Press, London (in press).

Shappell, S.B., Taylor, A.A., Hughes, H., Mitchell, J.R., Anderson, D.C. and Smith, C.W. (1990a). Comparison of antioxidant and nonantioxidant lipoxygenase inhibitors on neutrophil function. Implications for pathogenesis of myocardial reperfusion injury. J. Pharmacol. Exp. Ther. 252, 531–538.

Shappell, S.B., Toman, C., Anderson, D.C., Taylor, A.A., Entman, M.L. and Smith, C.W. (1990b). Mac-1 (CD11b/CD18) mediates adherence-dependent hydrogen peroxide production by human and canine neutrophils. J. Immunol. 144, 2702–2711.

Shingu, M. and Nobunaga, M. (1984). Chemotactic activity generated in human serum from the fifth component on hydrogen peroxide. Am. J. Pathol. 117, 210–216.

Simpson, P.J., Mickelson, J., Fantone, J.C., Gallagher, K.P. and Lucchesi, B.R. (1987). Iloprost inhibits neutrophil function in vitro and in vivo and limits experimental infarct size in canine heart. Circulation. Res. 60, 666–673.

Simpson, P.J., Todd, R.F., III, Fantone, J.C., Mickelson, J.R., Griffin, J.D. and Lucchesi, B.R. (1988). Reduction of experimental canine myocardial reperfusion injury by a monoclonal antibody (anti-Mol, anti-CD11b) that inhibits leukocyte adhesion. J. Clin. Invest. 81, 624–629.

Simpson, P.J., Todd, R.F., III, Michelson, J.K., Fantone, J.C., Gallagher, K.P., Lee, K.A., Tamura, Y., Cronin, M. and Lucchesi, B.R. (1990). Sustained limitation of myocardial reperfusion injury by a monoclonal antibody that alters leukocyte function. Circulation 81, 226–237.

Smiley, P.L., Stremler, K.E., Prescott, S.M., Zimmerman, G.A. and McIntyre, T.M. (1991). Oxidatively fragmented phosphatidylcholines activate human neutrophils through the receptor for platelet-activating factor. J. Biol. Chem. 266, 11104–11110.

Smith, C.W. (1992). In "Adhesion. Its Role in Inflammatory Disease" (ed J.M. Harlan and D.Y. Liu), pp. 85–115. W.H. Freeman and Company, New York.

Smith, C.W., Rothlein, R., Hughes, B.J., Mariscalco, M.M., Schmalstieg, F.C. and Anderson, D.C. (1988). Recognition of an endothelial determinant for CD18-dependent human neutrophil adherence and transendothelial migration. J. Clin. Invest. 82, 1746–1756.

Smith, C.W., Marlin, S.D., Rothlein, R., Toman, C. and Anderson, D.C. (1989). Cooperative interactions of LFA-1 and Mac-1 with intercellular adhesion molecule-1 in facilitating adherence and transendothelial migration of human neutrophils *in vitro*. J. Clin. Invest. 83, 2008–2017.

Smith, C.W., Entman, M.L., Lane, C.L., Beaudet, A.L., Ty, T.I., Youker, K., Hawkins, H.K. and Anderson, D.C. (1991a). Adherence of neutrophils to canine cardiac myocytes *in vitro* is dependent on intercellular adhesion molecule-1. J. Clin. Invest. 88, 1216–1223.

Smith, C.W., Kishimoto, T.K., Abbassi, O., Hughes, B.J., Rothlein, R., McIntyre, L.V., Butcher, E. and Anderson, D.C. (1991b). Chemotactic factors regulate lectin adhesion molecule 1 (LECAM-1)-dependent neutrophil adhesion to cytokine-stimulated endothelial cells *in vitro*. J. Clin. Invest. 87, 609–618.

Stahl, G.L., Terashita, Z.-I. and Lefer, A.M. (1988). Role of platelet activating factor in propagation of cardiac damage during myocardial ischemia. J. Pharmacol. Exp. Ther. 244, 898–904.

Suzuki, M., Onauen, W., Kiretys, P.R., Grisham, M.B., Meininger, C., Schelling, M.E., Granger, H.J. and Granger, D.N. (1989). Superoxide mediates reperfusion-induced leukocyte-endothelial cell interactions. Am. J. Physiol. H1740–H1745.

Svendsen, J.H., Bjerrum, P.J. and Haunso, S. (1991). Myocardial capillary permeability after regional ischemia and reperfusion in in vivo canine heart. Circulation Res. 68, 174–184.

Taylor, A.A., Gasic, A.C., Kitt, T.M., Shappell, S.B., Rui, J., Lena, M.L., Smith, C.W. and Mitchell, J.R. (1989). A specific leukotriene B$_4$ antagonist protects against myocardial ischemia–reflow injury. Clin. Res. 37, 528A (Abstract).

Todd, R.F., III, and Freyer, D.R. (1988). The CD11/CD18 leukocyte glycoprotein deficiency. Hematol. Oncol. Clinics North Am. 2, 13–28.

Todd, R.F., III, Arnaout, M.A., Rosin, R.E., Crowley, C.A., Peters, W.A. and Babior, B.M. (1984). Subcellular localization of the large subunit of Mol (Mol$_1$; formerly gp110), a surface glycoprotein associated with neutrophil adhesion. J. Clin. Invest. 74, 1280–1290.

Tsao, P.S. and Lefer, A.M. (1990). Time course and mechanism of endothelial dysfunction in isolated ischemic and hypoxic rat hearts. Am. J. Physiol. 28, H1660–H1666.

Vogt, W., von Zabern, I., Hesse, D., Nolte, R. and Haller, Y. (1987). Generation of an activated form of human C5 (C5b-like C5) by oxygen radical. Immunol. Lett. 14, 209–215.

Watson, S.R., Fennie, C. and Lanky, L.A. (1991). Neutrophil influx into an inflammatory site inhibited by a soluble homing receptor-IgG chimaera. Nature (Lond.) 349, 164–167.

Weisman, H.F., Barton, T., Leppo, M.K., Marsh, H.C., Jr, Carson, G.R., Concino, M.F., Boyle, M.P., Roux, K.H., Weisfeldt, M.L. and Fearon, D.T. (1990). Soluble human complement receptor type 1: *in vivo* inhibitor of complement suppressing post-ischemic myocardial inflammation and necrosis. Science 249, 146–151.

Williams, F.M., Collins, P.D., Nourshargh, S. and Williams, T.J. (1988). Suppression of ^{111}In-neutrophil accumulation in rabbit myocardium by MoA ischemic injury. J. Mol. Cell. Cardiol. 20(Suppl.), S33.

Wright, S.D., Lo, S.K. and Detmers, P.A. (1989). In "Leukocyte Adhesion Molecules: Structure, Function and Regulation" (ed T.A. Springer, D.C. Anderson, R. Rothlein and A.S. Rosenthal), pp. 190–207. Springer Verlag, New York.

Youker, K., Smith, C.W., Anderson, D.C., Miller, D., Michael, L.H., Rossen, R.D. and Entman, M.L. (1992). Neutrophil adherence to isolated adult cardiac myocytes: induction by cardiac lymph collected during ischemia and reperfusion. J. Clin. Invest. 89, 602–609.

Zimmerman, G.A., McIntyre, T.M., Mehra, M. and Prescott, S.M. (1990). Endothelial cell-associated platelet-activating factor: a novel mechanism for signaling intercellular adhesion. J. Cell Biol. 110, 529–540.

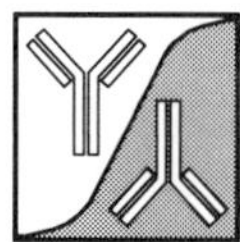

6. *Complement Activation in Cardiac Disease*

Roger D. Rossen

1. *Introduction*

Complement activation, mediated by attachment of antibodies to antigens expressed within cardiac tissue, has long been recognized as a cause of inflammation, and tissue injury within the heart. Antibody- and complement-mediated cell injury resulting in cardiac dysfunction has been implicated in rheumatic fever (Ghose and Mammen, 1977), Chagas disease (Hubsch *et al.*, 1976), myocarditis of various aetiologies (Wolfgram and Rose, 1989) and the post-cardiotomy and post-myocardial infarction syndromes (Kaplan and Frengley, 1969; Watanabe and Hyodo, 1970; Laguens *et al.*, 1988, 1989). It has also been suggested that antibodies reactive with antigens expressed by coronary vascular endothelium may participate in the development of the proliferative, atherosclerosis-like vascular lesions seen in chronic cardiac allograft rejection (Rossen *et al.*, 1971a,b; Anderson *et al.*, 1977; Forbes *et al.*, 1985; Palmer *et al.*, 1985; Hammond *et al.*, 1989; Oluwole *et al.*, 1989). A common feature of all of these conditions is the cognitive recognition of antigens expressed by endothelial cells, myocytes or other targets which results in the development of antibodies to cell surface membranes or subcellular organelles. Whether heart cell-reactive antibodies

develop in response to cross-reactive antigens expressed by an infectious agent, as occurs in rheumatic carditis (Ghose and Mammen, 1977; Kaplan, 1965; Kaplan and Frengley, 1969) or Chagas disease (Hubsch *et al.*, 1976; Laguens *et al.*, 1988, 1989) or to antigens released from cardiac tissue in the course of a virus infection or ischaemia (Kaplan and Frengley, 1969), it has been accepted for some time that interactions of antibodies with antigenic targets in the heart can activate complement, release chemotactic stimuli that attract leukocytes, and induce foci of inflammation which can alter the structure and function of the myocardium. Complement-dependent inflammation under such circumstances is not unexpected and is notable more for the site in which it occurs, and the disabilities which it causes, than for other unique characteristics.

More puzzling, perhaps, and initially less understandable is the intracardiac inflammation seen after ischaemic injury (Hill and Ward, 1969, 1971). This occurs in the apparent absence of heart-reactive antibodies (Pinckard *et al.*, 1980; Crawford *et al.*, 1988) and is apparently evoked by an intrinsic ability of ischaemic myocardium to activate the complement cascade (Schafer *et al.*, 1986). The purpose of this chapter is: (a) to summarize knowledge concerning the mechanisms responsible for

Immunopharmacology of the Heart
ISBN 0–12–200245–8

the release of subcellular constituents during an ischaemic episode that activates the complement cascade; (b) to review the consequences of intracardiac complement activation; and (c) to discuss mechanisms likely to regulate this process.

2. What Are the Characteristics of Ischaemic Heart Tissue that Make It a Particularly Efficient Nonimmune Activator of the Complement Cascade?

Hill and Ward were the first to recognize that ischaemic injury to muscle, particularly cardiac muscle, can activate the complement cascade. They suggested that injury resulted in the release of tissue lysosomal proteases which can generate a leukotactic substance originally thought to be a proteolytic fragment of the third component of complement (Hill and Ward, 1969). Subsequently this factor was recognized to be C5a (Weisman et al., 1990; Dreyer et al., 1989, 1991, 1992).

Prior treatment with cobra venom factor, a substance which effectively depletes C3 and later-acting complement components, significantly reduced complement-dependent, neutrophilic leukocyte (PMN) infiltrates in ischaemic myocardium during the first 48 h after coronary artery ligation (Hill and Ward, 1969, 1971). Pinckard et al. (1975) provided evidence to suggest that mitochondria from heart cells may be richly endowed with substances which trigger the early-acting components (C1, C4, C2 and C3) of the complement cascade. These investigations were the outgrowth of studies designed to identify factors which provoke formation of autoantibodies reactive with cardiac mitochondria following myocardial infarction (Pinckard et al., 1973). They showed that cardiac mitochondria contain substances which bind C1 and activate the classical pathway in the absence of detectable antibody; they also provided clinical evidence that myocardial cell necrosis causes significant decreases in serum C1, C4 and C3 during the 72 h immediately following a myocardial infarction. Subsequent investigations confirmed that although mitochondrial and other subcellular membranes may activate both the alternative and the classical complement pathways (Giclas et al., 1979), these subcellular constituents were capable of forming stable complexes only with C1 and C4, not with C3. Indeed cardiac mitochondria contain substances with a particularly high affinity for C1q, on the order of 1×10^9 to 1×10^{10} M^{-1} (Storrs et al., 1981, 1983). Proteins from cardiac mitochondria (Rossen et al., 1988; Kagiyama et al., 1989) and/or cardiolipin, a negatively charged phospholipid found principally in the inner mitochondrial membrane (Kovacsovics et al., 1985; Peitsch

et al., 1988), are thought to be the principal subcellular tissue components responsible for binding and activating C1.

The molecular mechanisms responsible for activating complement and generating complement-dependent leukocyte chemoattractive agents may be different in skeletal muscle. Using an isolated canine gracilis muscle to study ischaemia–reperfusion injury, Rubin et al. (1990) showed that upon reperfusion there was an 18% decrease in the level of factor B, a key component of the alternative pathway of complement activation, but no change in serum C4 levels.

Initially, it was unclear whether interventions designed to suppress inflammation provoked by complement activation could significantly reduce myocardial injury when implemented after a coronary occlusion, since irreversible myocardial cell damage is evident 20–25 min after the onset of ischaemia unless the tissue is reperfused (Sommers and Jennings, 1964; Herdson et al., 1965). Subsequent investigations have documented that reperfusion at later times will result in the salvage of part of the myocardium which otherwise would be irreversibly injured by coronary occlusion. However, delayed reperfusion creates problems in its own right, related principally to the influx of inflammatory cells and the release of their products in the heart muscle (Werns and Lucchesi, 1989). The studies of Maroko et al. (1978) were among the first to demonstrate the potential usefulness of interventions designed to limit influx of leukocytes into ischaemic myocardium, *after coronary occlusion*. They showed that injection of cobra venom factor 30 min after occlusion of the left anterior descending coronary artery in the dog, in quantities sufficient to reduce serum complement levels to less than 20% of normal, within 2–4 h thereafter, significantly reduced the loss of myocardial creatine phosphokinase and decreased the histological evidence of necrosis demonstrable after 24 h. A very similar reduction in myocardial necrosis associated with reduced leukocyte localization in ischaemic myocardium was seen in baboons treated with cobra venom factor 30 min after coronary artery ligation (Crawford et al., 1988). In untreated controls there was extensive evidence of tissue deposits of C3, C4 and C5; but in cobra venom factor-treated animals, only C4 deposition was evident. Weisman et al. (1990a,b) have further documented the intimate linkage between cardiac ischaemic injury, complement activation and localization of leukocytes in ischaemic myocardium by showing that intravascular infusion of sCR1 in rats undergoing myocardial ischaemia significantly reduces cardiac tissue localization of white blood cells and elements of the membrane attack complex of the complement cascade (C5b–C9) (Weisman et al., 1990a). Treatment with sCR1 in this experimental model of ischaemic coronary artery disease reduced the estimated quantity of infarcted myocardium by 44% (Weisman et al., 1990b).

Using ^{125}I-labelled human C1q, a subcomponent of the first component of complement, as a probe for sites in the heart that bind and activate complement, we found that experimental coronary occlusions lasting from 15 to 45 min in the dog were sufficient to provide a stimulus upon reperfusion for the accumulation of C1q in formerly ischaemic myocardium (Rossen *et al.*, 1985). In these studies, ischaemic myocardium was positively identified and defined by regional myocardial blood flow measurements, using radiolabelled microspheres. Ischaemic myocardial segments which accumulated high concentrations of C1q frequently generated stimuli that attracted ^{99}Tc-labelled autologous granulocytes, suggesting that sites which bound C1q were often sites for the assembly of the complement cascade and the generation of C5a. From such investigations it is now clear that there is a close temporal and physical relationship between complement activation and the accumulation of leukocytes in formerly ischaemic tissues, upon reperfusion. Moreover, these studies have demonstrated that stimuli which activate complement are released from formerly ischaemic myocardium certainly within 30 min of the onset of reperfusion (Rossen *et al.*, 1988) and possibly sooner (Dreyer *et al.*, 1989). Lymphatic fluids, collected from formerly ischaemic canine myocardium, after reperfusion, appear to contain a substance functionally homologous to C5a, as demonstrated by its chemotactic activity for granulocytes and its ability to induce a bipolar morphology in leukocytes resembling that induced in neutrophils when stimulated with zymosan-activated serum (Dreyer *et al.*, 1989). This chemotactic activity in canine cardiac lymphatic fluids is effectively abolished by adding antibodies to complement C5 to the post-reperfusion lymph (Dreyer *et al.*, 1992).

Canine cardiac lymph, collected after reperfusion of ischaemic myocardium, also contains soluble macromolecular complexes incorporating C1q and molecules originating from cardiac mitochondria and sarcoplasmic reticulum (Rossen *et al.*, 1988). Evidence for this assertion includes the fact that rabbit antisera, made against proteins precipitable from post-reperfusion lymph by anti-C1q, react not only with C1q but also with a 20 200 Da protein present in canine cardiac sarcoplasmic reticulum and three distinct constituents found in NP-40 lysates of isolated canine cardiac mitochondria. Indeed, detergent lysates of isolated canine cardiac mitochondria contain ten or more additional constituents, presumably protein in origin, which, *in vitro*, can bind C1 and activate the complement cascade (Kagiyama *et al.*, 1989). Continuing study of the complement-activating substances that may be isolated upon detergent lysis of canine cardiac mitochondria have demonstrated that some but not all contain cardiolipin (Rossen *et al.*, 1991).

It is likely that the mitochondrial proteins which contain cardiolipin are important substrates for assembly of the complete complement cascade during ischaemic injury. Cardiolipin efficiently activates C1 (Kovacsovics *et al.*, 1985; Peitsch *et al.*, 1988). But it is unlikely that cardiolipin, by itself, provides sites that permit covalent attachment of C4b and C3b and the assembly of the C3-converting enzymes necessary to complete activation of the complement cascade. Typically, sites which enable the activation of the complement cascade through the membrane attack complex bind hydrolysed C4b and C3b and protect the nascent converting enzyme from regulatory proteins that ordinarily interfere with its assembly and hasten its dissociation. Bindon *et al.* (1990) have shown that there is no necessary correlation between substrate binding of C1q and complement-mediated target cell lysis; their studies suggest that substrate characteristics may influence binding and enzymatic activities of the C1r$_2$C1s$_2$ tetramer as well as C4b and C3b. Surfaces of bacteria, and mitochondria, which resemble bacteria in many important respects (Gupta and Dudani, 1987; Youngleson *et al.*, 1989; Reizer *et al.*, 1991; Hartman *et al.*, 1992), appear to provide such enabling sites, whereas membranes of intact and healthy mammalian cells, as discussed below, generally do not.

Studies of patients following a spontaneous coronary artery occlusion indicate that sufficient complement is activated after a significant coronary artery ischaemic event to cause transient depletion of circulating complement components. Systemic complement activation under these circumstances is presumably due to release of phlogistic substances from the ischaemic myocardium. For example, during the initial 72 h following a spontaneous coronary artery occlusion, serum levels of the third and fourth complement components are significantly depressed in patients with documented myocardial infarctions (Pinckard *et al.*, 1975). The quantity of the terminal activation complex, C5b–C9, released into the circulation of such patients appears to be proportional to the severity of cardiac injury (Langlois and Gawryl, 1988). Sera of patients with acute myocardial infarctions also contain significantly elevated levels of the C1rC1s–C1 inhibitor complex, C5a, C3a and the C3bBbP activation complex of the alternative pathway. During the first 72 h following the onset of an ischaemic event, soluble macromolecular complexes containing C1q, presumably bound to subcellular components of myocardial origin, are found in sera of patients with documented myocardial infarctions but not in sera of patients hospitalized for anginal pain who fail to develop myocardial necrosis (Rossen *et al.*, 1988). The C1q binding complexes from these patients do not appear to contain anti-heart antibodies; thus it is unlikely that they represent circulating antigen–antibody complexes. Rather, as in the dog model, these C1q binding complexes are thought to contain subcellular myocardial tissue components bound to C1q.

3. Mechanisms Likely to Promote Release of Complement-fixing Cardiac Subcellular Organelles During Ischaemic Injury

Several events conspire to bring molecules associated with mitochondria in contact with circulating complement proteins during ischaemia. Onset of ischaemia, especially sudden onset of profound ischaemia, results in brisk metabolic changes, including a switch from aerobic metabolism to anaerobic glycolysis with the inevitable accumulation in cardiac cells of products of glycolytic metabolism, including lactate and alpha glycerol phosphate (Jennings *et al.*, 1990, 1991; Jennings and Reimer, 1991). Within 15–20 s of the onset of ischaemia, metabolites, produced in the switch from oxidation of mainly fatty acids to glycolysis (Reimer *et al.*, 1983), result in the accumulation of osmotically active products intracellularly and cause cardiac myocytes to swell (Jennings *et al.*, 1985). These changes ultimately damage the cytoskeleton of the myocyte (Sage and Jennings, 1988b). With total ischaemia, *in vitro*, ultra-structural studies demonstrate blebs created by separation of the sarcolemma from the myofibrils; when viewed by transmission electron microscopy, these blebs appear as spaces 3–5 sarcomeres long that may contain numerous mitochondria (Sage and Jennings, 1988a). Despite the osmotically induced swelling, the plasma membranes initially remain intact. However, after 10–60 min of ischaemia, breaks in the sarcolemma may be seen (Greve *et al.*, 1990). It is not clear whether the osmotically induced cell swelling causes the breaks in the plasma membrane (Steenbergen *et al.*, 1985), or whether additional events dependent upon prolonged ischaemia and the resulting energy deficiency are responsible for these disruptions in the limiting membrane. Such lesions are readily demonstrable in ultrastructural studies of canine myocardium after ischaemic intervals lasting more than 60 min (Steenbergen *et al.*, 1985; Sage and Jennings, 1988a). But sarcolemmal membrane interruptions and mitochondrial swelling may be evident after only 10 min of ischaemia in the cat (Greve *et al.*, 1990). Because of differences in experimental methods, it is not clear whether the differences in time of onset of these events reflect actual differences in sensitivity to ischaemic cardiac injury in dogs and cats.

From the standpoint of the present chapter, however, these observations provide a plausible explanation for the appearance of complement-fixing mitochondrial membrane components in extracellular fluids (cardiac lymph) collected after reperfusion of ischaemic myocardium (Rossen *et al.*, 1988; Kagiyama *et al.*, 1989). We suggest that cardiac mitochondria may be released through breaks in the sarcolemmal membrane during episodes of ischaemia lasting 10–15 min or longer; once

they are outside the cardiac myocyte, these mitochondria swell and burst, releasing subcellular components that activate the complement cascade. Indeed, after only 10 min of ischaemia, swelling may be apparent in sub-sarcolemmal mitochondria; with longer intervals of ischaemia, the fraction of mitochondria showing histo-logical evidence of injury increases and many begin to show damaged christae (Greve *et al.*, 1990). Eventually, mitochondria lose their normal ultrastructural character-istics as amorphous material appears within the limiting membranes. Nondescript vesicles may also be seen in the interstitial space between myocytes, possibly the ghosts of mitochondria which have burst, releasing their internal contents (Steenbergen *et al.*, 1985).

In addition to complement-fixing membrane consti-tuents from cardiac subsarcolemmal mitochondria, molecules derived from other subcellular structures, including the sarcoplasmic reticulum, may bind and activate C1 (Hansson *et al.*, 1987; Rossen *et al.*, 1988). Release of these complement-activating molecules results in the assembly of the membrane attack complex (C5b–C9) in proximity to the membranes of the damaged cardiac myocytes (Pinckard *et al.*, 1980; Stein *et al.*, 1985; Schafer *et al.*, 1986; Vlaicu *et al.*, 1988). Once the outer membrane has been breached, cytoskeletal proteins within the cell provide additional sites for complement activation and the generation of the anaphylatoxins, C3a and C5a. Figure 6.1 illustrates events which are thought to activate complement in the microenvironment of an ischaemic cardiac myocyte and summarize the likely consequences of that activation.

4. Effects of Intracardiac Complement Activation

C5a causes granulocyte aggregation and increased margination of leukocytes along the walls of blood vessels; it also provides chemotactic signals that attract and activate leukocytes to scavenge damaged cells (Hansson *et al.*, 1987). However, from studies of animals rendered profoundly neutropaenic by treatment with cyclophosphamide (Swenson *et al.*, 1987), it is clear that complement exerts profound physiological effects within the heart, even in the absence of circulating white cells. These effects are related to the ability of C3a and C5a to stimulate tissue mast cells and circulating basophils to release histamine, and induce production of vasoactive arachidonic acid metabolites. It is these secondary mediators which affect myocardial contractility, electrical conduction and vascular tone (del Balzo *et al.*, 1985, 1988, 1989, 1990; Regal, 1985; Swenson *et al.*, 1987; Schumacher *et al.*, 1991).

To study the physiological effects of the anaphylatoxins on cardiac function under conditions which minimized their chemotactic and leukocyte-activating effects, del

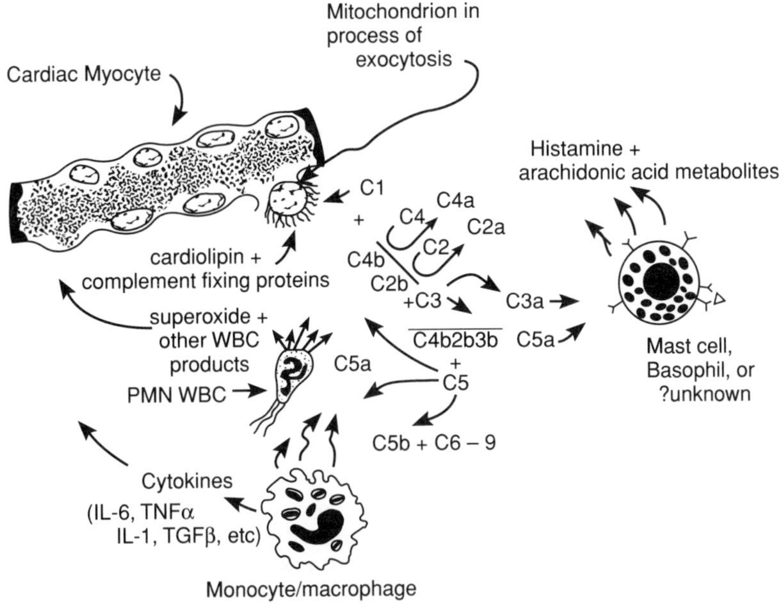

Figure 6.1 Hypothetical scheme of events which result in complement activation and the generation of an inflammatory response following ischaemic myocardial injury.

Balzo and colleagues (1985) administered purified human C3a into the aorta of guinea-pig hearts, isolated and perfused in a Langendorff apparatus. Infusion of $1–10\ \mu g$ C3a increased the sinus rate, prolonged the P–R interval, decreased the left ventricular contractile force and diminished coronary flow in a dose-dependent fashion. At higher doses of C3a ($24\ \mu g$), the decrease in contractility was preceded in three of five experiments by a brief, but significant increase in contractile force. Treatment of the putative C3a with carboxypeptidase B all but abolished these effects, confirming the identity of the polypeptide. The dose-dependent increase in heart rate associated with C3a infusion correlated significantly with the amount of histamine released from the heart; moreover, this tachycardia was suppressed by treatment with the histamine H_2 receptor antagonist, cimetidine. The changes in contractile force and in coronary artery flow, however, reflected the ability of C3a to stimulate arachidonic acid metabolism. For example, the effect of the high, $24\ \mu g$ dose of C3a on ventricular contractility was suppressed by treatment with the leukotriene receptor antagonist, FPL55712, whereas the reduction in coronary flow was decreased by pretreatment with indomethacin, a cyclo-oxygenase inhibitor.

While partially purified C5a also caused a positive chronotropic effect (Regal, 1985), it *increased* rather than decreased the force of contraction of isolated guinea-pig atrial muscle. Both the positive chronotropic and positive inotropic effects of C5a were significantly reduced on second and subsequent exposure to this peptide whereas repeated exposure to histamine did not cause tachyphylaxis (Regal, 1985). Only the positive inotropic

effect could be suppressed by pretreating the atria with the histamine H_2 receptor antagonist, metiamide.

Infusions of isolated human C3a caused a significant, albeit brief *increase* in coronary blood flow in pentobarbital anaesthetized dogs (Schumacher *et al.*, 1991) without any concurrent changes in arterial blood pressure, heart rate or left ventricular contractile force. Qualitatively similar results were obtained by these same investigators using isolated C5a or zymosan-activated serum as a source of anaphylatoxin. Comparable increases in coronary blood flow were observed following infusion of *human* C3a or zymosan-activated *canine* serum in the left circumflex coronary artery of fully conscious dogs, previously instrumented to obtain physiological measurements. Thus the effects of activated complement in this animal were similar, regardless of the species of origin of the anaphylatoxin.

To further identify the effects of complement on the heart during anaphylactic reactions, del Balzo and colleagues (1988) studied isolated guinea-pig atria or papillary muscles, presensitized with guinea-pig IgG1 antibodies specific for the antigen DNP. Cardiac muscle preparations from these animals were challenged with DNP, conjugated to bovine serum albumin. Human serum, diluted 1 : 1 with Tyrode's solution, provided a source of complement. Formation of antigen–antibody complexes, caused by reactions of the anti-DNP antibodies with DNP, consumed C3. Correlated with the consumption of C3 was the release of both C3a and histamine. The quantity of C3a produced was increased by administering an inhibitor of the enzyme, carboxypeptidase N, which normally degrades C3a to C3a des-Arg. Release of C3a des-Arg was accompanied by an increase in both the rate and the force of cardiac muscle contraction. These positive inotropic and chronotrophic effects were almost completely blocked by administration of an H_2 receptor antagonist. Thus the effects of activated complement on the rate and strength of muscle contraction were largely attributable to histamine released by C3a.

To compare the ability of complement-dependent anaphylatoxins to release secondary mediators of anaphylaxis, del Balzo *et al.* (1990) infused recombinant human C5a, purified C3a and a synthetic carboxy-terminal peptide, $C3a_{72-77}$ into the coronary circulation. The effects of all three peptides were qualitatively similar, except for the quantity required to produce a measurable effect, C5a being significantly more potent (0.01–0.1 nM) than C3a (0.1–1.0 nM) which, in turn, was significantly more potent than the C3a peptide (50–300 nM). In keeping with the response to these anaphylatoxins in other tissues, complete tachyphylaxis was observed following repeated treatment with any of the three peptides in the heart; however, cross-tachyphylaxis was incomplete. For example, the $C3a_{72-77}$ peptide desensitized the tissue to native C3a but not completely to C5a. Considering that inactivated C3a (C3a without its carboxy-terminal arginine) has no cardiac effects, the $C3a_{72-77}$ peptide appears to be the minimum structure required to trigger anaphylatoxin activity in the heart.

From studies of the effects of specific H_2 and H_1 histamine receptor antagonists, as well as the cyclo-oxygenase inhibitor, indomethacin, the leukotriene receptor antagonist, FPL55712, and the thromboxane synthetase inhibitor, CGS13080, del Balzo *et al.* (1990) deduced that the effects of the complement-dependent anaphylatoxins on cardiac function were caused by the combined actions of histamine, TXA_2 or other arachidonic acid metabolites, together with adenosine, released into the coronary artery circulation following infusion of the anaphylatoxins. For example, H_2 receptor blockade suppressed the tachycardia induced by C5a. This positive chronotropic effect was further reduced by the leukotriene receptor antagonist, suggesting that peptidoleukotrienes also contribute to this response. The initial brief increase in contractility following C3a or C5a administration was apparently caused by histamine acting through H_2 receptors, since it could be blocked by pretreatment with a selective H_2 antagonist. The more prolonged decrease in myocardial contractility which followed exposure to the anaphylatoxins may represent a complex interaction of histamine, acting upon H_1 receptors, together with cyclo-oxygenase and 5-lipoxygenase products of arachidonic acid metabolism. In isolated papillary muscles, H_1 blockade alone was sufficient to suppress the negative inotropic effect. But in whole heart preparations, complete abolition of the effect required combined treatment with H_1 antagonists, a cyclo-oxygenase

inhibitor, and the leukotriene receptor antagonist, FPL 55712.

Indomethacin blocked the coronary vasoconstrictor response to C5a, suggesting that the reduction in coronary artery blood flow caused by C5a triggered TXA_2 release (Allan and Levi, 1981). The vasoconstrictor response to C5a was exacerbated by treatment with histamine H_1 and H_2 receptor antagonists in the absence of indomethacin (del Balzo *et al.*, 1985, 1990). The net vasoconstrictor effect of H_1 and H_2 receptor antagonists in this model suggests that the vasospastic effect of TXA_2, evoked by C5a, is partly offset by vasodilatory effects of histamine acting upon the coronary artery circulation.

Atrioventricular conduction abnormalities associated with release of C3a and C5a in the heart have been attributed to adenosine (Belardinelli *et al.*, 1989). Adenosine is released into the coronary circulation during anaphylaxis (Genovese *et al.*, 1987; Heller and Regal, 1988); it delays atrioventricular nodal conduction, dilates coronary vessels, and decreases both contractility and automaticity. Histamine may contribute to the atrioventricular conduction defects which follow intracardiac C3a or C5a release, since combined treatment with H_1 and H_2 receptor antagonists also reduces conduction defects created by C5a administration (del Balzo *et al.*, 1990).

While these studies of the consequences of infusing C3a and/or C5a in the coronary artery circulation illuminate mechanisms likely to injure ischaemic heart muscle or myocardium targeted by infectious agents or auto-immune responses, they also provide information which could be used to minimize the complications of thrombolytic therapy. For example, it is clear that tissue plasminogen activator activates the complement cascade (Bennet *et al.*, 1987). Complement activation, in turn, generates C3a. C3a causes platelet aggregation, even when degraded to C3a des-Arg (Polley and Nachman, 1983). Both C3a and C5a stimulate production of PAF (Camussi *et al.*, 1977). In addition to its pro-coagulant activity (Luscher, 1987), PAF appears to have effects on the heart similar to those of the secondary mediators of anaphylaxis discussed above: it decreases left ventricular contractile force, reduces coronary blood flow, increases coronary vascular resistance and impairs atrioventricular conduction (Levi *et al.*, 1984). Since microenvironments created by atherosclerotic vascular lesions already provide sites that initiate thrombus formation and activate the complement cascade (Seifert *et al.*, 1988, 1990), we postulate that strategies that will reduce the capacity of thrombolytic therapy to activate complement may enhance blood flow through the reperfused vessel and reduce the chance of re-occlusion.

The mechanisms of activation and regulation shared by the complement and coagulation cascades has been illuminated by studies investigating the biocompatibility of membranes used in haemodialysis and pump oxygenators (Chenoweth, 1986; Schohn *et al.*, 1986; Schoen *et al.*, 1987; Moll *et al.*, 1990). While effective anticoagulant dosages of heparin may reduce the chance of re-occlusion and protect formerly ischaemic muscle (Hobson *et al.*, 1989), the concentrations of heparin commonly used for this purpose may not prevent activation of the complement cascade and the physiological effects associated with release of the complement-dependent anaphylatoxins (Chenoweth, 1986; Bengtson *et al.*, 1989; Utley, 1990; Mottaghy *et al.*, 1991). Thus while heparin may forestall re-occlusion, it does not necessarily abolish pro-inflammatory stimuli resulting from the combined effects of ischaemia, treatment with recombinant tissue plasminogen activator and the continuing pro-coagulant activity of the atherosclerotic lesions which initiated coronary artery thrombosis in the first place. Further research may demonstrate that certain additional therapies, including those used by del Bazo *et al.* (1990), to control the physiological effects of secondary mediator release following intra-coronary infusion of the anaphylatoxins may be useful adjuncts to thrombolytic therapy for coronary artery occlusion.

C5a, released following reperfusion of ischaemic myocardium, is a potent leukocyte chemotactic agent. The consequences attendant upon infiltration of formerly ischaemic myocardium with granulocytic leukocytes are extensively discussed by Entman and Smith (Chapter 5 of this volume). However, it is important to note that other cellular elements of the immune system are also likely to be attracted by C5a release to ischaemic myocardium, once perfusion is re-established. For example, C5a is also a potent chemotactic and activating agent for peripheral blood monocytes. It stimulates their maturation and the synthesis and release of cytokines, such as IL-1, IL-6, and TNF-α. C5a also induces monocytes to produce superoxide radicals and release lysosomal enzymes that can cause tissue injury (Antrum and Solomkin, 1986; Hansson *et al.*, 1987; Ohura *et al.*, 1987; Mrowietz and Christophers, 1988; Yancey *et al.*, 1989; Scholz *et al.*, 1990; Haeffner and Kazatchkine, 1991). Thus, in myocardial ischaemia, stimuli which induce infiltration of granulocytic leukocytes into the damaged tissue are also likely to induce tissue infiltration of comparatively long-lived monocytes, and regional accumulations of their products. While morphological studies have demonstrated that infiltrating monocytes mature into the tissue macrophages which persist for some time amidst injured myocytes following ischaemic injury, their role in the development of (McCluskey *et al.*, 1982) and the recovery from (Virmani *et al.*, 1990) the inflammatory response to ischaemia remains conjectural. On the other hand, in myocarditis, cardiac tissue infiltration of mononuclear leukocytes correlates significantly with the extent and intensity of tissue deposits of immunoglobulin and complement (Hammond *et al.*, 1987) consistent with the notion that these cells respond to and participate in the inflammatory response to intracardiac infection (Laguens *et al.*, 1989).

5. Influences which May Protect Cardiac Tissues Against the Powerful Effector Mechanisms Released by Complement Activation

Although myocardial ischaemia commonly provides a stimulus sufficient to cause transient depletion of circulating complement, it is clear that in the target organ, complement-dependent localization of neutrophilic leukocytes is limited to a subset of the myocardium which experienced severely reduced perfusion (Rossen *et al.*, 1985, 1988; Dreyer *et al.*, 1991). Characteristically host tissues are protected against injury by activated autologous complement activation (Hideshima *et al.*, 1990; Nose *et al.*, 1990; Zimmermann *et al.*, 1990; Davies, 1991). Indeed recent studies suggest that normal cardiac myocytes express DAF, one of the regulatory proteins which protects against bystander cell injury by activated complement (Zimmermann *et al.*, 1990). This protein is lacking on many but not all cell surfaces in ischaemic regions in myocardium of patients who have died of a myocardial infarction. The studies by Zimmerman *et al.* (1990) are among the first to document that somatic cells in the heart may be among the host tissues which enjoy some protection against tissue injury by proteins which regulate autologous complement activation.

Many of the molecular mechanisms responsible for limiting complement-mediated tissue injury have become evident in the past several years. Particular progress has been made in understanding the cell membrane-associated inhibitory proteins which protect host cells from complement-mediated injury (Davies, 1991). "Homologous restriction" is the term which is currently used to describe the apparent inefficiency of autologous complement, when activated, to cause injury to one's own cells (Lachmann, 1991). Molecules mediating homologous restriction include decay-accelerating factor (DAF or CD55), membrane cofactor protein (MCP or CD46), the 65 kDa homologous restriction factor (65 kD HRf), also known as C8 binding protein, and the membrane inhibitor of reactive lysis (MIRL or CD59) which is the 20 kDa homologous restriction factor (20 kD HRf). Endothelial cells and most peripheral blood cells in primates express these molecules; consequently these cells are relatively protected against bystander effects of complement activation. C3 converting enzymes critical for advancing the activation of complement beyond C3b are inhibited by DAF (Fujita *et al.*, 1987) and MCP (Seya and Atkinson, 1989), while the 20 kDa and 65 kDa HRfs interfere with the formation of an effective membrane attack complex (Lachmann, 1991).

Membrane-bound DAF is only one of the components involved in regulating complement activation which has "decay accelerating activity". The circulating plasma proteins, C4BP and factor H also interfere with formation of or dissociate the assembly of molecules which form a C3 converting enzyme. Another protein with "decay accelerating activity" includes complement receptor 1 (CR1) which serves to entrap circulating or cell-associated antigen–antibody complexes which have previously bound C3b. CR1 are particularly abundant on peripheral blood cell surfaces, including the red blood cells in primates. Because of their relative abundance, red cells provide a rich supply of cell membrane-associated molecules that can clear activated complement components. Several of the complement regulatory proteins, including CR1, factor H and MCP also enhance the inactivation of C3b or C4b by factor I-mediated proteolysis. The imaginative use of recombinant soluble CR1 to scavenge activated complement components generated in the course of myocardial ischaemia to protect against the injurious effects of complement-mediated inflammation has already been discussed (Weisman *et al.*, 1990b). Other than these studies, little

has been done to examine the potential cardioprotective effects of the complement regulatory proteins in ischaemic injury. Clearly, further examination of the potential ability of the complement regulatory proteins to influence the consequences of complement activation following ischaemic injury is warranted. For example, CD59, the low molecular weight (20 kDa) homologous restriction factor that appears to inhibit complement-mediated target cell injury by inhibiting insertion of C9 to form a poly C9 pore, was originally identified only on endothelia and cells of the peripheral blood (Hideshima *et al.*, 1990; Nose *et al.*, 1990; Sayama *et al.*, 1990). Subsequent studies indicate that this regulatory molecule is widely distributed in a number of somatic tissues (Meri *et al.*, 1991; Rooney *et al.*, 1991). CD59 also appears to control platelet activation by the complement C5b–C9 complex (Sims *et al.*, 1989). Thus in myocardial ischaemia and other conditions where intravascular complement activation generates C5b-7 complexes that can mediate bystander lysis, platelet CD59 may be an important regulator of the pro-coagulant effects of complement activation, mediated perhaps by C5b–C9 assembly on the platelet membrane surface, which results in secretion and activation of factor V from the platelet α granules (Wiedmer *et al.*, 1986a,b, 1987). Sims

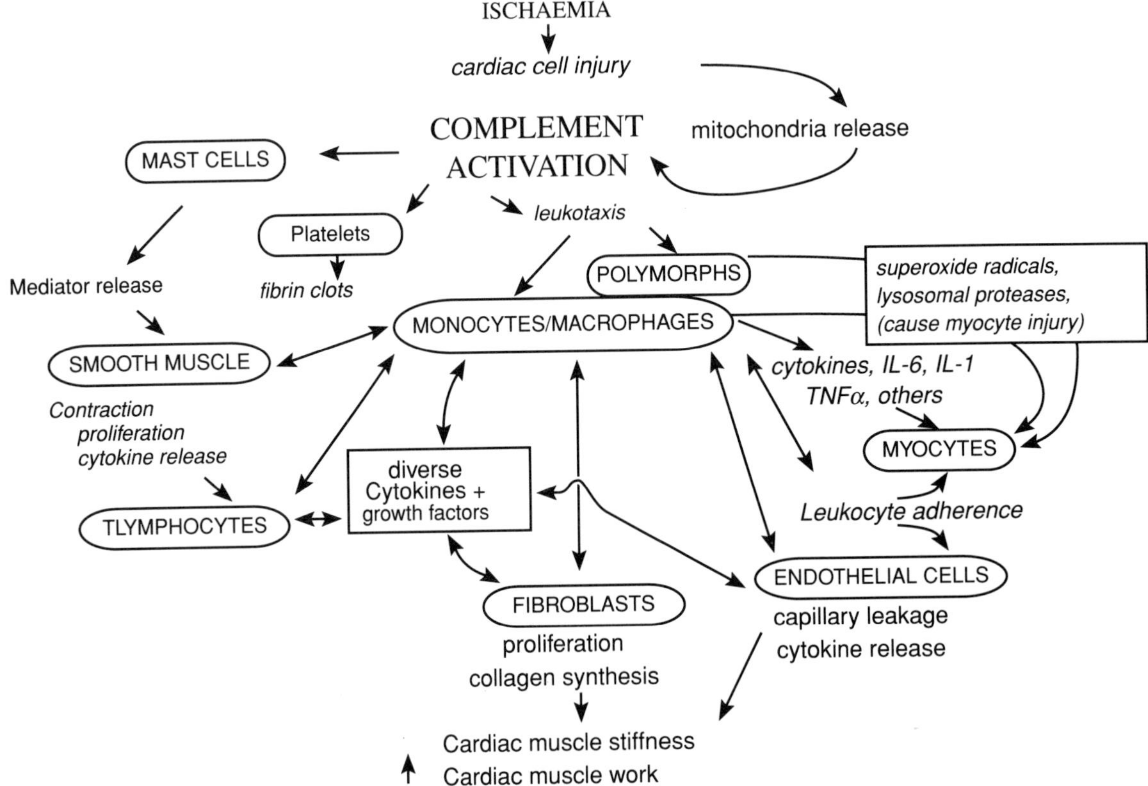

Figure 6.2 Cellular and molecular events, initiated by ischaemic myocardial injury, which promote acute inflammation and cause myocardial tissue injury. Also indicated are some of the mechanisms involved in the resolution of the inflammatory responses that facilitate healing and repair.

and Wiedmer (1991) report that the addition of antibodies that neutralize the effects of CD59 increases C9 deposition in platelet membranes under assault by C5b–C8 complexes and augments C5b–C9-mediated coagulation. Considering the likely effects of platelet activation in the events which predispose to myocardial ischaemia and coronary artery occlusions, investigation of the role of CD59, and other proteins that regulate the terminal activation pathway of the complement cascade, may provide useful insights that will lead to the development of agents which may further suppress the pro-inflammatory consequences of complement activation within myocardial tissues.

6. Summary

Complement activation, triggered by mitochondria or other subcellular elements of cardiac cells, initiates a complex series of cellular and molecular interactions, simplified and summarized in Fig. 6.2. Figure 6.2 attempts to illustrate not only the acute effects associated with complement activation in proximity to cardiac cells, but also attempts to foresee some of the long-term consequences associated with complement activation, including the functional changes in leukocytes, smooth muscle cells, endothelia, fibroblasts and the myocytes themselves, which ultimately impair myocardial function. Not shown are the many regulatory elements, discussed in the latter part of this chapter, which limit the extent of the interactions of these plasma proteins, and cellular elements so as to moderate tissue injury.

7. Acknowledgements

Dr Rossen's scholarly activities are supported by merit review research awards from the United States Department of Veterans Affairs, and USPHS–NIH grant awards HL41408, HL42550, AI28071 and AI33236.

8. References

Allan, G. and Levi, R. (1981). Thromboxane and prostacyclin release during cardiac immediate hypersensitivity reactions *in vitro*. J. Pharmacol. Exp. Ther. 217, 157–161.

Anderson, N.D., Wyllie, R.G. and Shaker, I.J. (1977). Pathogenesis of vascular injury in rejecting rat renal allografts. Johns Hopkins Med. J. 141, 135–147.

Antrum, R.M. and Solomkin, J.S. (1986). Monocyte dysfunction in severe trauma: evidence for the role of C5a in deactivation. Surgery 100, 29–37.

Belardinelli, L., Linden, J. and Berne, R.M. (1989). The cardiac effects of adenosine. Prog. Cardiovasc. Dis. 32, 73–97.

Bengtson, A., Millocco, I., Heideman, M. and Berggren, H. (1989). Altered concentrations of terminal complement complexes, anaphylatoxins, and leukotrienes in the coronary sinus during cardiopulmonary bypass. J. Cardiothorac. Anesth. 3, 305–310.

Bennet, W.R., Yawn, D.H., Migliore, P.J., Young, J.B., Pratt, C.M., Raizner, A.E., Roberts, R. and Bolli, R. (1987). Activation of the complement system by recombinant tissue plasminogen activator. J. Am. Coll. Cardiol. 10, 627–632.

Bindon, C.I., Hale, G. and Waldmann, H. (1990). Complement activation by immunoglobulin does not depend solely on C1q binding. Eur. J. Immunol. 20, 277–281.

Camussi, G., Mencia, H.J.M. and Benveniste, J. (1977). Release of platelet-activating factor and histamine. I. Effect of immune complexes, complement and neutrophils on human and rabbit mastocytes and basophils. Immunology 33, 523–534.

Chenoweth, D.E. (1986). Anaphylatoxin formation in extracorporeal circuits. Complement 3, 152–165.

Crawford, M.H., Grover, F.L., Kolb, W.P., McMahan, C.A., O'Rourke, R.A., McManus, L.M. and Pinckard, R.N. (1988). Complement and neutrophil activation in the pathogenesis of ischemic myocardial injury. Circulation. 78, 1449–1458.

Davies, K.A. (1991). Complement. Baillières Clin. Haematol. 4, 927–955.

del Balzo, U.H., Levi, R. and Polley, M.J. (1985). Cardiac dysfunction caused by purified human C3a anaphylatoxin. Proc. Natl. Acad. Sci. USA 82, 886–890.

del Balzo, U., Polley, M.J. and Levi, R. (1988). Activation of the third complement component (C3) and C3a generation in cardiac anaphylaxis: histamine release and associated inotropic and chronotropic effects. J. Pharmacol. Exp. Ther. 246, 911–916.

del Balzo, U., Polley, M.J. and Levi, R. (1989). Cardiac anaphylaxis. Complement activation as an amplification system. Circulation Res. 65, 847–857.

del Balzo, U., Sakuma, I. and Levi, R. (1990). Cardiac dysfunction caused by recombinant human C5A anaphylatoxin: mediation by histamine, adenosine and cyclooxygenase arachidonate metabolites. J. Pharmacol. Exp. Ther. 253, 171–179.

Dreyer, W.J., Smith, C.W., Michael, L.H., Rossen, R.D., Hughes, B.J., Entman, M.L. and Anderson, D.C. (1989). Canine neutrophil activation by cardiac lymph obtained during reperfusion of ischemic myocardium. Circulation Res. 65, 1751–1762.

Dreyer, W.J., Smith, C.W., Michael, L.H., West, M.S., Smith, C.W., Rothlein, R., Rossen, R.D., Anderson, D.C. and Entman, M.L. (1991). Neutrophil accumulation in ischemic canine myocardium. Circulation 84, 400–411.

Dreyer, W.J., Michael, L.H., Nguyen, T., Smith, C.W., Anderson, D.C., Entman, M.L. and Rossen, R.D. (1992). Kinetics of C5a release in cardiac lymph of dogs experiencing coronary artery ischemia-reperfusion injury. Circulation Res. 71, 1518–1524.

Forbes, R.D., Lowry, R.P., Gomersall, M. and Blackburn, J. (1985). Comparative immunohistologic studies in an adoptive transfer model of acute rat cardiac allograft rejection. Transplantation 40, 77–85.

Fujita, T., Inoue, T., Ogawa, K., Iida, K. and Tamura, N. (1987). The mechanism of action of decay-accelerating factor (DAF). DAF inhibits the assembly of C3 convertases by dissociating C2a and Bb. J. Exp. Med. 166, 1221–1228.

Genovese, A., Sakuma, I., Boykin, M.T., Levi, R. and Belardinelli, L. (1987). Cardiac anaphylaxis: adenosine release and negative dromotropic effect. Fed. Proc. 46, 932.

Ghose, T. and Mammen, M. (1977). Interaction *in vitro* between myocardial cells and autologous lymphocytes and sera from patients with rheumatic carditis. Chest 71, 730–734.

Giclas, P.C., Pinckard, R.N. and Olson, M.S. (1979) *In vitro* activation of complement by isolated human subcellular membranes. J. Immunol. 122, 146–151.

Greve, G., Rotevatn, S., Svendby, K. and Grong, K. (1990). Early morphologic changes in cat heart muscle cells after acute coronary artery occlusion. Am. J. Path. 136, 273–283.

Gupta, R.S. and Dudani, A.K. (1987). Mitochondrial binding of a protein affected in mutants resistant to the microtubule inhibitor podophyllotoxin. Eur. J. Cell Biol. 44, 278–285.

Haeffner, C.N. and Kazatchkine, M.D. (1991). Induction of interleukin-1 synthesis. A new criterion of biocompatibility of extracorporeal circuits. Presse Med. 20, 797–801.

Hammond, E.H., Menlove, R.L. and Anderson, J.L. (1987). Predictive value of immunofluorescence and electron microscopic evaluation of endomyocardial biopsies in the diagnosis and prognosis of myocarditis and idiopathic dilated cardiomyopathy. Am. Heart J. 114, 1055–1065.

Hammond, E.H., Yowell, R.L., Nunoda, S., Menlove, R.L., Renlund, D.G., Bristow, M.R., Gay, W.A.J., Jones, K.W. and O'Connell, J.B. (1989). Vascular (humoral) rejection in heart transplantation: pathologic observations and clinical implications: see comments. J. Heart Transplant. 8, 430–443.

Hansson, G.K., Lagerstedt, E., Bengtsson, A. and Heideman, M. (1987). IgG binding to cytoskeletal intermediate filaments activates the complement cascade. Exp. Cell Res. 170, 338–350.

Hartman, D.J., Hoogenraad, N.J., Condron, R. and Hoj, P.B. (1992). Identification of a mammalian 10-kDa heat shock protein, a mitochondrial chaperonin 10 homologue essential for assisted folding of trimeric ornithine transcarbamoylase *in vitro*. Proc. Natl. Acad. Sci. USA 89, 3394–3398.

Heller, L.J. and Regal, J.F. (1988). Effect of adneosine on histamine release and atrioventricular conduction during guinea pig cardiac anaphylaxis. Circulation Res. 62, 1147–1158.

Herdson, P.B., Sommers, H.M. and Jennings, R.B. (1965). A comparative study of the fine structure of normal and ischemic dog myocardium with special reference to early changes following temporary occlusion of a coronary artery. Am. J. Pathol. 46, 367–386.

Hideshima, T., Okada, N. and Okada, H. (1990). Expression of HRF20, a regulatory molecule of complement activation, on peripheral blood mononuclear cells. Immunology 69, 396–401.

Hill, J.H. and Ward, P.A. (1969). C3 Leukotactic factors produced by a tissue protease. J. Exp. Med. 130, 505–518.

Hill, J.H. and Ward, P.A. (1971). The phlogistic role of C3 leukotactic fragment in myocardial infarcts in rats. J. Exp. Med. 133, 885–900.

Hobson, R.W., II, Neville, R., Watanabe, B., Canady, J., Wright, J.G. and Belkin, M. (1989). Role of heparin in reducing skeletal muscle infarction in ischaemia–reperfusion. Microcirc. Endothelium Lymphatics 5, 259–276.

Hubsch, R.M., Sulzer, A.J. and Kagan, I.G. (1976). Evaluation of an autoimmune type antibody in the sera of patients with Chagas' disease. J. Parasitol. 62, 523–527.

Jennings, R.B. and Reimer, K.A. (1991). The cell biology of acute myocardial ischaemia. Ann. Rev. Med. 42, 225–246.

Jennings, R.B., Schaper, J., Hill, M.L., Steenbergen, C.J. and Reimer, K.A. (1985). Effects of reperfusion late in the phase of reversible ischemic injury. Changes in cell volume, electrolytes, metabolites and ultrastructure. Circulation Res. 56, 262–278.

Jennings, R.B., Murry, C.E., Steenbergen, C.J. and Reimer, K.A. (1990). Development of cell injury in sustained acute ischemia. Circulation 82 (Suppl. 3), II2–II20.

Jennings, R.B., Murry, C.E. and Reimer, K.A. (1991). Preconditioning myocardium with ischemia. Cardiovasc. Drugs Ther. 5, 933–938.

Kagiyama, A., Savage, H.E., Michael, L.H., Hanson, G., Entman, M.L. and Rossen, R.D. (1989). Molecular basis of complement activation in ischaemic myocardium: identification of specific molecules of mitochondrial origin that bind human C1q and fix complment. Circulation Res. 64, 607–615.

Kaplan, M.H. (1965). Autoantibodies to heart and rheumatic fever: the induction of autoimmunity to heart by streptococcal antigen cross-reactive with heart. Ann. N.Y. Acad. Sci. 124, 904–915.

Kaplan, M.H. and Frengley, J.D. (1969). Autoimmunity to the heart in cardiac disease. Current concepts of the relation of autoimmunity to rheumatic fever, postcardiotomy and postinfarction syndromes and cardiomyopathies. Am. J. Cardiol. 24, 459–473.

Kovacsovics, T., Tschopp, J., Kress, A. and Isliker, H. (1985). Antibody-independent activation of C1, the first component of complement, by cardiolipin. J. Immunol. 135, 2695–2700.

Lachmann, P.J. (1991). The control of homologous lysis. Immunol. Today 12, 312–315.

Laguens, R.P., Meckert, P.C. and Chambo, J.G. (1988). Antiheart antibody-dependent cytotoxicity in the sera of mice chronically infected with *Trypanosoma cruzi*. Infect. Immun. 56, 993–997.

Laguens, R.P., Cabeza, M.P.M. and Chambo, J.G. (1989). Immunologic studies on a murine model of Chagas disease. Medicine (Buenos Aires) 49, 197–202.

Langlois, P.F. and Gawryl, M.S. (1988). Detection of the terminal complement complex in patient plasma following acute myocardial infarction. 70, 95–105.

Levi, R., Burke, J.A., Guo, Z.G., Hattori, Y., Hoppens, C.M., McManus, L.M., Hanahan, D.J. and Pinckard, R.N. (1984). Acetyl glyceryl ether phosphorylcholine (AGEPC). A putative mediator of cardiac anaphylaxis in the guinea pig. Circulation Res. 54, 117–124.

Luscher, E.F. (1987). Activated leukocytes and the hemostatic system. Rev. Infect. Dis. 9 (Suppl. 5), S546–S552.

Maroko, P.R., Carpenter, C.B., Chiarello, M., Fishbein, M.C., Radvany, P., Kostman, J.D. and Hale, S.L. (1978). Reduction by cobra venom factor of myocardial necrosis after coronary artery occlusion. J. Clin. Invest. 61, 661–670.

McCluskey, E. R., Corr, P.B., Lee, B.I., Saffitz, J.E. and Needleman, P. (1982). The arachidonic acid metabolic capacity of canine myocardium is increased during healing of acute myocardial infarction. Circulation Res. 51, 743–750.

Meri, S., Waldmann, H. and Lachmann, P.J. (1991). Distribution of protectin (CD59), a complement membrane attack inhibitor, in normal human tissues. Lab. Invest. 65, 532–537.

Moll, S., De, M.P., Reber, G., Schifferli, J. and Leski, M. (1990). Comparison of two hemodialysis membranes, polyacrylonitrile and cellulose acetate, on complement and coagulation systems. Int. J. Artif. Organs 13, 273–279.

Mottaghy, K., Oedekoven, B., Poppel, K., Kovacs, B., Kirschfink, M., Bruchmuller, K., Kashefi, A. and Geisen, C. (1991). Heparin-coated versus non-coated surfaces for extracorporeal circulation. Int. J. Artif. Organs 14, 721–728.

Mrowietz, U. and Christophers, E. (1988). Modulation of human monocyte functions during acute bacterial infection. Scand. J. Immunol. 28, 139–146.

Nose, M., Katoh, M., Okada, N., Kyogoku, M. and Okada, H. (1990). Tissue distribution of HRF20, a novel factor preventing the membrane attack of homologous complement, and its predominant expression on endothelial cells in vivo. Immunology 70, 145–149.

Ohura, K., Katone, I.M., Wahl, L.M., Chenoweth, D.E. and Wahl, S.M. (1987). Co-expression of chemotactic ligand receptors on human peripheral blood monocytes. J. Immunol. 138, 2633–2639.

Oluwole, S.F., Tezuka, K., Wasfie, T., Stegall, M.D., Reemtsma, K. and Hardy, M.A. (1989). Humoral immunity in allograft rejection. The role of cytotoxic alloantibody in hyperacute rejection and enhancement of rat cardiac allografts. Transplantation 48, 751–755.

Palmer, D.C., Tsai, C.C., Roodman, S.T., Codd, J.E., Miller, L.W., Sarafian, J.E. and Williams, G.A. (1985). Heart graft arteriosclerosis. An ominous finding on endomyocardial biopsy. Transplantation 39, 385–388.

Peitsch, M.C., Tschopp, J., Kress, A. and Isliker, H. (1988). Antibody-independent activation of the complement system by mitochondria is mediated by cardiolipin. Biochem. J. 249, 495–500.

Pinckard, R.N., Olson, M.S., Kelley, R.E., DeHeer, D.H., Palmer, J.D., O'Rourke, R.A. and Goldfein, S. (1973). Antibody-independent activation of human C1 after interaction with heart subcellular membranes. J. Immunol. 110, 1376–1382.

Pinckard, R.N., Olson, M.S., Giclas, P.C., Terry, R., Boyer, J.T. and O'Rourke, P.A. (1975). Consumption of classical complement components by heart subcellular membranes in vitro and in patients after acute myocardial infarction. J. Clin. Invest. 56, 740–750.

Pinckard, R.N., O'Rourke, R.A., Crawford, M.H., Grover, F.S., McManus, L.M., Ghidoni, J.J., Storrs, S.B. and Olson, M.S. (1980). Complement localization and mediation of ischemic injury in baboon myocardium. J. Clin. Invest. 66, 1050–1056.

Polley, M.J. and Nachman, R.L. (1983). Human platelet activation by C3a and C3a des-arg. J. Exp. Med. 158, 603–615.

Regal, J.F. (1985). Effect of C5a on isolated guinea pig atria. Immunopharmacol. 9, 27–31.

Reimer, K.A., Hill, M.L. and Jennings, R.B. (1983). Prolonged depletion of ATP because of delayed repletion of the adenine nucleotide pool following reversible myocardial ischemic injury in dogs. Adv. Myocardiol. 4, 395–407.

Reizer, A., Pao, G.M. and Saier, M.H.J. (1991). Evolutionary relationships among the permeate proteins of the bacterial phosphoenolpyruvate: sugar phosphotransferase system. Construction of phylogenetic trees and possible relatedness to proteins of eukaryotic mitochondria. J. Mol. Evol. 33, 179–193.

Rooney, I.A., Davies, A., Griffiths, D., Williams, J.D., Davies, M., Meri, S., Lachmann, P.J. and Morgan, B.P. (1991). The complement-inhibiting protein, protectin (CD59 antigen), is present and functionally active on glomerular epithelial cells. Clin. Exp. Immunol. 83, 251–256.

Rossen, R.D., Butler, W.T., Reisberg, M.A., Brooks, D.K., Leachman, R.D., Milam, J.D., Mittal, K.K., Montgomery, J.R., Nora, J.J. and Rochelle, D.G. (1971a). Immunofluorescent localization of human immunoglobulin in tissues from cardiac allograft recipients. J. Immunol. 106, 171–180.

Rossen, R.D., Butler, W.T., Johnson, A.H. and Mittal, K.K. (1971b). Immunofluorescent and serologic studies of the humoral antibody response to cardiac allotransplantation in man. Transplant. Proc. III, 445–448.

Rossen, R.D., Swain, J.L., Michael, L.H., Weakley, S., Giannini, E. and Entman, M.L. (1985). Selective accumulation of the first component of complement and leukocytes in ischemic canine heart muscle. Circulation Res. 57, 119–130.

Rossen, R.D., Michael, L.H., Kagiyama, A., Savage, H.E., Hanson, G., Reisberg, M.A., Moake, J.N., Kim, S.H., Self, D., Weakley, S., Giannini, E. and Entman, M.L. (1988). Mechanism of complement activation after coronary artery occlusion: evidence that myocardial ischemia in dogs causes release of constituents of myocardial subcellular origin that complex with human C1q in vivo. Circulation Res. 62, 572–584.

Rossen, R.D., Schulmeier, G.A., Nowicki, S., Hall, B., Michael, L.H., Entman, M.L. and Baughn, R.E. (1991). Identification of C1q-binding molecules in cardiac mitochondria that contain cardiolipin. FASEB J. 5, A1104(#4252).

Rubin, B.B., Smith, A., Liauw, S., Isenman, D., Romaschin, A.D. and Walker, P.M. (1990). Complement activation and white cell sequestration in postischemic skeletal muscle. Am. J. Physiol. 259, H525–H531.

Sage, M.D. and Jennings, R.B. (1988a). Cytoskeletal injury and subsarcolemmal bleb formation in dog heart during in vitro total ischemia. Am. J. Pathol. 133, 327–337.

Sage, M.D. and Jennings, R.B. (1988b). Myocyte swelling and plasmalemmal integrity during early experimental myocardial ischemia in vivo. Scanning Microsc. 2, 477–484.

Sayama, K., Shiraishi, S., Shirakata, Y., Kobayashi, Y., Okada, N., Okada, H. and Miki, Y. (1990). Characterization of homologous restriction factor (HRF20) in human skin and leucocytes. Clin. Exp. Immunol. 82, 355–358.

Schafer, H., Mathey, D., Hugo, F. and Bhakdi, S. (1986). Deposition of the terminal C5b-9 complement complex in infarcted areas of human myocardium. J. Immunol. 137, 1945–1949.

Schoen, F.J., Clagett, G.P., Hill, J.D., Chenoweth, D.E., Anderson, J.M. and Eberhart, R.C. (1987). The biocompatibility of artificial organs. Asaio Trans. 33, 824–833.

Schohn, D.C., Jahn, H.A., Eber, M. and Hauptmann, G. (1986). Biocompatibility and hemodynamic studies during polycarbonate versus cuprophane membrane dialysis. Blood Purif. 4, 102–111.

Schulz, W., McClurg, M.R., Cardenas, G.J., Smith, M., Noonan, D.J., Hugli, T.E. and Morgan, E.L. (1990). C5a-mediated release of interleukin 6 by human monocytes. Clin. Immunol. Immunopathol. 57, 297–307.

Schumacher, W.A., Fantone, J.C., Kunkel, S.E., Webb, R.C. and Lucchesi, B.R. (1991). The anaphylatoxins C3a and C5a are vasodilators in the canine coronary vasculature *in vitro* and *in vivo*. Agents Actions 34, 345–349.

Seifert, P.S. and Kazatchkine, M.D. (1988). The complement system in atherosclerosis. Atherosclerosis 73, 91–104.

Seifert, P.S., Catalfamo, J.L. and Dodds, W.J. (1988). Complement C5a (desArg) generation in serum exposed to damaged aortic endothelium. Exp. Mol. Pathol. 48, 216–225.

Seifert, P.S., Hugo, F., Tranum-Jensen, J., Zahringer, U., Muhly, M. and Bhakdi, S. (1990). Isolation and characterization of a complement-activating lipid extracted from human atherosclerotic lesions. J. Exp. Med. 172, 547–557.

Seya, T. and Atkinson, J.P. (1989). Functional properties of membrane cofactor protein of complement. Biochem. J. 264, 581–588.

Sims, P.J. and Wiedmer, T. (1991). The response of human platelets to activated components of the complement system. Immunol. Today 12, 338–342.

Sims, P.J., Rollins, S.A. and Wiedmer, T. (1989). Regulatory control of complement on blood platelets. Modulation of platelet procoagulant responses by a membrane inhibitor of the C5b-9 complex. J. Biol. Chem. 264, 19228–19235.

Sommers, H.M. and Jennings, R.B. (1964). Experimental acute myocardial infarction. Histologic and histochemical studies of early myocardial infarcts induced by temporary or permanent occlusion of a coronary artery. Lab. Invest. 13, 1491–1503.

Steenbergen, C., Hill, M.L. and Jennings, R.B. (1985). Volume regulation and plasma membrane injury in aerobic, anaerobic, and ischemic myocardium *in vitro*. Circulation Res. 57, 864–875.

Stein, J.H., Osgood, R.W., Barnes, J.L., Reineck, H.J., Pinckard, R.N. and McManus, L.M. (1985). The role of complement in the pathogenesis of postischemic acute renal failure. Miner. Electrolyte Metab. 11, 256–261.

Storrs, S.B., Kolb, W.P., Pinckard, N.P. and Olson, M.S. (1981). Characterization of the binding of purified human C1q to heart mitochondrial membranes. J. Biol. Chem. 256, 10924–10929.

Storrs, S.B., Kolb, M.S. and Olson, M.S. (1983). C1q binding and C1 activation by various isolated cellular membranes. J. Immunol. 131, 416–422.

Swenson, L.J., Pantely, G.A., Anselone, C.G. and Bristow, J.D. (1987). Increased vascular resistance during complement-activated plasma infusion in swine. Am. J. Physiol. 253, H58–H65.

Utley, J.R. (1990). Pathophysiology of cardiopulmonary bypass: current issues. J. Card. Surg. 5, 177–189.

Virmani, R., Forman, M.B. and Kolodgie, F.D. (1990). Myocardial reperfusion injury. Histopathological effects of perfluorochemical. Circulation 81(Suppl. 3), IV57–IV68.

Vlaicu, R., Rus, H.G. and Niculescu, F. (1988). Immuno-histochemical localization of the terminal C5b-9 complement complexes associated to S-protein and macrophage immunoreactive deposits in human damaged myocardial areas. Med. Interne 26, 21–27.

Watanabe, N. and Hyodo, Y. (1970). Anti-heart reactant in rheumatic fever. Paediatr. Univ. Tokyo 18, 149–153.

Weisman, H.F., Bartow, T., Leppo, M.K., Boyle, M.P., Marsh, H.C.J., Carson, G.R., Roux, K.H., Weisfeldt, M.L. and Fearon, D.T. (1990a). Recombinant soluble CR1 suppressed complement activation, inflammation, and necrosis associated with reperfusion of ischemic myocardium. Trans. Assoc. Am. Physicians 103, 64–72.

Weisman, H.F., Bartow, T., Leppo, M.K., Marsh, H.C.J., Carson, G.R., Concino, M.F., Boyle, M.P., Roux, K.H., Weisfeldt, M.L. and Fearon, D.T. (1990b). Soluble human complement receptor type 1: *in vivo* inhibitor of complement suppressing post-ischemic myocardial inflammation and necrosis. Science 249, 146–151.

Werns, S.W. and Lucchesi, B.R. (1989). Myocardial ischemia and reperfusion: the role of oxygen radicals in tissue injury. Cardiovasc. Drugs Ther. 2, 761–769.

Wiedmer, T., Esmon, C.T. and Sims, P.J. (1986a). Complement proteins C5b-9 stimulate procoagulant activity through platelet prothrombinase. Blood 68, 875–880.

Wiedmer, T., Esmon, C.T. and Sims, P.J. (1986b). On the mechanism by which complement proteins C5b-9 increase platelet prothrombinase activity. J. Biol. Chem. 261, 14587–14592.

Wiedmer, T., Ando, B. and Sims, P.J. (1987). Complement C5b-9-stimulated platelet secretion is associated with a Ca^{2+}-initiated activation of cellular protein kinases. J. Biol. Chem. 262, 13674–13681.

Wolfgram, L.J. and Rose, N.R. (1989). Coxsackievirus infection as a trigger of cardiac autoimmunity. Immunol. Res. 8, 61–80.

Yancey, K.B., Lawley, T.J., Dersookian, M. and Harvath, L. (1989). Analysis of the interaction of human C5a and C5a des Arg with human monocytes and neutrophils: flow cytometric and chemotaxis studies. J. Invest. Dermatol. 92, 184–189.

Youngleson, J.S., Jones, D.T. and Woods, D.R. (1989). Homology between hydroxybutyryl and hydroxyacyl coenzyme A dehydrogenase enzymes from *Clostridium acetobutylicum* fermentation and vertebrate fatty acid beta-oxidation pathways. J. Bacteriol. 171, 6800–6807.

Zimmerman, A., Gerber, H., Nussenzweig, V. and Isliker, H. (1990). Decay-accelerating factor in the cardiomyocytes of normal individuals and patients with myocardial infarction. Virchows Arch. A Pathol. Anat. Histopathol. 417, 299–304.

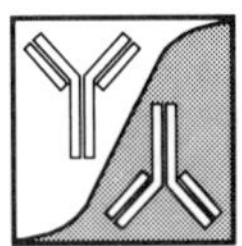

7. Prevention of Sudden Cardiac Death by Immunopharmacological Intervention

Michael J. Curtis

1. Abstract

Sudden cardiac death is a syndrome characterized by death attributable to a sudden loss of cardiac sufficiency. It is the largest single cause of death in North America and the European Economic Community. For a variety of reasons, in the majority of its victims, its immediate cause is believed to be ventricular arrhythmias, initiated as a consequence of ischaemic heart disease (cardiac ischaemia, infarction or reperfusion). The natural history of these arrhythmias is characterized by distinct phases of susceptibility, indicative of several different underlying pathological mechanisms. During acute myocardial ischaemia there may be at least two phases of susceptibility to arrhythmias. During the development of infarction there is another phase. If ischaemic but viable tissue is reperfused, a separate phase of so-called reperfusion-induced arrhythmias may occur. In this chapter, the potential for amelioration of sudden cardiac death by immuno-modulation is evaluated by consideration of: (1) the involvement of the immune system in the pathogenic mechanisms involved in each phase of arrhythmia; and (2) the anti-arrhythmic effects of drugs acting on the immune system. Evidence is presented that some, but not all, of the phases of arrhythmogenesis relevant to sudden cardiac death may be susceptible to vitiation by immunopharmacological intervention. Direct evidence of this from animal models of sudden cardiac death concerns mainly the more long-established components of the immune system (e.g. TXA_2). Detailed work on the role of more recently discovered substances (e.g. selectins) may be warranted.

2. Sudden Cardiac Death

Although relative vagueness in its definition has diminished gradually over the years as the facts have become better established, sudden cardiac death is a

Immunopharmacology of the Heart
ISBN 0–12–200245–8

syndrome, defined in most instances by postmortem findings. Its nomenclature was introduced for the purpose of categorizing the cause of death in an individual. Sudden cardiac death does not have a rigid scientific definition. However, it is an extremely important entity affecting more than 400 000 people each year in the USA alone (Gordon and Kannel, 1971; Lown, 1982).

Despite the caveats noted above, a definition of sudden cardiac death is necessary to allow this article to proceed. Clinically, it has been defined as death occurring within 1 h of the victim last being seen alive, and has been suggested to occur in approximately 30% of the total patient population with coronary artery disease (Armstrong *et al.*, 1971). It has been estimated that approximately 43% of deaths that occur during the 4 weeks following the onset of chest pain (the first tangible symptom other than death itself) do so during the first hour after the onset of pain (Armstrong *et al.*, 1971). In addition, it has been estimated that fewer than 10% of patients dying during the first hour after the onset of pain are seen by a physician (Oliver, 1982).

It is apparent from the above that the majority of sudden cardiac deaths occur in the absence of an unequivocal medical establishment of the immediate cause of death. The possibilities for this are believed to include, principally, cardiac arrhythmia, cardiac rupture and cardiac output failure. However, there are good reasons to suspect that the most common cause of death is ventricular fibrillation (VF), and that this VF is principally a consequence of acute myocardial ischaemia (Oliver, 1982; Campbell, 1983, 1984). In considerably fewer cases, reperfusion of ischaemic myocardium may also cause death (Tzivoni *et al.*, 1983). In addition, infarct-induced VF may constitute another basis for sudden cardiac death (in coronary artery disease patients surviving the acute phase of myocardial ischaemia). These suppositions are based on experiment; in all animal models of coronary artery disease, in studies using anaesthetized and conscious rat with a single permanent coronary obstruction (Curtis *et al.*, 1987) through to conscious dog with thrombus-induced ischaemia in the presence of a healed infarction (Black *et al.*, 1991), anaesthetized dog with myocardial infarction resulting from a two-stage obstruction of a coronary artery (Harris, 1950) or rat, pig and dog with reperfusion of an obstructed coronary artery (Manning and Hearse, 1984) the principal life-threatening event is VF. This information is summarized in Fig. 7.1

The figure illustrates that there are two principal approaches which can, in theory, be expected to have a major impact on sudden cardiac death: prevention of the development of coronary obstruction and direct prevention of VF itself. The development of coronary obstruction and its modulation by immunopharmaco-logical intervention is discussed in Chapter 4 of this book, by Darley-Usmar and Hassall. Although

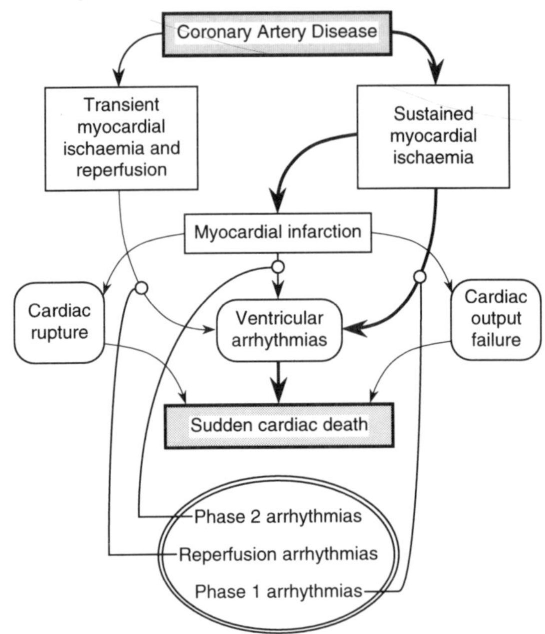

Figure 7.1 Causes of sudden cardiac death. The flow chart shows the linkage between coronary artery disease and sudden cardiac death. The relative importance of different causes of sudden cardiac death is indicated by the thickness of the arrows linking the settings to their adverse consequences. The thickest arrows indicate the most prevalent clinical condition relevant to sudden cardiac death. The major phases of arrhythmia are labelled by the words shown in the oval box at the bottom of the figure.

prevention of coronary obstruction and the resultant ischaemia, were it achievable, would constitute an effective means of prevention of sudden cardiac death, intervention aimed at amelioration of the consequences of coronary obstruction (such as VF) will remain a necessary approach to sudden cardiac death so long as ischaemic heart disease prevails (i.e. as long as the means of effective prevention of coronary obstruction remains elusive). The present chapter is therefore focused on the ability of immunopharmacological intervention to alleviate VF resulting from acute ischaemia, infarction and reperfusion, and the role of the immune system as a pharmacological target for prevention of this VF.

3. *Mechanism of Arrhythmogenesis in Ischaemia, Infarction and Reperfusion*

If immunopharmacological intervention is to modulate

sudden cardiac death by prevention of VF then it would be necessary that chemical components of the immune system play an important role in the mechanism of arrhythmogenesis in ischaemic heart disease. The present chapter is focused primarily on direct arrhythmogenic actions of immunologically relevant substances, as noted above. The most feasible means by which this would be possible is the production of serious disturbance of cellular electrophysiology by chemical mediators of the immune response cascades, leading to abnormal initiation and/or conduction of electrical activity in the heart. In order to evaluate whether this is the case, it is first necessary to consider what changes in syncytial and cellular electrophysiology take place during ischaemia, infarction and reperfusion.

In acute ischaemia, the involved tissue becomes depolarized during diastole, refractoriness becomes heterogeneous and conduction velocity is slowed. These changes facilitate focal (flow of injury current) and re-entrant syncytial mechanisms of arrhythmogenesis (Janse, 1986). Reperfusion-induced arrhythmias appear to be associated with a multiphasic profile of hyper-polarization and depolarization, variable changes in refractoriness, and the appearance of afterpotentials (Ferrier *et al.*, 1985, and others). In hearts with an established infarct, re-entry around a fixed anatomical obstruction (the infarct) appears to be an important mechanism (Janse, 1986), although automaticity originating in damaged but surviving Purkinje fibres also appears to be relevant (Wit, 1989).

As will become clear later, the settings which may be most relevant to immunopharmacological modulation, namely the period during which the infarct is evolving (either when coronary obstruction is maintained or following reperfusion subsequent to ischaemia sustained for 60 min or more), and the early moments of reper-fusion are least well categorized in terms of electro-physiological mechanisms of arrhythmogenesis.

The changes described above are important since any immunological entity or mediator relevant to arrhythmo-genesis must be capable of eliciting similar changes if it is to be regarded as a relevant cause of sudden cardiac death (and thus an appropriate target for immunopharmaco-logical intervention).

4.   *Natural History of Arrhythmogenesis in Ischaemic Heart Disease*

Arrhythmias occurring in ischaemic heart disease show characteristic profiles of occurrence and severity related to the lapse of time from the onset of ischaemia, the extent to which cells remain viable, and the timing of reper-fusion (if reperfusion occurs). In man, this timing (and its possible anatomical heterogeneity) is difficult to define with precision or accuracy owing to the difficulty in establishing the onset of ischaemia, the severity of ischaemia (in terms of degree of reduction of coronary flow) and the heterogeneity of the clinical condition (in terms of anatomical location and extent of involved tissue and the variable level of collateral anastomoses). In addition there is the complication of previous medication and new therapeutic intervention initiated after admis-sion to hospital which may obscure identification of the antecedents of arrhythmogenesis.

Deliberately objective clinical studies of the natural history of ischaemic heart disease in relation to arrhythmogenesis are scanty. Those that have been carried out have used a sometimes heterogeneous patient population. Campbell *et al.* (1981) evaluated 38 previously unmedicated patients admitted for "acute myocardial infarction" (an imprecise term which encompasses acute ischaemia and infarction) and related arrhythmia occurrence to the onset of symptoms (which were not defined in specific terms). It was found that primary VF (defined as that occurring in the absence of shock) manifested principally within the first 4 h after the onset of symptoms (it was not possible to achieve any greater temporal precision than this). VF was associated with the presence of ventricular premature beats (VPBs) of a particular variety in which the VPB occurred prior to completion of the previous T wave (the so-called R-on-T phenomenon). After 4 h, VF and R-on-T VPBs were almost absent. However, non-R-on-T VPBs persisted during the subsequent 12 h of assessment. No publica-tion contradicting this profile of arrhythmogenesis in patients with maintained ischaemia has yet been published in the intervening 12-year period, so it would appear that this profile has been accepted as representa-tive amongst clinical practitioners.

Animal studies have revealed a time–response profile grossly similar to this clinical profile, but the precision and accuracy are much greater owing to the ability in animal experiments for the investigator to control independent variables such as the exact onset of ischaemia, ischaemic zone size and collateral flow. Thus, Harris (1950) recognized that arrhythmias elicited by coronary obstruction (studied in anaesthetized dogs) may occur in two phases, with "phase-1" arrhythmias occur-ring during the first 30 min of ischaemia and "phase-2" arrhythmias beginning after 1–2 h of occlusion. More recently, in studies using the same species, Meesman (1982) found that phase-1 ventricular arrhythmias (particularly the least severe ones such as VPBs) have two components (1a and 1b) with associated peaks in suscept-ibility after around 4 min and 15 min after the onset of ischaemia. This subdivision of phase-1 arrhythmias into 1a and 1b has not yet been confirmed in conscious dogs (or in other species). In conscious rats, sustained abrupt and complete occlusion of a main coronary vessel leads to a well-defined monophasic profile of phase-1 arrhythmias with subdivision into 1a and 1b appearing to be unwarranted (Curtis, 1986).

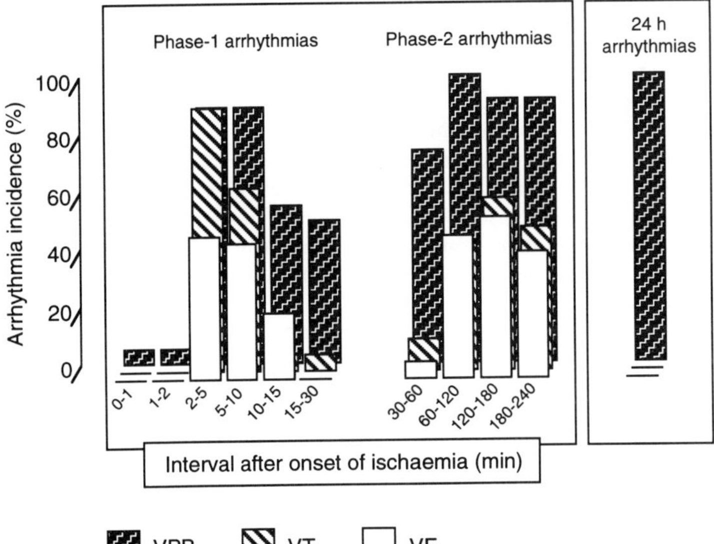

Figure 7.2 Ventricular arrhythmias in conscious rats (*n* = 18) with permanent left coronary occlusion. Two distinct phases of arrhythmias are seen (phases 1 and 2). The most severe arrhythmia (VF) shows the most pronounced biphasic profile, and the least severe arrhythmia (VPBs) the least. By 24 h after the onset of ischaemia, by which time an infarct is fully developed, VF and VT incidences have subsided to a negligible level, whereas VPBs persist. (Data adapted from Curtis, 1986.)

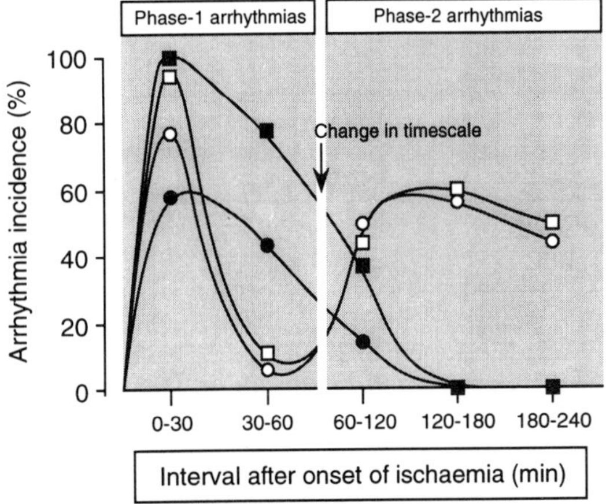

Figure 7.3 Ventricular arrhythmias in conscious rats (*n* = 18) and in isolated, crystalloid-perfused rat hearts (Langendorff mode), each with permanent left coronary occlusion. The x-axis shows a series of consecutive time bins and the y-axis shows percentage incidence per group of VT and VF (new episodes) occurring within each time bin. The data points are centred in the middle of each time bin. (Data adapted from Curtis, 1986, and Ravingerova and Curtis, 1991.) □, VT *in vivo*; ○, VF *in vivo*; ■, VT *in vitro*; ●, VF *in vitro*.

that in isolated rat hearts perfused with blood-free (crystalloid) modified Krebs buffer. The most salient feature here is that phase-2 arrhythmias are absent in the isolated crystalloid perfused heart. The other major point which can be identified from Fig. 7.3 is that phase-1 arrhythmias, although delayed slightly in onset and offset, are present in crystalloid-perfused hearts as well as *in vivo*.

5. *Phase-1, Phase-2 and Reperfusion Arrhythmias: Role of the Immune System*

Figure 7.3 raises several important points about arrhythmogenesis during phase-1 and phase-2 relevant to the focus of this chapter. First, all the conditions necessary for phase-1 arrhythmias *in vivo* appear to be present in the isolated crystalloid-perfused heart. It is important, then, to note that components of the immune response involving blood-borne elements (platelets, leukocytes, neutrophils, complement, etc.), absent in the latter preparation, would therefore appear to be unnecessary for phase-1 arrhythmogenesis. This is not to say that such elements may not be capable of contributing to arrhythmogenesis *in vivo*. However, since their involvement is not necessary *in vitro* (where arrhythmias show an equivalent malevolence and natural history to those occurring *in vivo* in the same species), it is difficult to contemplate that intervention designed to blunt the immune response

The natural history of arrhythmogenesis in the conscious rat with myocardial ischaemia is shown in Fig. 7.2. In Fig. 7.3, a comparison is made between the profile of phase-1 and phase-2 arrhythmias *in vivo* versus would be expected to be capable of producing significant attenuation of arrhythmogenesis during acute ischaemia (i.e. the period when phase-1 arrhythmias occur). Although this conclusion is not necessarily supported by drug intervention studies (see later), and may be specific to rat (it has not been confirmed in other species), by itself it would imply that immunopharmacological intervention would not be expected to have a major impact on this potentially important cause of sudden cardiac death.

The second point to note is that the converse to the above would appear to be true with regard to phase-2 arrhythmias. Figure 7.3 illustrates that isolation of the heart and perfusion in the absence of blood elements essentially abolishes phase-2 arrhythmias. One interpretation of this observation would be that phase-2 arrhythmias are critically dependent on blood elements (participants in the immune response) making phase-2 arrhythmias a manifestation of the immune response. This constitutes an immunological hypothesis for phase-2 arrhythmias which predicts that phase-2 arrhythmias are a target for immunopharmacological intervention. However, by way of a caveat, it should be noted that, compared with the *in vivo* setting, the isolated heart is devoid of functional innervation, so it could be argued that it is an intact and functioning cardiac nervous system (the neural hypothesis) which is critical for phase-2 arrhythmogenesis rather than a functioning immune system (the immunological hypothesis). In support of the neural hypothesis, it has been shown that the activity of the sympathetic nervous system and, in particular, the presence of β-adrenoceptor activation, plays an important role in the initiation of phase-2 arrhythmias in dogs (Patterson *et al.*, 1986), a finding supported by other independent data (Scherlag *et al.*, 1989). However, this dog data has yet to be confirmed in other species. It is a pity that research in the area of sudden cardiac death is so compartmentalized in terms of species and model chosen for study.

Reperfusion of ischaemic myocardium can be a potent stimulus for VF in animal experiments (Manning and Hearse, 1984). However, reperfusion is not regarded by clinicians as being important as a cause of sudden cardiac death (the reasons for this are discussed in detail in Curtis and Hearse, 1989a). Since reperfusion elicits VF most effectively after short durations of ischaemia (more brief than 30 min) and that this occurs reproducibly in crystalloid-perfused (blood-free) hearts (Curtis and Hearse, 1989a,b) and since reperfusion *in vivo* after sustained ischaemia produces little in the way of life-threatening arrhythmias (Euler *et al.*, 1988), it is difficult to argue a major role for the immune system in sudden cardiac death resulting from reperfusion, the logic behind this conclusion being equivalent to that expounded in more detail above in the case of ischaemia-induced phase-1 arrhythmias. However, for the sake of completeness, some inclusion of data on the subject of reperfusion

arrhythmias is given later, especially in relation to drug intervention studies. Reperfusion after sustained ischaemia produces arrhythmias which persist for several hours (Euler *et al.*, 1988) in contrast to reperfusion after a brief (<30 min) period of ischaemia (Curtis and Hearse, 1989a) and even though the severity of the former is low compared with the latter, their consideration is warranted in view of possible evidence of involvement of components of the immune system in their initiation, as will be discussed later.

In order to assess the question posed in the title of this chapter it is necessary to examine in more detail the evidence of a role for the immune system in sudden cardiac death. First, some criteria for approaching this issue are considered.

6. Proposed Criteria for Evaluating the Role of Components of the Immune System in Arrhythmogenesis

The identity and role of any substance (including components of the immune system) in arrhythmogenesis may be assessed using general criteria described recently (Curtis *et al.*, 1993) and reproduced in modified form (relevant to the immune system) in Table 7.1. As noted above, it is important to consider the possibility that any arrhythmogenic effects of a cell or chemical component of the immune response may be mediated by direct means or indirectly. The former requires modulation of the cardiac pumps and channels which determine resting membrane potential and action potential configuration. The latter involves effects such as vasoconstriction or

Table 7.1 Criteria for establishing that a component of the immune system functions as mediator of ventricular arrhythmias in ischaemic heart disease

1. Identification of the presence of (or activation of) the component in ischaemic or reperfused tissue.
2. Identification of potentially arrhythmogenic electrophysiological effects of the component at pathophysiologically relevant concentrations.
3. Identification of antiarrhythmic effects of drugs with specific actions on the accumulation of the component, concomitant with substantial alterations in the component's local accumulation.
4. Identification of pro-arrhythmic effects of drugs with specific actions to enhance the accumulation of the component, concomitant with selective and substantial alteration of local accumulation of the component.
5. Identification of pro-arrhythmic effects of drugs which selectively mimic actions of the component.
6. Demonstration that at pathophysiologically relevant local concentrations, administration of the component can mimic ischaemia-, infarction- or reperfusion-induced arrhythmias in the absence of ischaemia, infarction or reperfusion.

platelet aggregation (themselves leading to ischaemia) or release of secondary arrhythmogenic chemicals from the heart, endothelium or blood cells. Therefore, conclusions drawn from the use of isolated cells and tissues for exploring the electrophysiological actions of immunologically relevant cells and chemicals should be treated with circumspection.

Most of the data on the immune system and arrhythmogenesis falls into one of two types: demonstration of the presence of immunologically relevant material in the involved tissue and demonstration of antiarrhythmic properties of immunomodulatory interventions. Below, various chemical components of the immune system have been considered individually with respect to these factors.

Cells, such as neutrophils and platelets, also can participate in arrhythmogenesis and do so by one of two mechanisms. They may either contribute to coronary obstruction by virtue of their involvement in thrombosis or thrombogenesis (see elsewhere in this book), or they may release chemical substances which elicit cardiac electrophysiological disturbance. These cells cannot directly affect cellular electrophysiology by their own physical presence. Since a cellular electrophysiological change is a prerequisite for arrhythmogenesis then, in this chapter, it is the chemical products of immunologically relevant cells, plus chemicals with effects on such cells, which give their name to the following section headings. A framework for the pathogenesis of sudden cardiac death incorporating chemical components of the immune system is shown in Fig. 7.4.

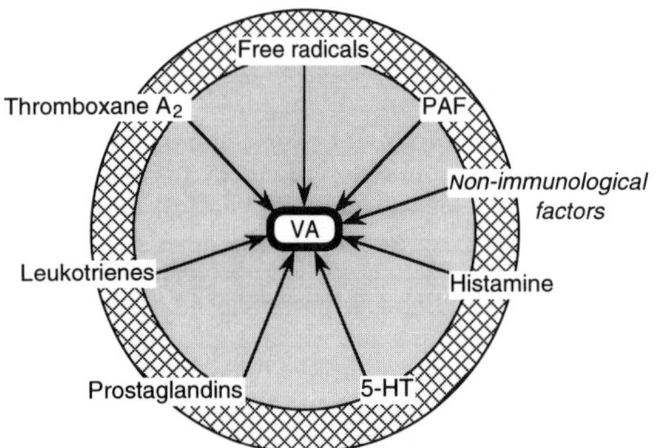

Figure 7.4 Immunopharmacologically relevant mediators of ventricular arrhythmias. The figure represents a working model for assessment of the role of immunological mediators of VAs in ischaemic heart disease. Grey shading represents a domain of direct arrhythmogenic effects and cross-hatched shading represents a domain of synergistic or antagonistic interactions between substances (in terms of modulation of accumulation or actions of one substance by another). Non-immunopharmacological factors refers to a variety of chemicals which have been suggested to play a role in sudden cardiac death, particularly during phase-1; these are discussed in detail elsewhere (Curtis et al., 1993).

7. PAF (and Platelets)

Platelet activating factor (PAF) is an important target for immunopharmacological intervention, and has diverse actions in the body in general (Braquet, 1989) and in the heart in particular (Tamargo et al., 1988). Its actions on myocytes are complex, some occurring directly, and some as a consequence of actions on platelets.

PAF is released during atrial pacing in patients with coronary artery disease (Montruccio et al., 1986) and elevated levels have been detected in the ischaemic myocardium of primate (Annable et al., 1985). PAF can elicit coronary vasoconstriction and negative inotropism (Benveniste et al., 1983; Sybertz et al., 1985; Pugsley et al., 1991), properties that have been attributed to the release and actions of secondary mediators relevant to the immune system, namely peptidoleukotrienes and TXA₂ (Piper and Stewart, 1986), and with direct actions on coronary vessels. On the basis of these observations, PAF's actions may represent potential targets for intervention aimed at prevention of sudden cardiac death.

Studies with several PAF antagonists (and PAF itself) support this notion. The PAF antagonists SRI-63,441 and BN52021 have been shown to protect dogs against reperfusion-induced arrhythmias, and PAF was found to be directly arrhythmogenic in the same species (Wainwright et al., 1989). Various PAF antagonists have been shown to inhibit ischaemia- and reperfusion-induced arrhythmias, including BN50739 (Koltai et al., 1991), CV-6209 (Stahl et al., 1988) and RP59227 (Auchampach et al., 1991).

The principal electrophysiological effect of low concentrations of PAF (10 pM to 100 nM) applied to the healthy heart is shortening of APD; this coincides with the appearance of ventricular arrhythmias (Tamargo et al., 1988). These effects suggest that PAF may function as a chemical mediator of ventricular arrhythmias. In support of this, 1 nM PAF was found to increase the incidence of VF in a model of low flow global ischaemia (Flores and Sheridan, 1990).

Platelets themselves (activated by PAF) may play a role in sudden cardiac death by virtue of their involvement in thrombosis (Braquet, 1989). In addition, some involvement of platelets in arrhythmogenesis independent of thrombosis is possible as a consequence of release of secondary mediators with electrophysiological actions. Exacerbation of prolongation of APD by platelets during low-flow global ischaemia can be antagonized by the PAF antagonist, BN52021; however, the exacerbation of arrhythmias by platelets in the same model is not antagonized (Flores and Sheridan, 1991). This suggests that platelets play a role in arrhythmogenesis independently of the presence of PAF.

For reasons discussed above, the presence of platelets is not necessary for phase-1 ischaemia-related arrhythmogenesis, so platelets are unlikely to represent an important target for immunopharmacological intervention in this setting, despite the evidence (discussed above) that platelets are capable of facilitating phase-1 arrhythmias. This is supported by evidence from Mullane and McGiff (1985) by the use of thrombocytopaenia induced by anti-platelet serum. During infarct development (and possibly during reperfusion after a sustained period of ischaemia), platelets may represent a more important therapeutic target since their accumulation in involved tissue is well documented (Romson *et al.*, 1982), and since phase-2 arrhythmias are absent in crystalloid-perfused (i.e. platelet-free) hearts (Fig. 7.3).

Thus, the relationship between PAF and platelets, coronary vascular calibre, action potential duration, secondary mediators and susceptibility to arrhythmias is complex and has potential relevance to sudden cardiac death; however, further work is needed to clarify this complexity. Thorough investigation of the role of PAF and platelets in phase-2 arrhythmogenesis and in reperfusion-induced arrhythmia development following sustained ischaemia is particularly warranted.

8. *Prostaglandins and Thromboxanes*

The principal site of cardiac prostaglandin (PG) generation is the coronary vasculature (Hsueh and Needleman, 1978), although Bolton *et al.* (1980) found that PG metabolites are also produced by cardiac myocytes. During hypoxia (De Deckere *et al.*, 1977; Karmazyn and Dhalla, 1983) or ischaemia and reperfusion (Kraemer *et al.*, 1976; Coker *et al.*, 1981; Moffat, 1987) there is a significant increase in PG production, especially of prostacyclin (PGI_2).

Some evidence exists that some PGs may function as chemical mediators of arrhythmogenesis in sudden cardiac death. Indeed, PGI_2 has been reported to potentiate the arrhythmogenic effects of coronary occlusion in rats (Au *et al.*, 1979). However, this finding is not consistent in that other investigators find protective effects of PGI_2. PGI_2 is a coronary vasodilator and has been suggested to play a role as an endogenous cardioprotective agent, and many studies have shown this and other PGs to be antiarrhythmic (e.g. Mest *et al.*, 1972, 1974). This has been reviewed by Parratt (1993). Nevertheless, it has been argued that the concept that PGs are protective is unlikely (Moffat, 1987; Karmazyn, 1989) on the grounds that protective effects of exogenously applied PGs occur at concentrations which are not physiologically relevant.

Evidence that endogenous PGs may represent a target for immunopharmacological intervention designed to reduce sudden cardiac death is provided by data from drug studies. Dix *et al.* (1982) demonstrated that 100 mg/kg sulphinpyrazone reduced the incidence of VF during acute ischaemia in cats. Karmazyn (1984) showed that the same drug was cardioprotective *in vitro* during reperfusion. Aspirin, meclofenamate and indomethacin

(all cyclo-oxygenase inhibitors) were found to decrease ventricular premature beats, VT and VF in rats *in vivo* during acute regional ischaemia (Fagbemi, 1984). These data suggest that cyclo-oxygenase products (which include PGs) play a role in arrhythmogenesis. However, the studies cited do not allow differentiation between the main subtypes of cyclo-oxygenase products, PGs and thromboxane (TX) which may have very different effects on susceptibility to arrhythmias. Au *et al.* (1979) found that PGE_2 inhibited (rather than promoted) ischaemia-induced arrhythmias in the rat whereas neither $PGF_{2\alpha}$ nor ibuprofen modified tissue electrophysiological responses to simulated ischaemia in isolated superfused canine Purkinje fibres (Moffat *et al.*, 1989); additionally, ibuprofen had no effect on oscillatory after potentials during simulated reperfusion despite an inhibition of PG production (Moffat *et al.*, 1989). At present it is difficult to draw any definitive conclusions concerning the role of PGs as a target for immunopharmacological intervention to prevent arrhythmias in acute ischaemia and reperfusion.

TX, on the other hand, possesses many actions including induction of platelet aggregation (Hamburg *et al.*, 1975) and constriction of coronary vascular smooth muscle (Ellis *et al.*, 1976) which suggest a potential involvement in arrhythmogenesis by virtue of a pro-ischaemic action. Evidence of an involvement in arrhythmogenesis is supported indirectly by data concerning antiarrhythmic effects of cyclo-oxygenase inhibitors (see above), blockers of TX synthesis and blockers of TX receptors (Coker and Parratt, 1985a). However, it is difficult to distinguish between a direct arrhythmogenic effect of TX and a pro-ischaemic effect from the above studies. In support of a pro-ischaemic mechanism, although platelets possess TX receptors (Gresele *et al.*, 1991), no cardiac TX receptor has yet been described, implying that if TX plays any role in arrhythmogenesis it is more likely to do so by promoting platelet aggregation or vasoconstriction, and thus exacerbating the depth or extent of ischaemia rather than by acting directly on cardiac cells to promote arrhythmogenesis.

TX does nevertheless show a profile of properties consistent with an involvement in arrhythmogenesis, even if this involvement is indirect. Its concentration is elevated in venous blood draining ischaemic tissue with levels correlating with the incidence of ventricular arrhythmias (Coker *et al.*, 1981). Arrhythmias can be elicited in the absence of ischaemia by administration of stable TX analogues with agonist properties in anaesthetized dogs (Mehta *et al.*, 1982). Such actions are not secondary to systemic haemodynamic changes because arrhythmias are also elicited by close coronary arterial injection of TX mimetics such as U46619 in dogs (Parratt and Wainwright, 1986b). Antiarrhythmic effects have been demonstrated in dogs by administration of the TX synthase blocker UK38485 during coronary ligation

(Coker, 1984) and by administration of the TX synthase blocker R-68070 in dogs with ischaemia induced by thrombogenic injury to coronary vascular endothelium (De Clerck *et al.*, 1989). In the latter study an anti-thrombotic effect is unlikely since the drug used also suppressed reperfusion-induced arrhythmias implying that part of the benefit, at least, was attributable to actions additional to inhibition of thrombosis (an action relevant during ischaemia but which cannot have any bearing on protection against reperfusion arrhythmias). Another TX synthase blocker, U-63557A, at a dose effective in reducing TXB_2 levels by >75%, reduced the incidence of ischaemia-induced VF in cats (O'Connor *et al.*, 1989). Likewise, dazmegrel reduced coronary venous TX levels and diminished the incidence of reperfusion-induced VF in dogs (Coker and Parratt, 1985b). Thus evidence exists for a role for TX in arrhythmogenesis during acute ischaemia and during reperfusion. A role for TX in mediating reperfusion-induced arrhythmias is also supported by evidence that the TX antagonists AH23848 (Coker and Parratt, 1985a), CGS-13080 (O'Connor *et al.*, 1989) and BM13,177 (Parratt and Wainwright, 1986a) all inhibit reperfusion-induced arrhythmias in various species. Thus TX is a good candidate as a mediator of arrhythmias and represents a potentially important target for immunopharmacological intervention.

One caveat that must not be overlooked, however, is the fact that phase-1 arrhythmias occur in crystalloid perfused hearts (see above) so, although TX may be a chemical mediator sufficient to elicit VF, its presence may not be necessary during the phase-1 period. It would be of interest to examine the benefits of TX synthase blockers and receptor blockers in crystalloid perfused hearts. In view of the participation of the immune system (and blood-borne elements in particular, see above) during the phase-2 period, it is tempting to hypothesize that the arrhythmogenic actions of TX are particularly important in phase-2 arrhythmogenesis; further studies carried out *in vivo* to explore this may be warranted.

9. *Leukotrienes*

Leukotrienes (LTs), derived from arachidonic acid and synthesized via action of 5-lipoxygenase, have not been demonstrated to accumulate specifically in ischaemic or reperfused myocardium. Nevertheless, the presence of LTs in cells which themselves accumulate during ischaemia (neutrophils, macrophages, etc.) and the protective effects of modulators of LT actions suggest that LTs may represent a target for immunopharmacological suppression of sudden cardiac death during phase-2. Although LTs disturb sinus rhythm (Terashita *et al.*, 1981) and elevate ST segment (Ezeamuzie and Assem, 1983), they are not reported to elicit ventricular arrhythmias when applied globally to the heart (Terashita

et al., 1981; Letts and Piper, 1982). Nevertheless, evidence showing that LT antagonists L660711 and L648051 attenuate ischaemia-induced VF (Beatch *et al.*, 1989) suggests that further evaluation of the arrhythmogenicity of regionally administered LTs may be warranted. Their effects when administered *regionally* have yet to be assessed, and the importance of regional cellular and chemical dysfunction (i.e. the presence of an involved and an uninvolved zone) in arrhythmogenesis, especially during the phase-1 period, is unequivocal (Curtis and Hearse, 1989b; Curtis, 1991; Ridley *et al.*, 1992).

The conclusions which may be drawn concerning the role of LTs are limited at present by scarcity of data. There are also some minor inconsistencies concerning the general cardiopathic relevance of LTs. For example, although LY-233569 (a 5-lipoxygenase inhibitor) and LY-255283 (a selective LTB_4 antagonist) have been shown to have no effect on arrhythmias induced by ischaemia and reperfusion in dogs in conjunction with a lack of effect on ST elevation, leukocyte infiltration or infarct size (Hahn *et al.*, 1990), another 5-lipoxygenase inhibitor (AA-861) and a LT antagonist (ONO-1078) both reduced infarct size in dogs subjected to ischaemia and reperfusion (Ito *et al.*, 1989). In rats, no suppression of reperfusion-induced arrhythmias occurred with L660711 or L648051 (Beatch *et al.*, 1989)

Part of the difficulty in assessing the importance of LTs as a target for immunopharmacological intervention against sudden cardiac death results from complications of data interpretation due to the potentially pro-ischaemic effects of some LTs. This may obscure any direct effect on the myocardium (a similar problem to that described for TX, above). These effects include coronary vasoconstriction (Letts and Piper, 1982; Burke *et al.*, 1983; Ezeamuzie and Assem, 1983; Roth and Lefer, 1983) and, in the case of LTB_4, induction of platelet aggregation and chemotaxis (Mayer, 1988). These effects may contribute to extension of the size of an involved zone and the depth of ischaemia within that zone, two factors which play extremely important roles in arrhythmogenesis (Meesman, 1982; Curtis and Hearse, 1989a). Further work on the role of LTs is indicated.

10. *5-HT*

5-HT is a chemical mediator of aspects of the immune response which is present in platelets and is found in large quantities within the heart. At least four receptor subtypes mediate the actions of 5-HT: $5-HT_1$, $5-HT_2$, $5-HT_3$ and $5-HT_4$ (Bradley *et al.*, 1986; Clarke *et al.*, 1989). 5-HT may stimulate the heart directly via cardiac receptors (Kaumann *et al.*, 1990) or indirectly via release of noradrenaline from sympathetic nerve terminals (Richardson *et al.*, 1985; Thandroyen *et al.*, 1985).

Drug studies have provided conflicting data on the role of 5-HT actions as a target for immunopharmacological intervention for prevention of sudden cardiac death. 5-HT has been shown to initiate after depolarizations in papillary muscle (Kaumann, 1983) suggesting a possible role in reperfusion arrhythmias, yet the 5-HT mimetic, 4-nitro-5-methoxytryptamine reduced the severity of reperfusion-induced arrhythmias in an animal model (Shabunina *et al.*, 1988). On the other hand, in the same model, 5-HT produced a dose-dependent increase in VT and VF, effects being ameliorated by the 5-HT$_2$ antagonist ritanserin (El Mahdy, 1990). Also, the 5-HT antagonists ritanserin, ketanserin and ICS 205,930 all reduced the incidence of reperfusion-induced VF (Williams *et al.*, 1985; Coker *et al.*, 1986; Coker and Ellis, 1987). These data suggest an involvement of 5-HT in reperfusion arrhythmias. However, high concentrations of ketanserin and ritanserin produce effects indicative of sodium or potassium channel blockade (Saman *et al.*, 1985; Coker and Ellis, 1987), so use of these drugs may not permit identification of a specific involvement of 5-HT in arrhythmogenesis. In more recent studies, the selective 5-HT$_2$ antagonists ICI 170,809 and ICI 169,369 were found to have no effect on the incidence of ischaemia-induced VF, but had a defibrillatory effect (facilitating spontaneous termination of VF) during reperfusion of ischaemic myocardium in anaesthetized rat (Ellis and Coker, 1992). The authors attributed the actions to non-cardiac effects (inhibition of platelet aggregation), although both drugs were also capable of reducing maximum driving frequency in isolated paced rat ventricles (effects reminiscent of those produced by ketanserin and ritanserin, mentioned above) and so their defibrillatory actions during reperfusion may have occurred via possible direct effects on cardiac sodium or potassium channels. Whatever the mechanism of action of ICI 170,809 and ICI 169,369, data with these drugs do not support an important direct role for 5-HT$_2$ agonism in arrhythmogenesis in the rat, either during acute ischaemia or during reperfusion.

In addition to its direct effects in cardiac tissue, 5-HT may have other actions which contribute to an involvement in sudden cardiac death. 5-HT provokes contraction of isolated coronary arteries in many species and thus may play a role in coronary artery vasospasm (Saxena and Villalon, 1991), thereby eliciting arrhythmias indirectly. In man, patients with Prinzmetal's angina (McFadden *et al.*, 1991) and patients with coronary endothelial damage (Golino *et al.*, 1991) are particularly susceptible to 5-HT-induced vasospasm. The receptor subtype involved in 5-HT-induced vasoconstrictor responses is not fully established, and evidence suggests a role for 5-HT$_1$ and 5-HT$_2$ subtypes, depending on species and degree, and nature of concomitant vascular injury (Kaumann, 1983; Ichikawa *et al.*, 1989; Chester *et al.*, 1990). Although in patients suffering from Prinzmetal's angina, ketanserin was found to be therapeutically ineffective (De Catarina *et al.*, 1984), the importance of 5-HT-induced vasospasm in sudden cardiac death cannot

be overlooked, and further studies, particularly with 5-HT$_1$ antagonists, are warranted.

11. *Histamine*

Cardiac histamine is located primarily in mast cells (Giotti *et al.*, 1966), is released from the hypoxic heart (Anrep *et al.*, 1936), and may thus play some role as an immunopharmacologically relevant mediator of sudden cardiac death (Harris, 1950). Histamine can elicit electrophysiological effects on cardiac APD and can promote enhanced normal automaticity, abnormal automaticity, delayed after depolarizations and triggered activity (Wit and Rosen, 1983; Wolff and Levi, 1986; Cameron *et al.*, 1990; Cerbai *et al.*, 1990). The latter effects are similar to changes occurring in acutely reperfused tissue (e.g. Ferrier *et al.*, 1985), so this introduces the possibility that histamine may contribute to reperfusion-induced VF.

Despite the fact that electrophysiological actions of histamine are more consistent with changes occurring during reperfusion rather than during ischaemia, Podzuweit *et al.* (1982) succeeded in eliciting ventricular arrhythmias by local administration of histamine into the ischaemic ventricle in pig *in vivo*, and Dai (1987) has shown that ischaemia-induced ventricular arrhythmia onset and severity can be altered by rhodanine and aminoguanidine (agents that alter histamine synthesis and catabolism). To complicate matters further, Dai and Ogle (1990) have also shown that mast cells are surprisingly not the primary source of any histamine-mediated ischaemia-induced arrhythmias.

The subclass of receptor-mediating supposed arrhythmogenic actions of histamine has been examined by use of H$_1$ and H$_2$ antagonists. Although the H$_1$ blockers promethazine, mepyramine and chlorpheniramine have all been shown to inhibit ischaemia- and reperfusion-induced ventricular arrhythmias in isolated perfused rat hearts (Rochette *et al.*, 1988), the authors regarded the effect observed as being a consequence of H$_1$ receptor-independent actions, so involvement of H$_1$ receptors is unproven. On the other hand, the H$_2$ antagonists SK&F-93479 and ranitidine were found to prevent ischaemia-induced VF in rats and dogs *in vivo* with effects attributable to specific H$_2$ blockade rather than to ancillary effects such as Class-I actions (Dai, 1984, 1987) suggesting a role for H$_2$ receptors. However, Rochette *et al.* (1988) found that neither ranitidine nor cimetidine (up to 10^{-4}M) were capable of reducing ischaemia- or reperfusion-induced ventricular arrhythmias in isolated perfused rat heart. Also, Wolff *et al.* (1984) found a slight exacerbation of ischaemia-induced arrhythmias by ranitidine in dogs. Thus, studies with H$_1$ and H$_2$ antagonists have so far failed to clarify the role of histamine as a target for immunopharmacological intervention in prevention of sudden cardiac death and more work would appear to be required.

12. Free Radicals (•R)

In the setting of reperfusion, data concerning the role of •Rs in arrhythmogenesis are in greater abundance than that for any other putative arrhythmogenic factor (Manning and Hearse, 1984; Woodward and Zakaria, 1985; Bernier *et al.*, 1986; etc.). However, few studies directly address the question of whether •Rs produced by cells relevant to the immune system (and its functions) play a role in reperfusion arrhythmias. There are several lines of argument which would not support such a mechanism. Importantly, reperfusion arrhythmias elicited after a brief duration of ischaemia occur with considerable malevolence in hearts perfused *in vitro* in the absence of blood components (Manning and Hearse, 1984) and arguments against an immunological involvement (of any type) are equivalent to those set out (above) in the case of phase-1 ischaemia-induced arrhythmias. Furthermore, the severity of late-onset arrhythmias (elicited by reperfusion *in vivo* after ischaemia of a more sustained duration and which have a greater potential relevance to immunological mechanisms owing to the time-dependent accumulation of •R-producing cells such as neutrophils; Romson *et al.*, 1983) is much less than that associated with brief ischaemia, and little or no VF is encountered (Euler *et al.*, 1988); moreover the anti-radical agents allopurinol and *N*-t-butyl-a-phenylnitrone are ineffective against such arrhythmias (Parratt and Wainwright, 1987). •Rs produced by the immune system therefore remain a possible target for immuno-pharmacological intervention, but more work is needed to determine the extent to which such intervention is capable of limiting sudden cardiac death.

13. Future Considerations

Other substances with relevance to the immune system, not yet discussed, may also play a role in sudden cardiac death, although an insufficiency of data means that their consideration will be a matter for the future, rather than a subject that may be adequately addressed at present.

The importance of adhesion molecules in atherogenesis and thrombosis continues to be an area of intensive investigation; possible modulation of the rate and extent of cardiac ischaemia by molecules such as P-selectin, ICAM-1, ELAM-1, etc. is considered by Lefer in Chapter 3 of this book. At present, there is no direct evidence for participation in (or modulation of) sudden cardiac death at the cardiocyte level by adhesion molecules, and no data on the direct modulation of cardiac electrophysiology by them, with the single and important exception of PAF (see above). Instead, actions in the heart are primarily focused on the coronary vasculature.

Table 3.1 in Chapter 3 shows that maximal expression of P-selectin and PAF occurs within 10 min of a stimulatory event (such as acute myocardial ischaemia), so this permits possible involvement of these substances as mediators of phase-1 sudden cardiac death, according to the criteria shown in Table 7.1 of the present chapter. Indeed, as discussed above, objective evidence supports a role for PAF as a mediator of VF in sudden cardiac death. Although no data on the ability of P-selectin to modulate VF are presently available, the rapidity of its mobilization, emphasized earlier by Entman *et al.* (in Chapter 5) may be reason enough to warrant a preliminary investigation of its possible role in arrhythmogenesis relevant to sudden cardiac death. On the other hand, maximal expression of the remainder of the known adhesion molecules requires at least 4 h following a noxious stimulus such as ischaemia (maximal expression of ATHERO-ELAM requires more than 18 h) so, even if these substances were found to have an ability to modulate VF (e.g. when infused into a coronary artery), such an effect would be relevant only to late phase-2 arrhythmias (which are not as life-threatening as phase-1 arrhythmias). For this reason, most adhesion molecules are unlikely to function as targets for immunopharmacological intervention by virtue of an ability to affect cardiac rhythm directly themselves, although they may represent important therapeutic targets in the context of atherogenesis and infarction (as described earlier in Chapters 3 and 5), conditions which may give rise to sudden cardiac death.

14. Conclusion

Overall, immunopharmacological intervention is unlikely to have a major impact on sudden cardiac death, via direct modulation of the arrhythmogenic processes occurring during ischaemia, infarction and reperfusion. The most clinically important arrhythmias relevant to sudden cardiac death occur during the phase-1 period (if animal studies can be regarded as reliable indicators of clinical ischaemic heart disease) and arrhythmias here do not require the presence of an intact immune system, since they occur in isolated crystalloid-perfused hearts. However, in certain specific settings (e.g. the infarct development phase of ischaemic heart disease), modulation of the accumulation or actions of certain chemical mediators of the immune system may provide benefit to a specific cohort of potential sudden cardiac death victims.

15. Acknowledgements

Work by the author which contributed to parts of this review were funded by the British Heart Foundation, British Columbia and Yukon Heart and Stroke Foundation, and the Canadian Heart Foundation. The author would like to express his thanks to Dr Susan J. Coker, who read a draft of the manuscript and made some helpful suggestions with regard to its content.

16. References

Annable, C.R., McManus, L.M., Carey, K.D. and Pinckard, R.N. (1985). Isolation of platelet activating factor (PAF) from ischaemic baboon myocardium. Fed. Proc. 44, 1271.

Anrep, G.V., Barsoum, G.S. and Talaat, M. (1936). Liberation of histamine by the heart muscle. J. Physiol. 86, 431–451.

Armstrong, A., Duncan, B., Oliver, M.F., Julian, D.G., Donald, K.W., Fulton, M., Lutz, W. and Morrison, S.L. (1971). Natural history of acute coronary heart attacks. A community study. Br. Heart J. 34, 67–80.

Au, T.L.S., Collins, G.A., Harvie, C.J. and Walker, M.J.A. (1979). The actions of prostaglandins I_2 and E_2 on arrhythmias produced by coronary occlusion in the rat and dog. Prostaglandins 18, 707–720.

Auchampach, J.A., Muruyama, M., Cavero, I. and Gross, G.J. (1991). Blocking platelet activating factor receptors with RP59227 decreases myocardial infarct size and the incidence of ventricular fibrillation. J. Molec. Cell. Cardiol. 23 (Suppl. 3), S52.

Beatch, G.N., Courtice, I.D. and Salari, H. (1989). A comparative study of the antiarrhythmic properties of eicosanoid inhibitors, free radical scavengers and potassium channel blockers on reperfusion-induced arrhythmias in the rat. Proc. Western Pharmacol. Soc. 32, 285–289.

Benveniste, J., Boullet, C., Brink, C. and Labat, C. (1983). The actions of PAF-acether (platelet activating factor) on guinea pig isolated heart preparations. Br. J. Pharmacol. 80, 81–83.

Bernier, M., Hearse, D.J. and Manning, A.S. (1986). Reperfusion-induced arrhythmias and oxygen-derived free radicals. Studies with "anti-free radical" interventions and a free radical-generating system in the isolated perfused rat heart. Circ. Res. 58, 331–340.

Black, S.C., Chi, L., Mu, D.-X. and Lucchesi, B.R. (1991). The antifibrillatory actions of UK-68-798, a class-III antiarrhythmic agent. J. Pharmacol. Exp. Ther. 258, 416–423.

Bolton, H.S., Chanderbahn, R., Bryant, R.W., Bailey, J.M., Weglicki, W.B. and Vahouny, G.V. (1980). Prostaglandin synthesis by adult heart myocytes. J. Molec. Cell. Cardiol. 11, 1287–1298.

Bradley, P.B., Engel, G., Feniuk, W., Fozard, J.R., Humphrey, P.P.A., Middlemiss, D.N., Mylecharane, E.J., Richardson, B.P. and Saxena, P.R. (1986). Proposals for the classification and nomenclature of functional receptors for 5-hydroxytryptamine. Neuropharmacology 25, 563–576.

Braquet, P. (1989) (editor) "Ginkgolides: Chemistry, Biology, Pharmacology and Clinical Perspectives," vol. II. JR Prous Science Publishers, Barcelona.

Burke, J.A., Levi, R., Guo, Z.-G. and Corey, E.J. (1983). Leukotrienes C_4, D_4 and E_4: effects on human and guinea pig cardiac preparations. J. Pharmacol. Exp. Ther. 221, 235–241.

Cameron, J.S., Switgart, C.R., Shin, G.S., Katz, D. and Bassett, A.L. (1990). Enhanced susceptibility to histamine-induced arrhythmias in spontaneously hypertensive rats. J. Cardiovasc. Pharmacol. 15, 626–632.

Campbell, R.W.F. (1983). Treatment and prophylaxis of ventricular arrhythmias in acute myocardial infarction. Am. J. Cardiol. 52, 55C–59C.

Campbell, R.W.F. (1984). Prophylactic antiarrhythmic therapy in acute myocardial infarction. Am. J. Cardiol. 44, 8E–10E.

Campbell, R.W.F., Murray, A. and Julian, D.G. (1981). Ventricular arrhythmias in the first 12 hours of acute myocardial infarction: natural history study. Br. Heart J. 46, 351–357.

Cerbai, E., Amerini, S. and Mugelli, A. (1990). Histamine and abnormal automaticity in barium- and strophanthidin-treated sheep Purkinje fibers. Agents Actions 31, 1–10.

Chester, A.H., Martin, G.R., Bodelsson, M., Arnelko-Nobin, B., Tajkarimi, S., Tornebrandt, K. and Yacoub, M.H. (1990). 5-Hydroxytryptamine receptor profile in healthy and diseased human epicardial coronary arteries. Cardiovasc. Res. 24, 932–937.

Clarke, D.E., Craig, D.A. and Fozard, J.R. (1989). The 5-HT$_4$ receptor: naughty but nice. Trends Pharmacol. Sci. 10, 385–386.

Coker, S.J. (1984). Further evidence that thromboxane exacerbates arrhythmias: effects of UK38485 during coronary artery occlusion and reperfusion in anaesthetized greyhounds. J. Molec. Cell. Cardiol. 16, 633–641.

Coker, S.J. and Parratt, J.R. (1985a). AH23848, a thromboxane antagonist, suppresses ischaemic and reperfusion-induced arrhythmias in anaesthetised greyhounds. Br. J. Pharmacol. 86, 259–264.

Coker, S.J. and Parratt, J.R. (1985b). Relationships between the severity of myocardial ischaemia, reperfusion-induced ventricular fibrillation, and the late administration of dazmegrel or nifedipine. J. Cardiovasc. Pharmacol. 7, 327–334.

Coker, S.J. and Ellis, A.M. (1987). Ketanserin and ritanserin can reduce reperfusion-induced but not ischaemia-induced arrhythmias in anaesthetised rats. J. Cardiovasc. Pharmacol. 10, 479–484.

Coker, S.J., Parratt, J.R., Ledingham, I. McA. and Zeitlin, I.J. (1981). Thromboxane and prostacyclin release from ischaemic myocardium in relation to arrhythmias. Nature (Lond.) 291, 323–324.

Coker, S.J., Dean, H.G., Kane, K.A. and Parratt, J.R. (1986). The effects of ICS 205-930, a 5-HT antagonist, on arrhythmias and catecholamine release during canine myocardial ischaemia and reperfusion. Eur. J. Pharmacol. 127, 211–218.

Curtis, M.J. (1986). The actions of calcium antagonists on arrhythmias and other responses to myocardial ischaemia in the rat. Ph.D. thesis, University of British Columbia, Vancouver, B.C.

Curtis, M.J. (1991). The rabbit dual coronary perfusion model: a new method for assessing the pathological relevance of individual products of the ischaemic milieu: role of potassium in arrhythmogenesis. Cardiovasc. Res. 25, 1010–1022.

Curtis, M.J. and Hearse, D. J. (1989a). Ischemia-induced and reperfusion-induced arrhythmias differ in their sensitivity to potassium: implications for the mechanism of initiation versus the mechanism of maintenance of ventricular fibrillation. J. Molec. Cell. Cardiol. 21, 21–40.

Curtis, M.J. and Hearse, D.J. (1989b). Reperfusion-induced arrhythmias are critically dependent on occluded zone size: relevance to the mechanism of arrhythmogenesis. J. Molec. Cell. Cardiol. 21, 625–637.

Curtis, M.J., MacLeod, B.A. and Walker, M.J.A. (1987). Models for the study of arrhythmias in myocardial ischaemia and infarction: the use of the rat. J. Molec. Cell. Cardiol. 19, 399–419.

Curtis, M.J., Pugsley, M.K. and Walker, M.J.A. (1993). Endogenous chemical mediators of arrhythmogenesis in ischaemic heart disease. Cardiovasc. Res. 27, 703–719.

Dai, S. (1984). Effects of SK&F 93479 on experimentally induced ventricular arrhythmias in dogs, rats and mice. Agents Actions 15, 131–136.

Dai, S. (1987). Ventricular histamine concentrations and arrhythmias during acute myocardial ischaemia in rats. Agents Actions 21, 66–71.

Dai, S. and Ogle, C.W. (1990). Ventricular histamine concentrations and mast cell counts in the rat heart during acute ischaemia. Agents Actions 29, 138–143.

De Catarina, R., Carpeggiani, C. and Abbate, A.L. (1984). A double blind, placebo-controlled study of ketanserin in patients with Prinzmetal's angina. Circulation 69, 889–894.

De Clerk, F., Beetins, J., van der Water, A., Vercammen, E. and Janssen, P.A. (1989). R68070: thromboxane A2 synthetase inhibition and thromboxane A2/prostaglandin endoperoxide receptor blockade combined in one molecule. ii. Pharmacological effects *in vivo* and *ex vivo*. Thromb. Haemost. 61, 43–49.

De Deckere, E.A.M., Nugteren, D.H. and Ten Hoor, F. (1977). Prostacyclin is the major prostaglandin released from the isolated perfused rabbit and rat heart. Nature (Lond.) 268, 160–163.

Dix, R.K., Kelliher, G.J., Jurkiewicz, N. and Bryan Smith, J. (1982). Effect of sulfinpyrazone on ventricular arrhythmias, prostaglandin synthesis, and catecholamine release following coronary occlusion in the cat. J. Cardiovasc. Pharmacol. 4, 1068–1076.

El Mahdy, S.A. (1990). 5-Hydroxytryptamine (serotonin) enhances ventricular arrhythmias induced by acute coronary artery ligation in rats. Res. Comm. Chem. Pathol. Pharmacol. 68, 383–386.

Ellis, A.M. and Coker, S.J. (1992). Contribution of antiplatelet activity to the effects of 5-HT$_2$ receptor antagonists on reperfusion-induced arrhythmias in anaesthetized rats. Eur. J. Pharmacol. 219, 97–104.

Ellis, E.F., Oelz, O., Roberts, J.L., II, Payne, N.A., Sweetman, D.J., Nies, A.S. and Oates, J.A. (1976). Coronary artery smooth muscle contraction by a substance released from platelet: evidence that it is thromboxane A2. Science 193, 1135–1137.

Euler, D.E., Murdock, D.K. and Scanlon, P.J. (1988). Effect of the duration of coronary occlusion, on delayed reperfusion arrhythmias. J. Electrophysiol. 2, 383–390.

Ezeamuzie, I.C. and Assem, E.S.K. (1983). Effects of leukotrienes C4 and D4 on guinea pig heart and the participation of SRS-A in the manifestations of guinea pig cardiac anaphylaxis. Agents Actions 13, 182–187.

Fagbemi, S.O. (1984). The effect of aspirin, indomethacin and sodium meclofenamate on coronary artery ligation arrhythmias in anaesthetised rats. Eur. J. Pharmacol. 97, 283–287.

Ferrier, G.R., Moffat, M.P. and Lukas, A. (1985). Possible mechanisms of ventricular arrhythmias elicited by ischemia following reperfusion: studies on isolated canine ventricular tissue. Circulation Res. 54, 184–194.

Flores, N.A. and Sheridan, D.J. (1990). Electrophysiological and arrhythmogenic effects of platelet activating factor during normal perfusion, myocardial ischaemia and reperfusion in the guinea pig. Br. J. Pharmacol. 101, 734–738.

Flores, N.A. and Sheridan, D.J. (1991). Alteration of the cellular electrophysiological effects of myocardial ischaemia by platelet activating factor antagonism. J. Molec. Cell. Cardiol. 23 (Suppl. V), 101.

Giotti, A., Guidotti, A., Mannaioni, P.F. and Ziletti, L. (1966). The influences of adrenotropic drugs and noradrenaline on the histamine release in cardiac anaphylaxis *in vitro*. J. Physiol. 184, 924–941.

Golino, P., Piscione, F., Willerson, J.T., Cappelli-Bigazzi, M., Focaccio, A., Villari, B., Indolfo, C., Russolillo, E., Condorelli, M. and Chiariello, M. (1991). Divergent effects of serotonin on coronary artery dimensions and blood flow in patients with coronary atherosclerosis and control patients. N. Engl. J. Med. 324, 641–648.

Gordon, I. and Kannel, W.B. (1971). Premature mortality from coronary heart disease: Framingham study. JAMA 215, 1617–1625.

Gresele, P., Deckmeyn, H., Nenci, G.G. and Vermylin, J. (1991). Thromboxane synthase inhibitors, thromboxane receptor antagonists and dual blockers in thrombotic disorders. Trends Pharm. Sci. 12, 158–163.

Hahn, R.A., MacDonald, B.R., Simpson, P.J., Potts, B.D. and Parcli, C.J. (1990). Antagonism of leukotriene B$_4$ receptors does not limit canine myocardial infarct size. J. Pharmacol. Exp. Ther. 252, 58–66.

Hamburg, M., Ovensson, J. and Samuelsson, B. (1975). Thromboxanes: a new group of biologically active compounds derived from prostaglandin endoperoxides. Proc. Natl Acad. Sci. USA 72, 2294–2998.

Harris, A.S. (1950). Delayed development of ventricular ectopic rhythms following experimental coronary occlusion. Circulation 1, 1318–1328.

Hsueh, W. and Needleman, P. (1978). Sites of lipase activation and prostaglandin synthesis in isolated, perfused rabbit hearts and hydronephrotic kidneys. Prostaglandins 16, 661–681.

Ichikawa, Y., Yokoyama, M., Akita, H. and Fukuzaki, H. (1989). Constriction of a large coronary artery contributes to serotonin-induced myocardial ischaemia in the dog with pliable coronary stenosis. J. Am. Coll. Cardiol. 14, 440–459.

Ito, T., Toki, Y., Heida, N., Okumura, K., Hashimoto, H., Ogawa, K. and Satake, T. (1989). Protective effects of a thromboxane synthase inhibitor, a thromboxane antagonist, a lipoxygenase inhibitor, and a leukotriene C$_4$, D$_4$ antagonist on myocardial injury caused by acute myocardial infarction in the canine heart. Jpn Circ. J. 53, 1115–1121.

Janse, M.J. (1986). In "The Heart and Cardiovascular System" (ed H.A. Fozzard), pp. 1203–1238. Raven Press, New York.

Karmazyn, M. (1984). A direct protective effect of sulphinpyrazone on ischaemic and reperfused rat hearts. Br. J. Pharmacol. 93, 331–336.

Karmazyn, M. (1989). Synthesis and relevance of cardiac eicosanoids with particular emphasis on ischaemia and reperfusion. Can. J. Physiol. Pharmacol. 67, 912–921.

Karmazyn, M. and Dhalla, N.S. (1983). Physiological and pathophysiological aspects of cardiac prostaglandins. Can. J. Physiol. Pharmacol. 61, 1207–1225.

Kaumann, A.J. (1983). A classification of heart serotonin receptors. Naunyn Schmiedeberg's Arch. Pharmacol. 322 (Suppl.), R42.

Kaumann, A.J., Snaders, L., Brown, A.M., Murray, K.J. and Brown, M. (1990). A 5-hydroxytryptamine receptor in human atrium. Br. J. Pharmacol. 100, 879–885.

Koltai, M., Tosaki, A., Hosford, A., Esanu, A. and Braquet, P. (1991). Effect of BN50739, a new platelet activating factor antagonist, on ischaemia-induced ventricular arrhythmias in isolated working rat hearts. Cardiovasc. Res. 25, 391–397.

Kraemer, R.J., Phernetton, T.M. and Folts, J.D. (1976). Prostaglandin-like substances in coronary venous blood following myocardial ischaemia. J. Pharmacol. Exp. Ther. 199, 611–619.

Letts, L.G. and Piper, P.J. (1982). The actions of leukotrienes C4 and D4 on guinea pig isolated hearts. Br. J. Pharmacol. 76, 169–176.

Lown, B. (1982). Management of patients at high risk of sudden death. Am. Heart J. 103, 689–697.

Manning, A.S. and Hearse, D.J. (1984). Reperfusion-induced arrhythmias: mechanisms and protection. J. Molec. Cell. Cardiol, 16, 497–518.

Mayer, M. (1988). Leukotrienes and other lipoxygenase metabolites of arachidonic acid: their role in edema, ischemia and demyelination of the central nervous system. Acta Univers. Palacki Olomuc. 119, 181–186.

McFadden, E.P., Clarke, J.G., Davies, G.J., Kaski, J.C., Haider, A.W. and Maseri, A. (1991). Effect of intracoronary serotonin on coronary vessels in patients with stable angina and patients with variant angina. N. Engl. J. Med. 324, 648–654.

Meesman, W. (1982). In "Early Arrhythmias Resulting from Myocardial Ischaemia" (ed J.R. Parratt), pp. 93–112. Macmillan, Basingstoke.

Mehta, J., Nichols, W., Mehta, P. and Conti, C.R. (1982). Thromboxane and prostacyclin in systemic and coronary vascular beds following endoperoxide analog infusion. Am. J. Cardiol. 49, 1014.

Mest, H.-J., Schror, K. and Forster, W. (1972). Antiarrhythmic properties of PGE2. Adv. Biosci. 9, 395–400.

Mest, H.-J., Mentz, P. and Forster, W. (1974). Effects of prostaglandins on experimental arrhythmias. Pol. J. Pharmacol. Pharm. 26, 151–156.

Moffat, M.P. (1987). Concentration-dependent effects of prostacyclin on the response of the isolated guinea pig heart to ischaemia and reperfusion: possible involvement of the slow inward current. J. Pharmacol. Exp. Ther. 242, 292–299.

Moffat, M.P., Ferrier, G.R. and Karmazyn, M. (1989). A direct role of endogenous prostaglandins in reperfusion-induced cardiac arrhythmias. Can. J. Physiol. Pharmacol. 67, 772–779.

Montruccio, G., Camussi, G., Tetta, G., Emmanuelli, G., Orzan, F., Libero, L. and Brusca, A. (1986). Intravascular release of platelet activating factor during atrial placing. Lancet 2, 293.

Mullane, K.M. and McGiff, J.C. (1985). Platelet depletion and infarct size in an occlusion–reperfusion model of myocardial ischemia in anaesthetised dogs. J. Cardiovasc. Pharmacol. 7, 733–738.

O'Connor, K.M., Frehling, T.D. and Kowey, P.R. (1989). The effect of thromboxane inhibition on vulnerability to ventricular fibrillation in the acute and chronic feline infarction models. Am. Heart J. 117, 848–853.

Oliver, M.F. (1982). Risk of correcting the risk of coronary disease and strokes with drugs. N. Engl. J. Med. 306, 297–299.

Parratt, J.R. (1993). Endogenous myocardial protective (anti-arrhythmic) substances. Cardiovasc. Res. (in press).

Parratt, J.R. and Wainwright, C.L. (1986a). The effects of the thromboxane antagonist BM13.177 on ischaemia and reperfusion induced arrhythmias in dogs. Br. J. Pharmacol. 88, 219P.

Parratt, J.R. and Wainwright, C.L. (1986b). Ventricular arrhythmias induced by local injections of vasoconstrictors following coronary occlusion. Br. J. Pharmacol. 88, 397P.

Parratt, J.R. and Wainwright, C.L. (1987). Failure of allopurinol and a spin trapping agent, N-t-butyl-a-phenyl nitrone to modify significantly ischaemia and reperfusion arrhythmias. Br. J. Pharmacol. 91, 49–59.

Patterson, E., Scherlag, B.J. and Lazzara, R. (1986). Mechanism of prevention of sudden death by nadolol: differential actions upon triggers and arrhythmia substrate after myocardial infarction in the dog. J. Am. Coll. Cardiol. 8, 1365–1371.

Piper, P.J. and Stewart, A.G. (1986). Antagonism of vaso-constriction induced by platelet activating factor in guinea pig perfused isolated hearts by selective PAF receptor antagonists. Br. J. Pharmacol. 88, 595–605.

Podzuweit, T. (1982). In "Early Arrhythmias Resulting from Myocardial Ischaemia" (ed J.R. Parratt), pp. 171–198. Macmillan, Basingstoke.

Pugsley, M.K., Penz, W.P., Walker, M.J.A. and Wong, T.M. (1992). Antiarrhythmic effects of U-50,488H in rats subjected to coronary artery occlusion. Eur. J. Pharmacol. 212, 15–19.

Ravingerova, T. and Curtis, M.J. (1991). Arrhythmogenesis and recovery of coronary flow during reperfusion after sustained (4 h) regional ischaemia in the absence of blood. J. Molec. Cell. Cardiol. 23 (Suppl. V), P158.

Richardson, B.P., Engel, G., Donatsch, P. and Stadler, P.A. (1985). Identification of serotonin M-receptor subtypes and their specific blockade by a new class of drugs. Nature (Lond.) 316, 126–131.

Ridley, P.D., Yacoub, M.H. and Curtis, M.J. (1992). A modified model of global ischaemia: application to the study of syncytial mechanisms of arrhythmogenesis. Cardiovasc. Res. 269, 313–322.

Rochette, L., Yao-Kouame, J., Bralet, J. and Opie, L.H. (1988). Effects of promethazine on ischemic and reperfusion arrhythmias in rat heart. Fund. Clin. Pharmacol. 2, 385–397.

Romson, J.L., Hook, B.G., Rigot, V.H., Schork, M.A., Swanson, D.P. and Lucchesi, B.R. (1982). Effects of ibuprofen on accumulation of Indium 111-labeled platelets and leukocytes in experimental myocardial infarction. Circulation 66, 1002–1010.

Romson, J.L., Hook, B.G., Ruikel, S.L., Abrams, E.D., Schork, M.A. and Lucchesi, B.R. (1983). Reduction of the extent of ischemic myocardial injury by neutrophil depletion in the dog. Circulation 67, 1016–1023.

Roth, D.M. and Lefer, A.M. (1983). Studies on the mechanisms of leukotriene induced coronary artery constriction. Prostaglandins 26, 573–582.

Saman, S., Thandroyen, F. and Opie, L.H. (1985). Serotonin and the heart: effects of ketanserin on myocardial function, heart rate and arrhythmias. J. Cardiovasc. Pharmacol. 7 (Suppl. 7), S70–S75.

Saxena, P.M. and Villalon, C.M. (1991). 5-Hydroxy-tryptamine, a chameleon in the heart. Trends Pharmacol. Sci. 12, 223–227.

Scherlag, B.J., Patterson, E.S., Berbari, E.J. and Lazzara, R. (1989). In "Adrenergic System and Ventricular Arrhythmias in Myocardial Infarction" (ed J. Brachmann and A. Schömig), pp. 299–312. Springer Verlag, Heidelberg.

Shabunina, E.V., Petrunin, I.A., Vinograd, L.K.H., Manukhina, E.B. and Meerson, F.Z. (1988). Prevention of arrhythmia in acute ischaemia in conscious animals using a serotonin analogue. Biull. Eksp. Biol. Med. 106, 410–412.

Stahl, G.L., Terashita, Z.I. and Lefer, A.M. (1988). Role of platelet activating factor in propagation of cardiac damage during myocardial ischemia. J. Pharm. Exp. Ther. 244, 898–904.

Sybertz, E.J., Watkins, R.W., Maum, T., Pula, K. and Rivelli, M. (1985). Cardiac, coronary and peripheral vascular effects of acetyl ether phosphorylcholine in the anaesthetised dog. J. Pharm. Exp. Ther. 232, 156–162.

Tamargo, J., Delgado, C., Diez, J. and Delpon, E. (1988). In "Ginkgolides, Chemistry, Biology, Pharmacology and Clinical Perspectives", Vol. 1 (ed P. Braquet), pp. 417–431. J.R. Prous Science Publishers, Barcelona.

Terashita, Z.-I., Fukui, H., Hirata, M., Terao, S., Ohkawa, S., Nishikawa, K. and Kikuchi, S. (1981). Coronary vasoconstriction and PGI_2 release by leukotrienes in isolated guinea pig hearts. Eur. J. Pharmacol. 73, 357–361.

Thandroyen, F.T., Saman, S. and Opie, L.H. (1985). In "Serotonin and the Cardiovascular System" (ed P.M. Vanhoutte), pp. 87–93. Raven Press, New York.

Tzivoni, D., Keren, A., Granot, H., Gottlieb, S., Benhorin, J. and Stern, S. (1983). Ventricular fibrillation caused by myocardial reperfusion in Prinzmetal's angina. Am. Heart J. 105, 323–325.

Wainwright, C.L., Parratt, J.R. and Bigaud, M. (1989). The effect of PAF antagonists on arrhythmias and platelets during acute myocardial ischaemia and reperfusion. Eur. Heart J. 10, 235–243.

Williams, F.M., Rothaul, A.L., Kane, K.A. and Parratt, J.R. (1985). Antiarrhythmic and electrophysiological effects of ICS 205-930, an antagonist of 5-hydroxytryptamine at peripheral receptors. J. Cardiovasc. Pharmacol. 7, 550–555.

Wit, A.L. (1989). In "Adrenergic System and Ventricular Arrhythmias in Myocardial Infarction" (ed J. Brachmann and A. Schömig), pp. 275–296. Springer-Verlag, Heidelberg.

Wit, A.L. and Rosen, M.R. (1983). Pathophysiological mechanisms of cardiac arrhythmias. Am. Heart J. 106, 798–811.

Wolff, A.A. and Levi, R. (1986). Histamine and cardiac arrhythmias. Circulation Res. 58, 1–16.

Wolff, A.A., Levi, R., Chenouda, A.A. and Fisher, V.J. (1984). Ventricular arrhythmias parallel cardiac histamine release after coronary artery occlusion in the dog: effects of ranitidine. Circulation 70 (Suppl. II), 225.

Woodward, B. and Zakaria, M.N.M. (1985). Effects of some free radical scavengers on reperfusion-induced arrhythmias in the isolated rat heart. J. Molec. Cell. Cardiol. 173, 485–499.

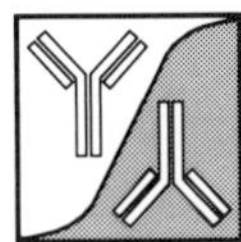

8. Inflammatory Mediators and the Stunned Myocardium

Garrett J. Gross

1. Introduction

1.1 STUNNED MYOCARDIUM

Brief periods of coronary artery occlusion (5–20 min) have been shown to produce prolonged (3–72 h) abnormalities in regional myocardial contractile function, tissue blood flow, cellular ultrastructure and tissue adenine nucleotide content in the absence of irreversible tissue injury. This phenomenon has been termed the "stunned" myocardium (Braunwald and Kloner, 1982). A number of mechanisms (and chemical mediators) have been proposed to be partially responsible for the prolonged abnormalities in mechanical and metabolic function observed, and include: production of oxygen-derived free radicals, disturbances in intracellular calcium homeostasis, structural alterations of the extracellular collagen matrix and dysfunction of cardiac sympathetic nerves (Bolli, 1990). There is also evidence that the response to these brief episodes of ischaemia and reperfusion may activate an inflammatory response; several studies have suggested a role for products of the arachidonic acid cascade and activated neutrophils in the pathogenesis of myocardial stunning (Wynsen *et al.*, 1988; Mullane and Engler, 1991). Thus the purpose of the present chapter is to summarize the evidence for and

against a role for products of the inflammatory response in the aetiology of stunned myocardium.

2. Role of Neutrophils

Many studies have convincingly demonstrated a role for oxygen-derived free radicals in myocardial stunning (Bolli, 1990). Since neutrophils are an important source of oxygen radicals and proteases (Mullane and Engler, 1991), these cells certainly have the potential to be a contributing factor in the pathogenesis of stunned myocardium; however, studies implicating a clear-cut role for these cells in stunning are equivocal.

It is clear that nonactivated or resting neutrophils do not affect cardiac function *in vivo* or *in vitro*. However, several studies have shown that activated neutrophils produce an impairment in contractile function in isolated papillary muscles (Kraemer *et al.*, 1990, 1991). In agreement, several *in vivo* studies in dogs have demonstrated that by removing neutrophils from the blood by Leukopak Filters, post-ischaemic dysfunction could be attenuated in hearts subjected to 15 min of coronary artery occlusion followed by 1–3 h of reperfusion (Engler and Covell, 1987; Westlin and Mullane, 1989). In

Immunopharmacology of the Heart
ISBN 0–12–200245–8

contrast, other investigators using a different neutrophil filter (Imugard IG 500; Jeremy and Becker, 1989), an antiserum to canine neutrophils (O'Neill *et al.*, 1989) and an F(ab[1])2 fragment of Anti-Mol (904) monoclonal antibodies (Schott *et al.*, 1989) to prevent neutrophil adherence to endothelial cells or myocytes found no effect on myocardial stunning. The reasons for the divergent results obtained in these studies is not clear, although there is evidence to suggest that Leukopak filters, in addition to removing neutrophils, also remove other white blood cells and platelets, activate the complement system and provoke adenosine release from red blood cells (Becker, 1991); therefore, results obtained with these filters are open to question. In addition, the brief duration of ischaemia (5–15 min) necessary to produce myocardial stunning does not appear to be sufficiently long to produce activation of neutrophils or their tissue infiltration. Taken together, these results suggest that activated neutrophils are not likely to be an important source of oxygen-derived free radicals sufficient to contribute to stunned myocardium. A radical-independent contribution of neutrophils to stunning cannot be excluded, but this is unsubstantiated.

3. *Role of Arachidonic Acid Pathway*

3.1 CYCLO-OXYGENASE

Non-steroidal anti-inflammatory drugs such as aspirin and indomethacin are potent inhibitors of cyclo-oxygenase and have been proposed to be useful in the prevention or treatment of certain cardiovascular disorders, most notably those associated with platelet aggregation and thrombosis (Lewis *et al.*, 1983). Indeed, several animal and clinical studies suggest that cyclo-oxygenase inhibitors may be beneficial in attenuating irreversible ischaemia-induced myocardial injury (Lepran *et al.*, 1981; Lewis *et al.*, 1983). Only one study has examined the effect of cyclo-oxygenase inhibitors in stunned myocardium. In this study, Farber and Gross (1990) investigated the effect of pretreatment with indomethacin on the recovery of regional systolic segment shortening in anaesthetized dogs subjected to 15 min of coronary artery occlusion (ischaemia) followed by 3 h of reperfusion. Indomethacin resulted in a marked reduction in TXB_2 and 6-keto-$PGF_{1\alpha}$, the stable hydration products of TXA_2 and PGI_2, respectively. However, there was no beneficial effect of indomethacin on the recovery of regional myocardial function compared with non-treated control animals ($5 \pm 6\%$ versus $12 \pm 11\%$ systolic shortening as a percentage of control values, see Fig. 8.1). Although these findings do not conclusively exclude a role for the cyclo-oxygenase pathway in the pathogenesis of stunned myocardium, the results do indicate that the TXA_2/PGI_2 balance would be more important than specific levels of TXA_2 or PGI_2 if the pathway does have any relevance, and that inhibition of both PGI_2 and TXA_2 synthesis equally is not beneficial. It is possible that the formation of other

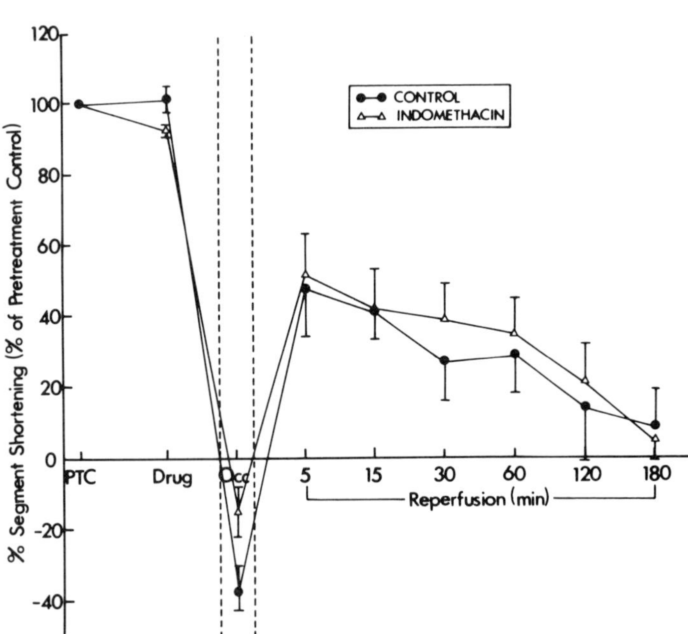

Figure 8.1 Percentage segment shortening (%SS) expressed as a percentage of the pretreatment control (PTC) value in the ischaemic–reperfused area at various times. Time course values for control dogs and animals pretreated with indomethacin (5.0 mg kg^{-1}, i.v.). Each point is the mean ± SEM. No significant differences were observed between groups. (Adapted from Farber and Gross, 1990.)

arachidonic acid metabolites may have been inhibited by indomethacin in the above study and this may have masked any possible beneficial or detrimental effect of blocking a specific component of this pathway downstream of the cyclo-oxygenase enzyme. Thus the role of cyclo-oxygenase products in stunning is not fully resolved.

3.2   THROMBOXANE (TXA2)

Since the discovery of a strongly vasospastic substance which is released during anaphylaxis, and the subsequent identification of the main component as TXA2, there have been a number of studies, performed using TXA2 synthesis inhibitors or receptor blockers, which suggest a role for TXA2 as a pathophysiological mediator of thrombosis, vasospasm, cardiac arrhythmias and myocardial infarction (see Chapters 3, 5 and 7). Recently, the involvement of TXA2 in stunned myocardium has been investigated in several laboratories.

Using the model of myocardial stunning described above in which anaesthetized dogs were subjected to 15 min of coronary artery occlusion and 3 h of reperfusion, Farber and Gross (1990) found that inhibition of TXA2 synthesis with dazmegrel resulted in a marked improvement in regional systolic shortening (66 ± 8% versus 12 ± 11%, see Fig. 8.2) compared with control animals. These data suggest that TXA2 may play an important role in myocardial stunning. However, one cannot exclude the possibility that the beneficial effects of TXA2 synthesis inhibition are due to the formation of PGI2 or other vasodilator prostaglandins and not specifically the result of a decrease in TXA2 *per se*.

To further test the importance of TXA2 in stunned myocardium, Farber and Gross (1990) pretreated anaesthetized dogs with either 0.5 or 10 mg/kg of a specific TXA2 receptor antagonist, BM 13.505. These doses of BM 13.505 markedly attenuated the increases in systolic and diastolic blood pressure produced by the specific TXA2 receptor agonist, U46619, but did not affect the synthesis of either TXA2 or PGI2. In contrast to the results obtained with the TXA2 synthesis inhibitor, dazmegrel, TXA2 receptor blockade with BM 13.505 had no effect on the recovery of regional wall motion following 15 min of coronary artery occlusion plus reperfusion (see Fig. 8.3). In contrast, Grover and Fulmor (1988) found that a different TXA2 receptor antagonist, SQ 29,548, enhanced the recovery of regional systolic shortening (60% versus 25%) in stunned myocardium of dogs subjected to 15 min of coronary artery occlusion followed by 5 h of reperfusion. These beneficial effects of SQ 29,548 could be achieved with administration prior to coronary artery occlusion and with administration immediately prior to reperfusion. The reasons for the differences in the study of Grover and Fulmor (1988) and that of Farber and Gross (1990) are not readily apparent but may be related to the experimental model, dose of drug or to differences in the

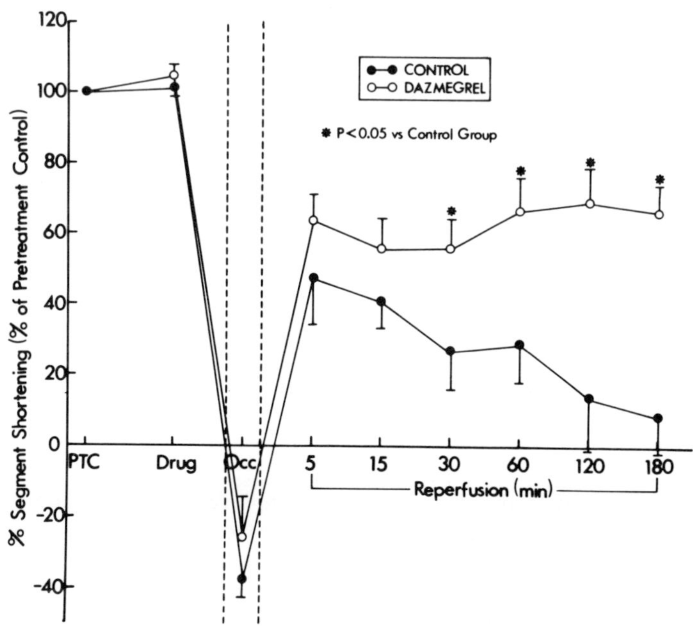

Figure 8.2 Percentage segment shortening (%SS) expressed as a percentage of the pretreatment control (PTC) value in the ischaemic–reperfused area at various times. Time course values for control dogs and animals pretreated with dazmegrel (3.0 mg kg^{-1}, i.v.). Each point is the mean ± SEM. *$P < 0.05$ versus the control group. (Adapted from Farber and Gross, 1990.)

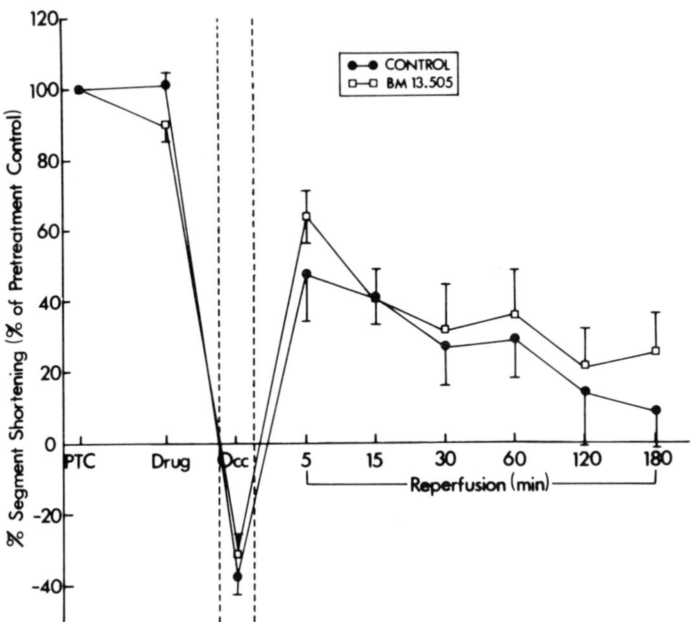

Figure 8.3 Percentage segment shortening (%SS) expressed as a percentage of the pretreatment control (PTC) value in the ischaemic–reperfused area at various times. Time course values for control dogs and animals pretreated with BM 13.505 (10 mg kg^{-1}, i.v.). Each point is the mean ± SEM. No significant differences were observed between groups. (Adapted from Farber and Gross, 1990.)

ancillary properties of the specific TXA$_2$ receptor antagonist studied.

It has been well documented that, although a TXA$_2$ synthetase inhibitor such as dazmegrel selectively blocks the formation of TXA$_2$ via cyclic endoperoxides, it may also reorient these endoperoxides to form increased amounts of prostaglandins such as PGI$_2$, PGE$_2$, PGF$_{2\alpha}$ or PGD$_2$ (Aiken *et al.*, 1981). This reorientation of cyclic endoperoxides has been termed "endoperoxide steal" and may be responsible for protective effects observed following administration of TXA$_2$ synthesis inhibitors which may be independent of any decrease in TXA$_2$ formation. Therefore, to eliminate the possibility of effects due to endoperoxide steal and to define more clearly the role of TXA$_2$ in myocardial stunning, Farber and Gross (1990) studied a group of dogs which were pretreated with indomethacin to prevent endogenous effects due to prostaglandins such as PGI$_2$, and then given the TXA$_2$ synthetase inhibitor dazmegrel prior to myocardial stunning. Indomethacin pretreatment completely blocked the previously noted dazmegrel-induced increase in PGI$_2$ (Farber *et al.*, 1988a) and also inhibited the protective effect of dazmegrel on the recovery of postischaemic function in stunned myocardium (see Fig. 8.4). These results, along with those obtained with the specific TXA$_2$ receptor blocker, BM 13.505, provide strong evidence that TXA$_2$ does not act as a mediator of stunned myocardium and that the protection observed following TXA$_2$ synthesis inhibition is most likely due to

endoperoxide shunting to vasodilatory prostaglandins such as PGI$_2$.

3.3 PROSTAGLANDINS

If the hypothesis that endogenous PGI$_2$ or other vasodilatory prostaglandins mediate the beneficial effects observed in stunned myocardium following TXA$_2$ synthesis inhibition is correct, exogenous administration of PGI$_2$ or an analogue should mimic these cardioprotective actions.

3.3.1 Iloprost

No reports have been published concerning the effects of PGI$_2$ *per se* in stunned myocardium. However, several studies have been performed with the stable PGI$_2$ analogue, iloprost. Initially, van der Giessen *et al.* (1986) administered iloprost to anaesthetized pigs which had been subjected to 20 min of coronary artery occlusion followed by 2 h of reperfusion. Although iloprost did not improve wall motion during ischaemia, it resulted in an enhanced recovery of regional wall motion during reperfusion.

More recently, Farber *et al.* (1988b) administered iloprost either prior to or during the occlusion period or at reperfusion in anaesthetized dogs subjected to 15 min of coronary artery occlusion followed by 3 h of reperfusion. Iloprost, administered prior to coronary artery occlusion (see Fig. 8.5), produced a dose-dependent

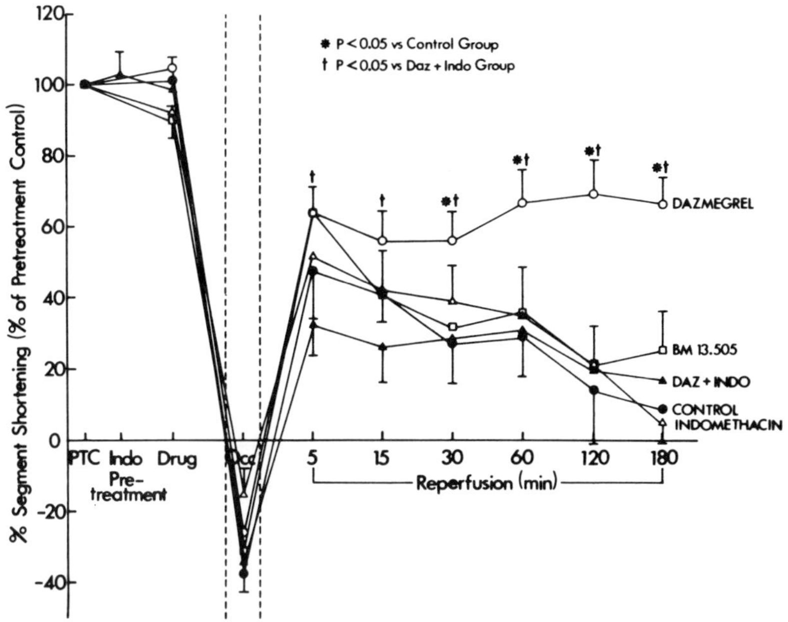

Figure 8.4 Percentage segment shortening (%SS) expressed as a percentage of the pretreatment control (PTC) value in the canine ischaemic–reperfused area at various times. Time course values for control, indomethacin (INDO), dazmegrel (DAZ) and DAZ + INDO pretreated dogs. Each point is the mean ± SEM. *P < 0.05 versus the control group. (Adapted from Farber and Gross, 1990.)

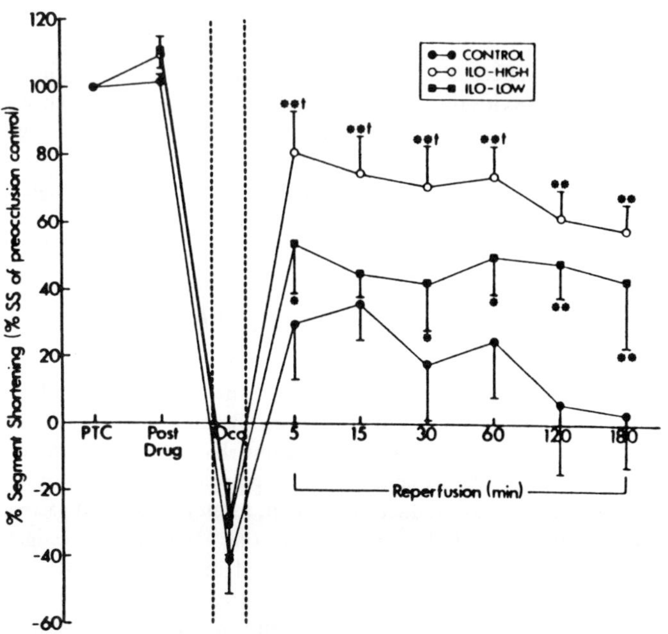

Figure 8.5 Percentage segment shortening (%SS) expressed as a percentage of the pretreatment control (PTC) value in the canine ischaemic–reperfused area at various times. Time course values for pretreatment with vehicle, iloprost at 0.05 μg^{-1} kg^{-1} min^{-1} (ILO-LOW) and iloprost at 0.1 μg^{-1} kg^{-1} min^{-1} (ILO-HIGH). Each point is the mean ± SEM. *P < 0.01 and P < 0.05 versus the control group. (Reprinted by permission from the American Heart Association; see Farber et al., 1988b.)

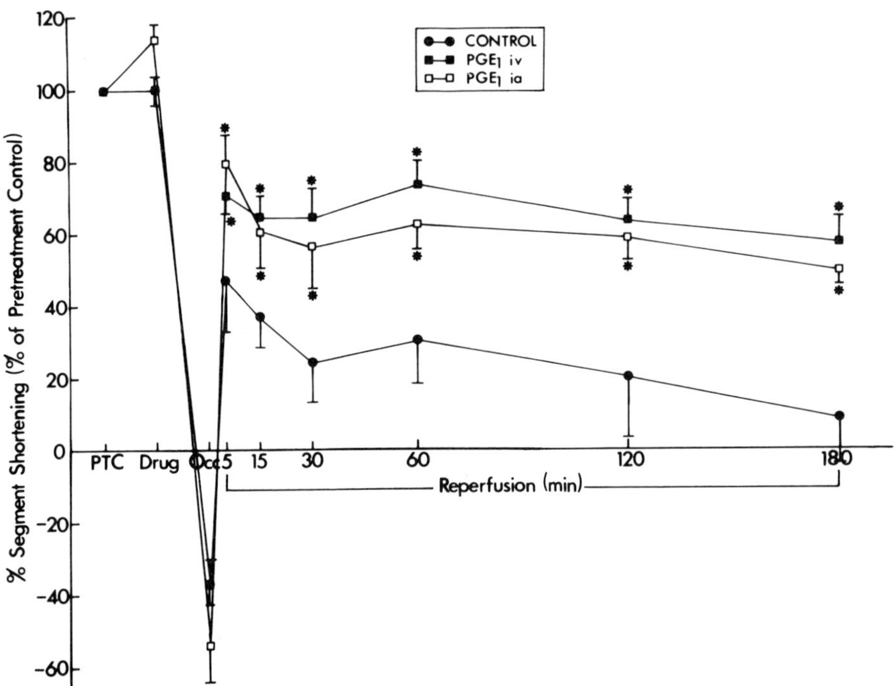

Figure 8.6 Percentage segment shortening (%SS) expressed as a percentage of the pretreatment control (PTC) value in the canine ischaemic–reperfused area at various times. Time course values for pretreatment with vehicle, PGE$_1$ (1 μg^{-1} kg^{-1} min^{-1}, i.v.). Each point is the mean $\pm$ SEM. *$P < 0.05$ versus the control group. (Reprinted by permission from the American Heart Association; see Farber and Gross, 1989.)

increase in regional systolic shortening (%SS) compared with the control group. Iloprost also enhanced the recovery of regional wall motion when it was given at reperfusion; however, the beneficial effect was not as great as with administration prior to occlusion.

3.3.2 Prostaglandin E$_1$ (PGE$_1$)

To determine whether the beneficial effect of iloprost could be mimicked by other vasodilatory prostaglandins, we investigated the effect of PGE$_1$ in stunned myocardium of anaesthetized dogs (Farber and Gross, 1989). PGE$_1$ was administered either intravenously or intra-atrially to avoid pulmonary metabolism. Postischaemic recovery of regional systolic shortening in the ischaemic–reperfused area was significantly enhanced at all times during reperfusion by intravenous or intra-atrial infusion of PGE$_1$ (see Fig. 8.6). The mechanism(s) by which iloprost or PGE$_1$ produce these salutary actions are unknown, but may be partially related to a reduction in afterload during occlusion, an inhibition of platelet or neutrophil function or a cytoprotective action at myocardial cell membranes.

3.4 LEUKOTRIENES

The role of lipoxygenase activation in the aetiology of stunned myocardium has been investigated by O'Neill *et al.* (1987) using anaesthetized dogs subjected to

15 min of coronary artery occlusion followed by 4 h of reperfusion. Nafazatrom, a potent lipoxygenase inhibitor, was administered prior to coronary artery occlusion and at reperfusion. Although the treated animals showed less dyskinesis during the occlusion period compared with controls, the recovery of regional wall motion during reperfusion was not improved by nafazatrom pretreatment. Thus, these data suggest that leukotrienes do not contribute to the pathogenesis of stunned myocardium.

4. PAF

PAF is a phospholipid which has been shown to be a potent mediator of several types of inflammation and shock (Braquet *et al.*, 1987). The results of several studies suggest that PAF is a mediator of the damage produced during prolonged periods of ischaemia and reperfusion in brain, kidney, heart and gastrointestinal tract. However, only one study has investigated the effect of PAF in stunned myocardium (Maruyama *et al.*, 1990). These investigators pretreated anaesthetized dogs subjected to 15 min of coronary artery occlusion followed by 3 h of reperfusion with either one or the other of two structurally unrelated PAF antagonists, BN 52021 or CV-3988. Neither compound had any beneficial effect on the recovery of regional segment shortening following reperfusion, although each compound elicited a significant reduction in myocardial infarct size (Maruyama *et al.*, 1990). These data suggest that PAF does not appear to be an important mediator of myocardial stunning.

5. *Corticosteroids*

Many chemical mediators released during myocardial ischaemia and reperfusion are similar to those released during an acute inflammatory response (Wynsen *et al.*, 1988). Even brief periods of myocardial ischaemia have been associated with oxygen radical production, membrane lipid peroxidation (Rao *et al.*, 1983) and activation of phospholipases. All of these processes have been shown to be detrimental to the heart and to produce myocardial cell damage. Since glucocorticoids are known to inhibit the inflammatory response and have been shown to decrease myocardial infarct size (Bernauer, 1985), there is a rationale for studying these compounds in stunned myocardium. The evidence supporting a potential beneficial effect of steroids in this model is summarized below.

5.1 METHYLPREDNISOLONE

Recently, Wynsen *et al.* (1988) studied the effect of methylprednisolone sodium succinate (MSS) on the

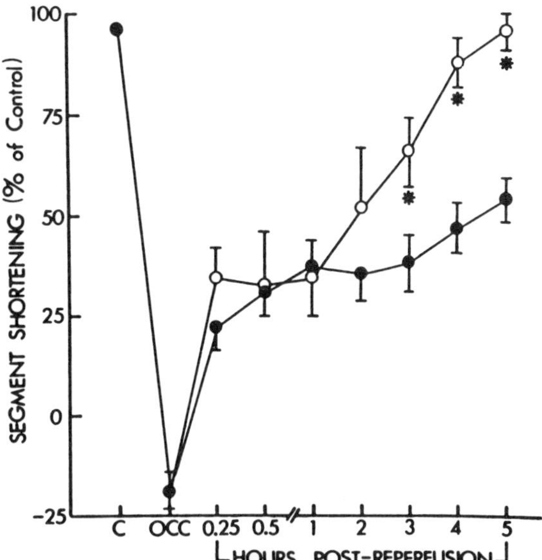

Figure 8.8 Time course of recovery of segment shortening (% of control) in postischaemic–reperfused myocardium over time. Open circles represent the group treated with MSS (20 mg kg^{-1}, i.v.) immediately prior to reperfusion. Closed circles represent the control group. All values are the mean ± SEM. $^*P < 0.05$ compared to control group. (Reprinted by permission from the American Heart Journal; see Wynsen *et al.*, 1988.)

recovery of regional systolic shortening (%SS) in conscious, chronically instrumented dogs subjected to 15 min of coronary artery occlusion and a 5 h period of reperfusion. Animals receiving MSS 90 min prior to occlusion showed a marked improvement in the recovery of %SS in the ischaemic–reperfused region throughout the entire reperfusion period (Fig. 8.7), whereas dogs given MSS at reperfusion demonstrated a delayed but marked improvement in %SS at 3, 4 and 5 h post-reperfusion (Fig. 8.8). Methylprednisolone given 15 min post-reperfusion had no beneficial effect. In addition, sodium succinate administration 90 min prior to occlusion also had no effect. These results could not be attributed to any haemodynamic effects of the steroid but were postulated to be the result of a steroid-induced increase in the release and/or synthesis of a PLA$_2$ inhibitory protein. However, no direct evidence for such an effect was obtained.

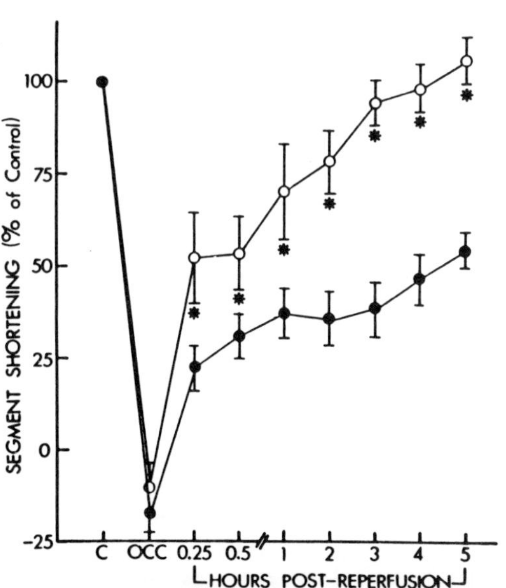

Figure 8.7 Time course of recovery of segment shortening (% of control) in postischaemic–reperfused myocardium. Open circles represent the dogs treated with MSS (20 mg kg^{-1}, i.v.) 90 min prior to coronary artery occlusion (OCC). Closed circles represent the control group. All values are the mean ± SEM. $^*P < 0.05$ compared to control group. (Reprinted by permission from the American Heart Journal; see Wynsen *et al.*, 1988.)

6. *Conclusions*

Most of the evidence presented in the present chapter does not support an important role for the inflammatory response in myocardial stunning following brief periods of ischaemia and reperfusion. The lack of effect of neutrophil inhibition or depletion, cyclo-oxygenase or lipoxygenase inhibitors and TXA_2 receptor and PAF receptor antagonists suggest that these cells, systems and chemical mediators are not involved in myocardial stunning, although the positive effect observed with methylprednisolone suggests that some component of the immune response may be involved. On the other hand, the positive results obtained with exogenous administration of TXA_2 synthesis inhibitors and several vasodilatory prostaglandins such as PGI_2 and PGE_1 suggest that pharmacological modification of the TXA_2/PGI_2 balance may be an effective means to attenuate myocardial stunning and may offer promise for future clinical intervention.

7. References

Aiken, J.W., Shebuski, R.J., Miller, O.V. and Gorman, R.R. (1981). Endogenous prostacyclin contributes to the efficiency of a thromboxane synthetase inhibitor for preventing coronary artery thrombosis. J. Pharmacol. Exp. 219, 299–308.

Becker, L.C. (1991). Do neutrophils contribute to myocardial stunning? Cardiovasc. Drugs Ther. 5, 909–914.

Bernauer, W. (1985). Inhibiting effect of dexamethasone on evolving myocardial necrosis in coronary-ligated rats, with and without reperfusion. Pharmacology 31, 328–336.

Bolli, R. (1990). Mechanism of myocardial "stunning". Circulation 82, 723–738.

Braquet, P., Touqui, J., Shen, T.Y. and Vargaftig, B.B. (1987). Perspectives in platelet activating factor research. Pharmacol. Rev. 39, 97–145.

Braunwald, E. and Kloner, R.A. (1982). The stunned myocardium: prolonged, postischemic ventricular dysfunction. Circulation 66, 1146–1149.

Engler, R. and Covell, J.W. (1987). Granulocytes cause reperfusion ventricular dysfunction after 15-minute ischemia in the dog. Circulation Res. 61, 20–28.

Farber, N.E. and Gross, G.J. (1989). Prostaglandin E_1 attenuates postischemic contractile dysfunction after brief coronary occlusion and reperfusion. Am. Heart J. 118, 17–24.

Farber, N.E. and Gross, G.J. (1990). Prostaglandin redirection by thromboxane synthetase inhibition: attenuation of myocardial stunning in the canine heart. Circulation 81, 369–380.

Farber, N.E., Pieper, G.M. and Gross, G.J. (1988a). Lack of involvement of thromboxane A_2 in postischemic recovery of stunned canine myocardium. Circulation 78, 450–461.

Farber, N.E., Pieper, G.M., Thomas, J.P. and Gross, G.J. (1988b). Beneficial effects of iloprost in the stunned canine myocardium. Circulation Res. 62, 204–215.

Grover, G.J. and Fulmor, I.E. (1988). Thromboxane A_2 antagonist and diltiazem-induced enhancement of contractile function: the effect of timing of treatment. J. Pharmacol. Exp. Ther. 247, 445–452.

Jeremy, R.W. and Becker, L.C. (1989). Neutrophil depletion does not prevent myocardial dysfunction after brief occlusion. J. Am. Coll. Cardiol. 13, 1155–1163.

Kraemer, R., Seligmann, B. and Mullane, K.M. (1990). Polymorphonuclear leukocytes reduce cardiac function in vitro by release of H_2O_2. Am. J. Physiol. 258, H1847–H1855.

Kraemer, R., Smith, C.W. and Mullane, K.M. (1991). Activated human polymorphonuclear leukocytes reduce rabbit papillary muscle function: role of CD18 glycoprotein adhesion complex. Cardiovasc. Res. 25, 172–175.

Lepran, I., Koltai, M. and Szekeres, L. (1981). Effect of non-steroidal antiinflammatory drugs in experimental myocardial infarction in rats. Eur. J. Pharmacol. 69, 235–238.

Lewis, H.D., Jr, Davis, J.W., Archibald, D.G., Steinke, W.E., Smitherman, T.C., Doherty, J.E., Schnaper, H.W., LeWinter, M.M., Linares, E., Pouget, J.M., Sabharwal, S.C., Chesler, E. and De Mots, H. (1983). Protective effects of aspirin against acute myocardial infarction and death in men with unstable angina. Results of a Veterans Administration Cooperative Study. N. Engl. J. Med. 309, 396–403.

Maruyama, M., Farber, N.E., Vercellotti, G.M., Jacob, H.S. and Gross, G.J. (1990). Evidence for a role of platelet activating factor in the pathogenesis of irreversible but not reversible myocardial injury after reperfusion in dogs. Am. Heart J. 120, 510–520.

Mullane, K. and Engler, R. (1991). Proclivity of activated neutrophils to cause postischemic cardiac dysfunction: participation in stunning? Cardiovasc. Drugs Ther. 5, 915–924.

O'Neill, P.G., Charlat, M.L., Han-soeb, K., Pocius, J., Michael, L.H., Hartley, C.J., Zhu, W.X., Roberts, R. and Bolli, R. (1987). Lipoxygenase inhibitor, nafazatrom fails to attenuate postischemic ventricular dysfunction. Cardiovasc. Res. 21, 755–760.

O'Neill, P.G., Charlat, M.L., Michael, L.H., Roberts, R. and Bolli, R. (1989). Influence of neutrophil depletion on myocardial function and flow after reversible ischemia. Am. J. Physiol. 256, H341–H351.

Rao, P.S., Cohen, M.V. and Mueller, H.S. (1983). Production of free radicals and lipid peroxides in early experimental myocardial ischemia. J. Mol. Cell. Cardiol. 15, 713–716.

Schott, R.J., Nao, B.S., McClanahan, T.B., Simpson, P.J., Stirling, M.C., Todd, R.F., III and Gallagher, K.P. (1989). F(ab')2 fragments of anti-mol (904) antibodies do not prevent myocardial stunning. Circulation Res. 65, 1112–1124.

Van der Giessen, W.J., Schoutsen, B., Tijssen, J.G.P. and Verdouw, P.D. (1986). Iloprost (ZK 36374) enhances recovery of regional myocardial function during reperfusion after coronary artery occlusion in the pig. Br. J. Pharmacol. 87, 23–27.

Westlin, W. and Mullane, K.M. (1989). Alleviation of myocardial stunning by leukocyte and platelet depletion. Circulation 80, 1828–1836.

Wynsen, J.C., Preuss, K.C., Gross, G.J., Brooks, H.L. and Warltier, D.C. (1988). Steroid-induced enhancement of functional recovery of postischemic, reperfused myocardium in conscious dogs. Am. Heart J. 116, 915–925.

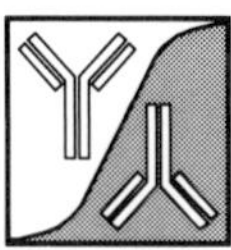

9. Immunotherapy and Reperfusion Injury in the Heart

Tsutomu Yamazaki, Yoshinori Seko, Ryozo Nagai and Yoshio Yazaki

1. Introduction

During the past few years, myocardial reperfusion therapy, such as percutaneous transluminal coronary angioplasty (PTCA) or percutaneous transluminal coronary recanalization (PTCR), has been widely performed in the management of acute myocardial infarction. It is well known, however, that leukocyte infiltration occurs in myocardium subjected to ischaemia followed by reperfusion (I/R) and causes damage to the myocardium contributing to myocardial reperfusion injury. In Fig. 9.1, we show a schematic representation of infarct size expansion as a result of myocardial reperfusion injury. As the ischaemic interval increases, the amount of the viable myocardium in the area at risk (horizontal line zone) decreases. The oblique line zone represents the additional cell death which is associated with reperfusion and can be salvaged by appropriate interventions.

Evidence has accumulated that progressive capillary plugging by leukocytes causing capillary no-reflow (Engler *et al.*, 1983; Engler, 1987; Schmid-Schonbein, 1987; Mehta *et al.*, 1988; Reynolds and McDonagh, 1989) and superoxide radical formation (Jolly *et al.*, 1984; Simpson and Lucchesi 1987) play a role in this reperfusion injury. Arachidonic acid metabolites

(Simpson *et al.*, 1987, 1988a), antibodies (Romson *et al.*, 1983) or filters (Engler *et al.*, 1986; Engler and Schmid-Schonbein, 1986) which either prevent leukocyte activation or deplete leukocytes from circulating blood have been reported to reduce myocardial reperfusion injury. A specific means of blocking leukocyte adhesion to vascular endothelial cells would therefore be a useful tool with which to study and perhaps to block leukocyte-mediated myocardial reperfusion injury.

Cell-adhesion molecules are associated with leukocyte–endothelial cell recognition, and the integrin $\beta 2$ family is composed of three structurally and functionally related glycoprotein heterodimers, each of which consists of an immunologically distinct, higher molecular weight α subunit (CD11) that is non-covalently associated with a common, lower molecular weight β subunit (CD18) (Sanchez-Madrid *et al.*, 1983; Kishimoto *et al.*, 1987; Patarroyo *et al.*, 1990) (Fig. 9.2). These three antigens are lymphocyte function-associated antigen-1 (LFA-1, CD11a/CD18), Mac-1 (the iC3b complement receptor, CD11b/CD18) and p150,95 (CD11c/CD18) and are discussed in detail in Chapter 5. All three antigens are expressed on the surface of leukocytes (Sanchez-Madrid *et al.*, 1983). It is increasingly evident that the cell-adhesion molecules of this integrin $\beta 2$ family participate in leukocyte adhesion to vascular endothelial cells and

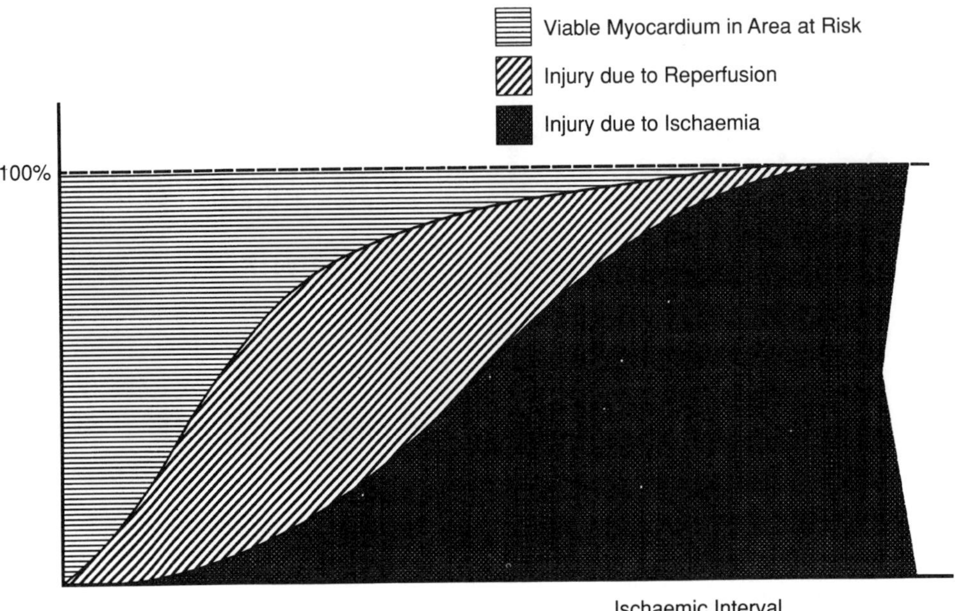

Figure 9.1 Schematic representation of infarct size expansion as a result of myocardial reperfusion injury.

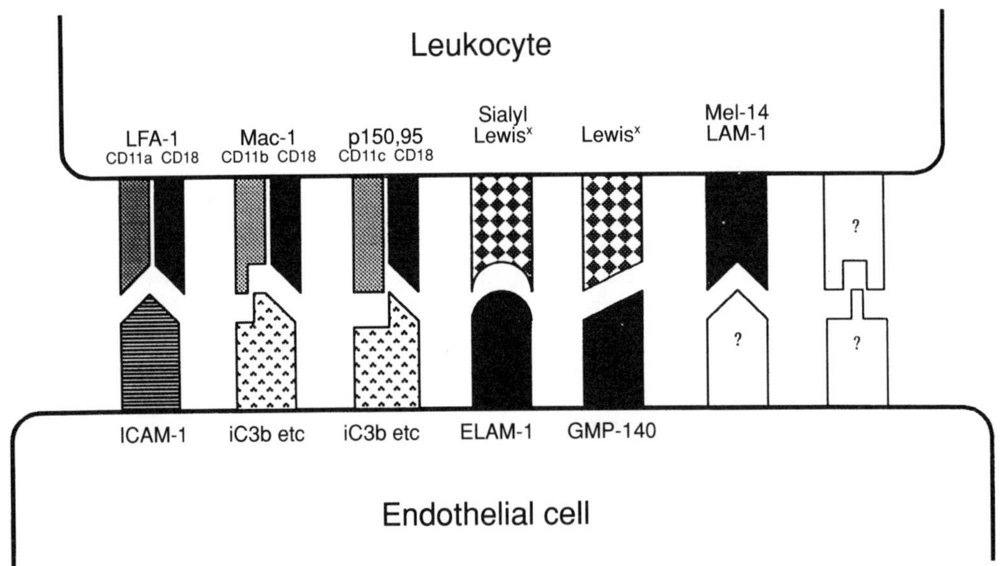

Figure 9.2 Cell-adhesion molecules on the surface of the leukocyte and the endothelial cell.

play a central role in leukocyte infiltration and, therefore, myocardial injury. Intercellular adhesion molecule-1 (ICAM-1, CD54), which is a ligand for LFA-1 (Marlin and Springer, 1987; Makgoba *et al.*, 1988), is expressed weakly on the luminal surface of vascular endothelial cells (Dustin *et al.*, 1986), and this molecule is increased at the site of inflammation (Dustin *et al.*, 1988; Rothlein *et al.*, 1988). However, it is not known whether this expression can be enhanced by I/R.

Recently, Simpson *et al.* (1988, 1990) indicated that the administration of anti-CD11b antibody reduced myocardial infarct size in dogs subjected to 90 min of ischaemia followed by 6–72 h of reperfusion, and Williams *et al.* (1990) showed that the administration of anti-CD 18 antibody reduced the accumulation of leukocytes in the area at risk of the reperfused myocardium. However, because all previous studies were performed with antibodies against other species, the true roles of cell-adhesion molecules in myocardial reperfusion injury are still unclear (Hogg, 1989). Thus, we have administered species-specific antibodies against CD11a, CD11b+c, CD18 and ICAM-1 molecules in a rat I/R model to clarify the roles of these cell-adhesion molecules in myocardial reperfusion injury.

This chapter is primarily focused on the results of our studies. Little data is as yet available on the effects of such antibodies on other types of cardiac injury so, to a certain extent, the infarction studies we have performed represent a pilot investigation of the potential of such intervention (which may have relevance to other aspects of cardiac disease, such as cardiac preservation in cardiac transplantation, in which immunopharmacological intervention may be indicated).

2.   *Materials and Methods*

2.1   MONOCLONAL ANTIBODIES

The generation of anti-rat ICAM-1, clone 1A29 (Tamatani and Miyasaka, 1990) (mouse IgG1), anti-rat CD11a, clone WT.1 (Tamatani *et al.*, 1991) (mouse IgG2a), anti-rat CA11b+c, clone OX42 (Robinson *et al.*, 1986) (mouse IgG2a) and anti-rat CD18, clone WT.3 (Tamatani *et al.*, 1991) (mouse IgG1) was by standard methods. OX42 was a gift from Dr Alan F. Williams (University of Oxford, England). Each antibody recognizes a functional epitope and inhibits leukocyte adhesion to endothelial monolayer *in vitro*. They were used as a 2 mg ml^{-1} solution in sterile saline.

2.2   SURGICAL PROCEDURES

A standard technique for producing myocardial ischaemia was used (see Chapter 2 of this book for rationale and references to full methodology). Wistar rats were anaesthetized and ventilated with a respirator. After left lateral thoracotomy, a 6-0 silk stitch was placed near the intramyocardial location of the left coronary artery. Coronary artery occlusion and recanalization were performed by fastening or unfastening the stitch. After recanalization, the thoracotomy incision was closed and the rats were allowed to recover from surgical anesthesia.

2.3   IMMUNOHISTOCHEMICAL STUDY

After 30 min of ischaemia followed by reperfusion for 0–144 h, individual rat hearts were excised. Sham-operated rat hearts were also excised. Freshly dissected left ventricles were frozen in OCT compound in liquid nitrogen. Cryostat sections of 6 μm thickness were prepared, air-dried and fixed in 100% acetone. The sections were incubated with WT.1, OX42, WT.3 or 1A29, biotinylated anti-mouse IgG and avidin-biotinylated peroxidase complex. After that, the sections

were counterstained with haematoxylin. The evaluation of the staining was conducted in a blind fashion.

2.4   ANTIBODY TREATMENT STUDY

Rats were assigned in a random blind fashion to either an antibody treatment group or control antibody (mouse whole IgG) treatment group. Each antibody (5 mg kg^{-1} of body weight) was administered intravenously 5 min before coronary artery occlusion. Each antibody treatment group was comprised of eight rats. After 48 h of reperfusion, the left ventricle of each heart was excised and cut transversely into six sections from the apex to the base. The tissue samples were incubated in triphenyl tetrazolium chloride. The apical side of each slice was then photographed. From these photographs, the area of infarction in each slice was determined by computerized planimetry. The infarct size was determined by adding all of the areas of infarction for each left ventricle, and expressed as a percentage of the left ventricle. This method for measurement of myocardial infarction is described in Chapter 2.

2.5   STATISTICAL ANALYSES

Statistical comparisons between treatment groups and the control group were made with Wilcoxon's two-sample rank-sum test (Mann–Whitney test) with *P*-values corrected by the Bonferroni method (Wallenstein *et al.*, 1980).

3.   *Results and Discussion*

3.1   LEUKOCYTE INFILTRATION INTO THE REPERFUSED MYOCARDIUM

It is known that leukocytes begin to infiltrate into the myocardia about 4–12 h after the onset of ischaemia (Cotran *et al.*, 1989). However, we found that leukocytes began to infiltrate into the I/R myocardium within 2 h after the onset of reperfusion despite the brief duration of ischaemia (30 min), and we confirmed that most of the infiltrating leukocytes were positive for CD11a, CD11b+c and CD18.

Leukocytes circulate within the blood vessels, and do not interact with the endothelial surface of the vasculature under normal physiological conditions (Lucchesi, 1987). Lo *et al.* (1989) demonstrated that leukocytes stimulated by phorbol dibutyrate, tumour necrosis factor or complement 5a adhered to endothelial cells, and that anti-CD11a/CD18 and anti-CD11b/CD18 monoclonal antibody each inhibited the adhesion. Therefore, we think that in the early phase of I/R, leukocytes are activated by cytokines or by unknown mechanisms, then adhere to vascular endothelial cells in a manner dependent on the cell-adhesion molecules of integrin $\beta2$ family, and finally accumulate in the myocardial tissue.

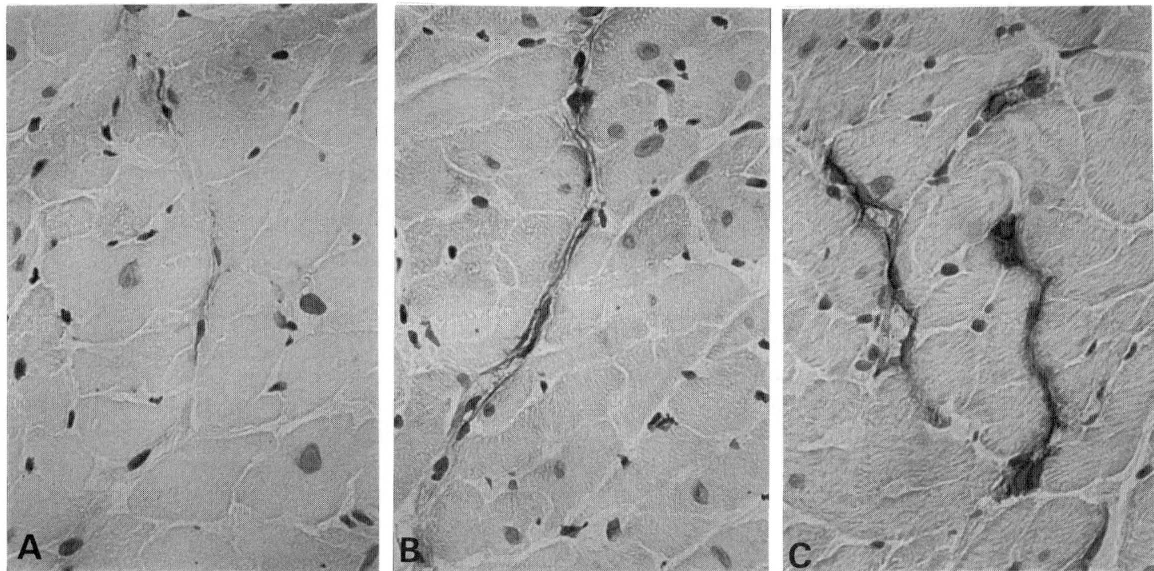

Figure 9.3 Immunohistochemical study of ICAM-1 expression. Panel A: Left ventricular myocardia of the sham-operated rat (× 100). Expression was weak on the luminal surface of capillary endothelial cells. Panel B and C: left ventricular myocardia after 30 min of ischaemia followed by 8 h (B) and 24 h (C) of reperfusion (× 100). ICAM-1 was expressed clearly on capillary and venous endothelial cells.

3.2 EXPRESSION OF ICAM-1 ON VASCULAR ENDOTHELIAL CELLS

There was only weak staining for ICAM-1 on the luminal surface of vascular endothelial cells in the left ventricle of the sham-operated rats (Fig. 9.3A). During 8–96 h after the start of reperfusion, however, ICAM-1 was expressed clearly on capillary and venous endothelial cells (Fig. 9.3B, 9.3C). This expression became weak again by 144 h after the start of reperfusion.

It is known that ICAM-1 is expressed weakly on the luminal surface of vascular endothelial cells and its expression is up regulated at the site of inflammation (Dustin *et al.*, 1988). Recently, Arnaout *et al.* (1988)

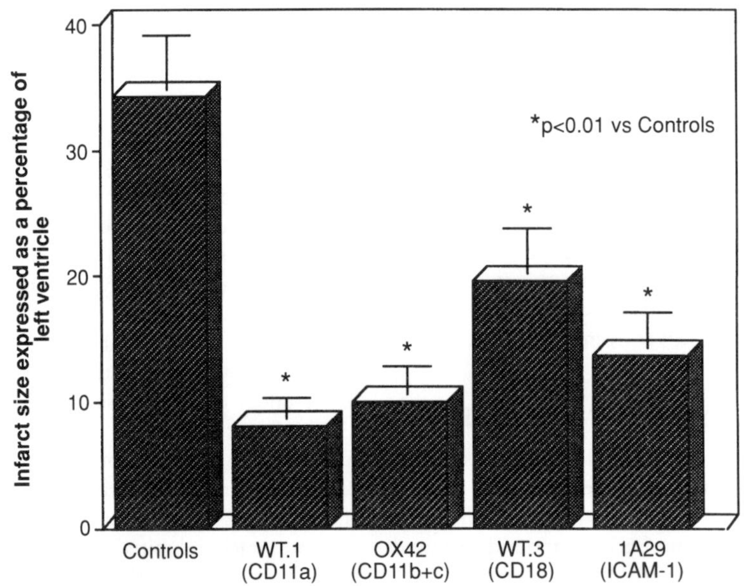

Figure 9.4 Mean (standard error) effects of antibody treatment on myocardial reperfusion injury. Mean myocardial infarct size expressed as a percentage of the left ventricle was significantly smaller (*P* < 0.01) in each monoclonal antibody treatment group as compared with the control group. However, among the monoclonal antibody treatment groups, there was no significant change.

reported that the pretreatment of endothelial cells with interleukin-1 induced leukocyte adhesion which was dependent on all three subunits of the integrin $\beta2$ family. They pointed out that ICAM-1 might mediate the binding of leukocytes to the endothelial cells. However, it has not been known whether the expression of ICAM-1 can be up-regulated by I/R.

For the first time, in the present study, we have demonstrated that I/R enhances the expression of ICAM-1 on the luminal surface of capillary and venous endothelial cells.

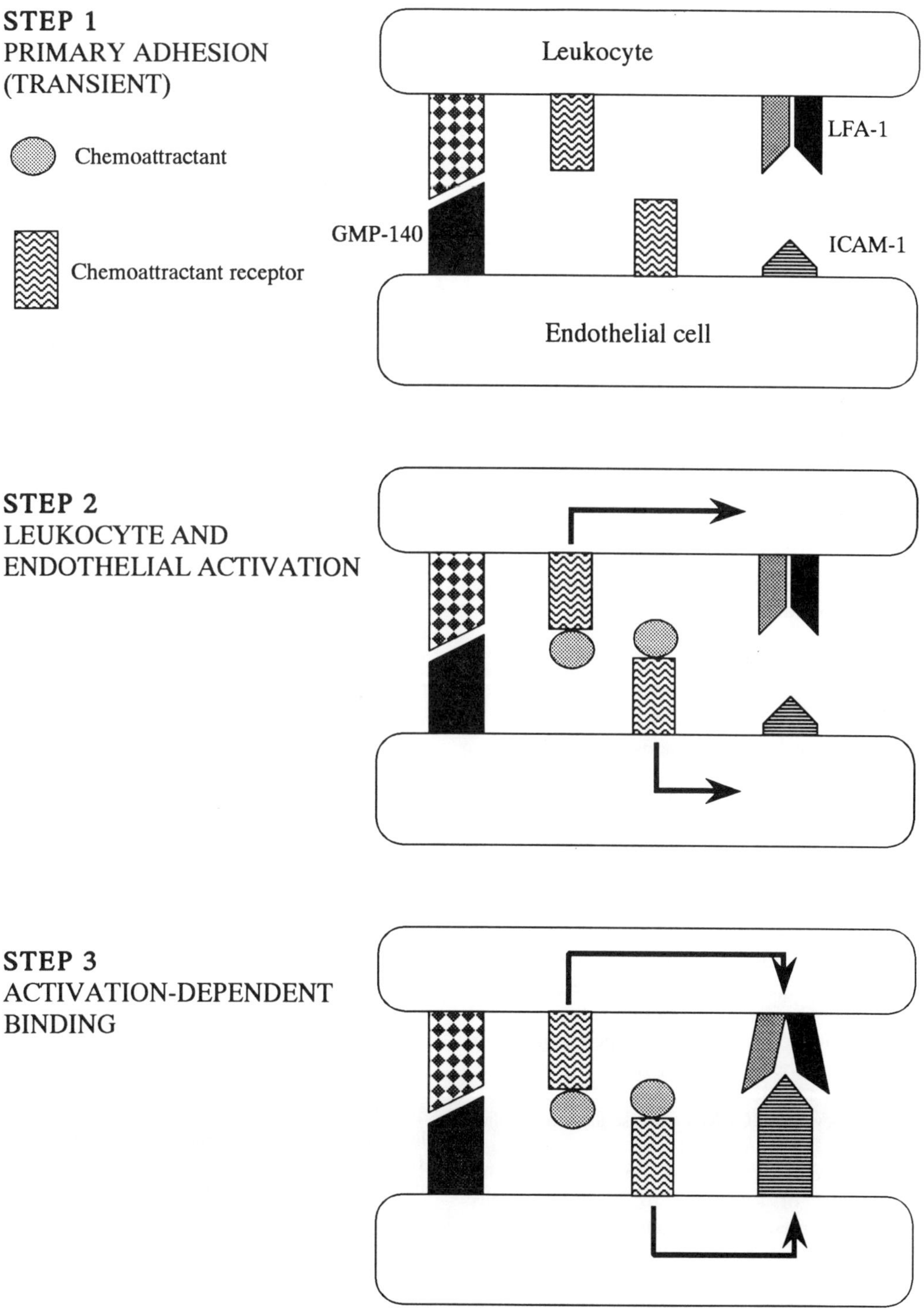

Figure 9.5 Model of leukocyte-endothelial cell recognition as an active, three-step process.

3.3   ANTIBODY TREATMENT STUDY

Forty rats (eight rats in each group) were randomized to receive either monoclonal antibody or non-specific mouse IgG in order to compare the effects of monoclonal antibodies against cell-adhesion molecules on the myocardial infarct size, which developed after 30 min of ischaemia followed by 48 h of reperfusion. There was a significant decrease in the myocardial infarct size which developed in the monoclonal antibody treatment groups. Each infarct size expressed as a percentage of the left ventricle was smaller (WT.1: $8.1 \pm 1.7\%$; OX42: $10.1 \pm 2.1\%$; WT.3: $19.6 \pm 3.6\%$; 1A29: $13.8 \pm 2.7\%$) when compared to the control group (mouse whole IgG: $34.3 \pm 4.2\%$; $P < 0.01$; Fig. 9.4).

Previous studies have indicated that the administration of monoclonal antibodies directed against the functional epitopes of CD11b (Simpson *et al.*, 1988b, 1990) or CD18 (Williams *et al.*, 1990) reduced I/R induced myocardial injury (Simpson *et al.*, 1988b, 1990; Williams *et al.*, 1990). Our present study confirms these data and extends them to show that individual monoclonal antibodies against CD11a or ICAM-1 are also effective in reducing I/R-induced myocardial injury.

We administered monoclonal antibodies by intravenous bolus before 30 min of ischaemia was induced and this treatment was effective in reducing myocardial necrosis. This suggests that the blocking of leukocyte adhesion in the early phase of reperfusion is of significance for the reduction of myocardial reperfusion injury.

4.   *Conclusions*

In the present study, we have demonstrated that leukocyte infiltration into the reperfused myocardium begins within 2 h after reperfusion and that I/R stimuli to the myocardia significantly enhances the expression of ICAM-1 on the luminal surface of capillary and venous endothelial cells. The fact that infiltration occurred 2 h after the start of reperfusion, despite only 30 min of ischaemia, indicates that tissue had been reperfused, and that it was this that caused the infiltration, since 4 h of sustained ischaemia (no reperfusion) is necessary for infiltration in the absence of reperfusion (Hogg, 1989). We have also indicated that myocardial reperfusion injury can be reduced by the administration of the antibodies which block cell-adhesion molecules (CD11a, b, c or CD18) on leukocytes or ICAM-1 on vascular endothelial cells.

Recently, Lawrence *et al.* (1991) indicated that the primary adhesion of leukocytes to endothelial cells through granule membrane protein-140 (GMP-140, CD62) is a prerequisite for the adhesion strengthening through ICAM-1. We also clarified that hypoxia–reperfusion is associated with the enhanced expression of GMP-140 on endothelial cells (Yamazaki *et al.*, 1992).

Finally, we have explored a general model in which leukocyte–endothelial cell interaction is viewed as an active process requiring three sequential events (Fig. 9.5). This model is an alteration of the scheme depicted by Butcher (1991). Within minutes after the start of reperfusion, the interaction is initiated by binding of GMP-140 on the endothelial cell to the ligand on the leukocyte. This initial adhesion is loose, transient and reversible, but followed by secondary events, such as specific chemoattractant or cell contact-mediated signals, which can trigger ICAM-1 expression on the endothelial cell. These three steps result in strong sustained attachment, completing the process of interaction and leading to reperfusion injury.

5.   *References*

Arnaout, M.A., Lanier, L.L. and Faller, D.V. (1988). Relative contribution of the leukocyte molecules Mo1, LFA-1, and p150,95 (LeuM5) in adhesion of granulocytes and monocytes to vascular endothelium is tissue- and stimulus-specific. J. Cell Physiol. 137, 305–309.

Butcher, E.C. (1991). Leukocyte–endothelial cell recognition: three (or more) steps to specificity and diversity. Cell 67, 1033–1036.

Cotran, R.S., Mumar, V. and Robbins, S.L. (1989). "Pathologic Basis of Disease" W.B. Saunders Company, Philadelphia.

Dustin, M.L., Rothlein, R., Bhan, A.K., Dinarello, C.A. and Springer, T.A. (1986). Induction by IL-1 and interferon-gamma: tissue distribution, biochemistry, and function of a natural adherence molecule (ICAM-1). J. Immunol. 137, 245–254.

Dustin, M.L., Staunton, D.E. and Springer, T.A. (1988). Supergene families meet in the immune system. Immunol. Today 9, 213–215.

Engler, R. (1987). Consequences of activation and adenosine-mediated inhibition of granulocytes during myocardial ischemia. FASEB J. 46, 2407–2412.

Engler, R.L. and Schmid-Schonbein, G.W. (1986). Granulocytes as active participants in acute myocardial ischemia and infarction. Am. J. Cardiovasc. Pathol. 1, 15–26.

Engler, R.L., Schmid-Schonbein, G.W. and Pavelec, R.S. (1983). Leukocyte capillary plugging in myocardial ischemia and reperfusion in the dog. Am. J. Pathol. 111, 98–111.

Engler, R.L., Dahlgren, M.D., Morris, D.D., Peterson, M.A. and Schmid-Schonbein, G.W. (1986). Role of leukocytes in response to acute myocardial ischemia and reflow in dogs. Am. J. Physiol. 251, H314–H322.

Hogg, N. (1989). The leukocyte integrins. Immunol. Today 10, 111–114.

Jolly, S.R., Kane, W.J., Bailie, M.B., Abrams, G.D. and Lucchesi, B.R. (1984). Canine myocardial reperfusion injury: its reduction by the combined administration of superoxide dismutase and catalase. Circulation Res. 54, 277–285.

Kishimoto, T.K., O'Connor, K., Lee, A., Roberts, T.M. and Springer, T.A. (1987). Cloning of the beta subunit of the leukocyte adhesion proteins: homology to an extracellular matrix receptor defines a novel supergene family. Cell 48, 681–690.

Lawrence, M.B. and Springer, T.A. (1991). Leukocytes roll on a selectin at physiologic flow rates: distinction from and prerequisite for adhesion through integrins. Cell 65, 859–873.

Lo, S.K., Detmers, P.A., Levin, S.M. and Wright, S.D. (1989). Transient adhesion of neutrophils to endothelium. J. Exp. Med. 169, 1779–1793.

Lucchesi, B.R. (1987). Role of neutrophils in ischemic heart disease: pathophysiologic role in myocardial ischemia and coronary artery reperfusion. Cardiovasc. Clin. 18, 35–48.

Makgoba, M.W., Sanders, M.E., Luce, G.E.G., Dustin, M.L., Springer, T.A., Clark, E.A., Mannoni, P. and Shaw, S. (1988). ICAM-1 a ligand for LFA-1-dependent adhesion of B, T and myeloid cells. Nature (Lond.) 331, 86–88.

Marlin, S.D. and Springer, T.A. (1987). Purified intercellular adhesion molecule-1 (ICAM-1) is a ligand for lymphocyte function-associated antigen 1 (LFA-1). Cell 51, 813–819.

Mehta, J.L., Nichols, W.W. and Mehta, P. (1988). Neutrophils as potential participants in acute myocardial ischaemia: relevance to reperfusion. J. Am. Coll. Cardiol. 11, 1309–1316.

Patarroyo, M., Prieto, J., Rincon, J., Timoner, T., Lundberg, C., Lindbom, L., Asjo, B. and Gahmberg, C.G. (1990). Leukocyte-cell adhesion: a molecular process fundamental in leukocyte physiology. Immunol. Rev. 114, 67–108.

Reynolds, J.M. and McDonagh, P.F. (1989). Early in reperfusion, leukocytes alter perfused coronary capillarity and vascular resistance. Am. J. Physiol. 256, H982–H989.

Robinson, A.P., White, T.M. and Mason, D.W. (1986). Macrophage heterogeneity in the rat as delineated by two monoclonal antibodies MRC OX-42, the latter recognizing complement receptor type 3. Immunol. 57, 239–247.

Romson, J., Hook, B.G., Kunkel, S.L., Abrams, G.D., Schork, M.A. and Lucchesi, B.R. (1983). Reduction of the extent of ischemic myocardial injury by neutrophil depletion in the dog. Circulation 67, 1016–1023.

Rothlein, R., Czajkowski, M., O'Neill, M.M., Marlin, S.D., Malnolfi, E. and Merluzzi, V.J. (1988). Induction of intercellular adhesion molecule 1 on primary and continuous cell lines by pro-inflammatory cytokines: regulation by pharmacologic agents and neutralizing antibodies. J. Immunol. 141, 1665–1669.

Sanchez-Madrid, F., Nagy, J.A., Robbins, E., Simon, P. and Springer, T.A. (1983). A human leukocyte differentiation antigen family with distinct a-subunits and a common b-subunit: the lymphocyte function-associated antigen (LFA-1), the C3bi complement receptor (OKM1/Mac-1), and the p150,95 molecule. J. Exp. Med. 158, 1785–1803.

Schmid-Schonbein, G.W. (1987). Capiliary plugging by granulocytes and the no-reflow phenomenon in the microcirculation. FASEB J. 46, 2397–2401.

Simpson, P.J. and Lucchesi, B.R. (1987). Free radicals and myocardial ischemia and reperfusion injury. J. Lab. Clin. Med. 110, 13–30.

Simpson, P.J., Mickelson, J., Fantone, J.C., Gallagher, K.P. and Lucchesi, B.R. (1987). Reduction of experimental canine myocardial infarct size with prostaglandin E1: inhibition of neutrophil migration and activation. J. Pharmacol. Exp. Ther. 244, 619–624.

Simpson, P.J., Fantone, J.C., Mickelson, J.K., Gallagher, K.P. and Lucchesi, B.R. (1988a). Identification of a time window for therapy to reduce experimental canine myocardial injury: suppression of neutrophil activation during 72 hours of reperfusion. Circulation Res. 63, 1070–1079.

Simpson, P.J., Todd, R.F., Fantone, J.C., Mickelson, J.K., Griffin, J.D. and Lucchesi, B.R. (1988b). Reduction of experimental canine myocardial reperfusion injury by a monoclonal antibody (anti-Mo1, anti-CD11b) that inhibits leukocyte adhesion. J. Clin. Invest. 81, 624–629.

Simpson, P.J., Todd, R.F., Mickelson, J.K., Fantone, J.C., Gallagher, K.P., Lee, K.A., Tamura, Y., Cronin, M. and Lucchesi, B.R. (1990). Sustained limitation of myocardial reperfusion injury by a monoclonal antibody that alters leukocyte function. Circulation 81, 226–237.

Tamatani, T. and Miyasaka, M. (1990). Identification of monoclonal antibodies reactive with the rat homolog of ICAM-1, and evidence for a differential involvement of ICAM-1 in the adherence of resting versus activated lymphocytes to high endothelial cells. Internat. Immunol. 2, 165–171.

Tamatani, T., Kotani, M. and Miyasaka, M. (1991). Characterization of the rat leukocyte integrin, CD11/CD18, by the use of LFA-1 subunit-specific monoclonal antibodies. Eur. J. Immunol. 21, 627–633.

Wallenstein, S., Zucker, C.L. and Fleiss, J.L. (1980). Some statistical methods useful in circulation research. Circulation Res. 47, 1–9.

Williams, F.M., Collins, P.D., Tanniere-Zeller, M. and Williams, T.J. (1990). The relationship between neutrophils and increased microvascular permeability in a model of myocardial ischaemia and reperfusion in the rabbit. Br. J. Pharmacol. 100, 729–734.

Yamazaki, T., Seko, Y., Okumura, K., Nagai, R. and Yazaki, Y. (1992). Expression of GMP-140 on the surface of vascular endothelial cells is the initial and essential response to ischemia–reoxygenation. J. Molec. Cell. Cardiol. 24, (Suppl. I), S.91.

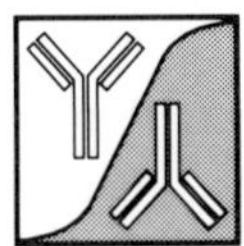

10. Immunopharmacology of Heart Transplantation

Š. Nyulassy, J. Slezák and T. Ravingerová

1. Introduction

Since the first successful kidney transplantation in 1954, the idea of transplantation of other organs and tissues has become increasingly popular. During the following decades the number of kidney grafts has grown and the first barriers were overcome by the concerted action of experimental and clinical investigation. The main milestones of the 1960s, the discovery of the major histocompatibility complex and the introduction of azathioprine, a new type of immunosuppressive, extended substantially the limitations of transplantation and led to the first successful heart transplantation in 1967. Despite numerous problems, development in the 1970s and 1980s has changed the context of heart transplantation. What was once an experimental and life-saving procedure has been transformed into a life-enhancing and technologically advanced form of therapy. Improved skills and handling of immunosuppressive drugs has resulted in a decline of infectious complications and a marked reduction in mortality rates. In the 1970s, anti-lymphocyte globulin treatment was introduced. The 1980s was characterized as the era of cyclosporin and monoclonal antibodies; these immunosuppressive agents have substantially improved the success rate of kidney transplantation and have also contributed to advances in

Immunopharmacology of the Heart
ISBN 0-12-200245-8

heart (and other organ) transplantation (Dec *et al.*, 1991; Sablinski *et al.*, 1991).

The first successful heart transplantation performed in 1967 led to a proliferation of the procedure in the following years. Nowadays several thousand hearts are transplanted yearly to patients in the end-stage of heart failure, a condition which is completely refractory to medical treatment and which may originate from coronary heart disease, cardiomyopathy, rheumatic heart disease, congenital heart disease or benign cardiac tumours.

In spite of surgical success in heart transplantation, the major problem has been rejection. This is usually a matter of balance between histocompatibility and immuno-suppression. It is this which represents the most important aspect of the immunopharmacology of heart transplantation, and which constitutes the focus of this chapter.

2. *Immunology of Rejection*[*]

In spite of the introduction of very efficient immuno-suppressive strategies (see later) the central remaining problem in organ transplantation is allograft rejection, and this is a major factor which threatens survival of patients after cardiac transplantation.

The immune system of the recipient is challenged by the histocompatibility antigens present on the surface of the donor cells. The antigens of the major histocompati-bility complex play the most important role, although ABO antigens are also relevant because the recipient's natural preformed antibodies react against them, in the case of an ABO incompatibility, hyperacute rejection threatens the graft (Dausset and Rapaport, 1966; Rapaport and Dausset, 1987).

2.1 ACUTE REJECTION

2.1.1 Immunology

The primary stimulus is transmitted to the recipient's immune system by the so-called "passenger" cells, i.e. antigen presenting cells (APCs) from the graft (Fig. 10.1). These cells express the I and II class antigens on their surfaces, antigens which are encountered by the lymphocytes of the recipient. Further stimuli triggering lymphocyte activation include production of interleukin-1 (IL-1). This leads to activation of T_H cells, a key moment in the immune response against the graft. Once activated, these cells start releasing IL-2, an essential co-factor in the activation of both CD8-positive T cells and B cells (Waldman, 1990). As a consequence of exposure

to antigen and the action of interleukins, a clonal proliferation and maturation of alloantigen-reactive cells takes place (Bradley *et al.*, 1992). Effector T cells develop, which migrate from the lymphoid tissue via the blood to the graft and cause damage at the antigen-containing sites. The effector T cells are of the CD4 and CD8 subclasses. The CD4 (helper/inducer) cells destroy the target cells through recognizing the HLA-A,B (class I) antigens, whereas the CD8 (suppressor/cytotoxic) cells recognize the class II histocompatibility antigens (HLA-D region coded). The activation of T cells is followed by the release of further lymphokines. Particularly important is interferon (IFNγ), which is known to increase the expression of HLA antigens on the cell membranes of the engrafted organ, thus enhancing the graft vulnerability to the effector cells (Wallach *et al.*, 1982). Furthermore, monocytes are activated. The action of each (specific T_C lymphocytes and activated monocytes) leads to the destruction of the target cells of the graft.

In addition, B cells are activated both locally and systemically, and antibodies are produced. The latter react with the respective antigens on the cell surface and damage is mediated by the activation of the complement system or by antibody-dependent cellular cytotoxicity (ADCC).

This sequence of events results in total destruction of the graft (Garovoy *et al.*, 1990).

2.1.2 Pathology

The morphological sequence of events in acute rejection of the heart has been described in detail previously (Cachera *et al.*, 1968; Kosek *et al.*, 1968; Slezák and Hubka, 1970). Macroscopically, hearts in the first days after implantation have a patchy fibrinous pericardial exudate, they are swollen, and cross-sections reveal randomly distributed pale areas of necrotic tissue and areas of haemorrhages in the ventricles.

Microscopically, the general trend in the development of cellular rejection and graft necrosis progresses in untreated grafts very fast over the first 7 days following transplantation and usually the heart fails within 7 to 14 days. Under these circumstances, survival beyond 14 days without treatment is quite unusual. In treated cases, development of cellular rejection stabilizes and/or decreases within 3 days after initial increase with possible episodes of acute rejection (late acute rejection).

The initial findings of diffuse interstitial infiltration of lymphocytes, histiocytes, polymorphonuclear leuko-cytes, and erythrocyte extravasation in the myocardium of a cardiac allograft are encountered during the first 24 h.

---

[*] Rejection – the true meaning of the word is the destruction of the graft by immune mechanisms, graft failure or graft loss. In this chapter this term will be used (as it is often done in transplantation literature) to describe either the process of the immune reaction against the graft, or the reversible deterioration of graft function due to immune mechanisms, in other words, rejection crisis or rejection episode.

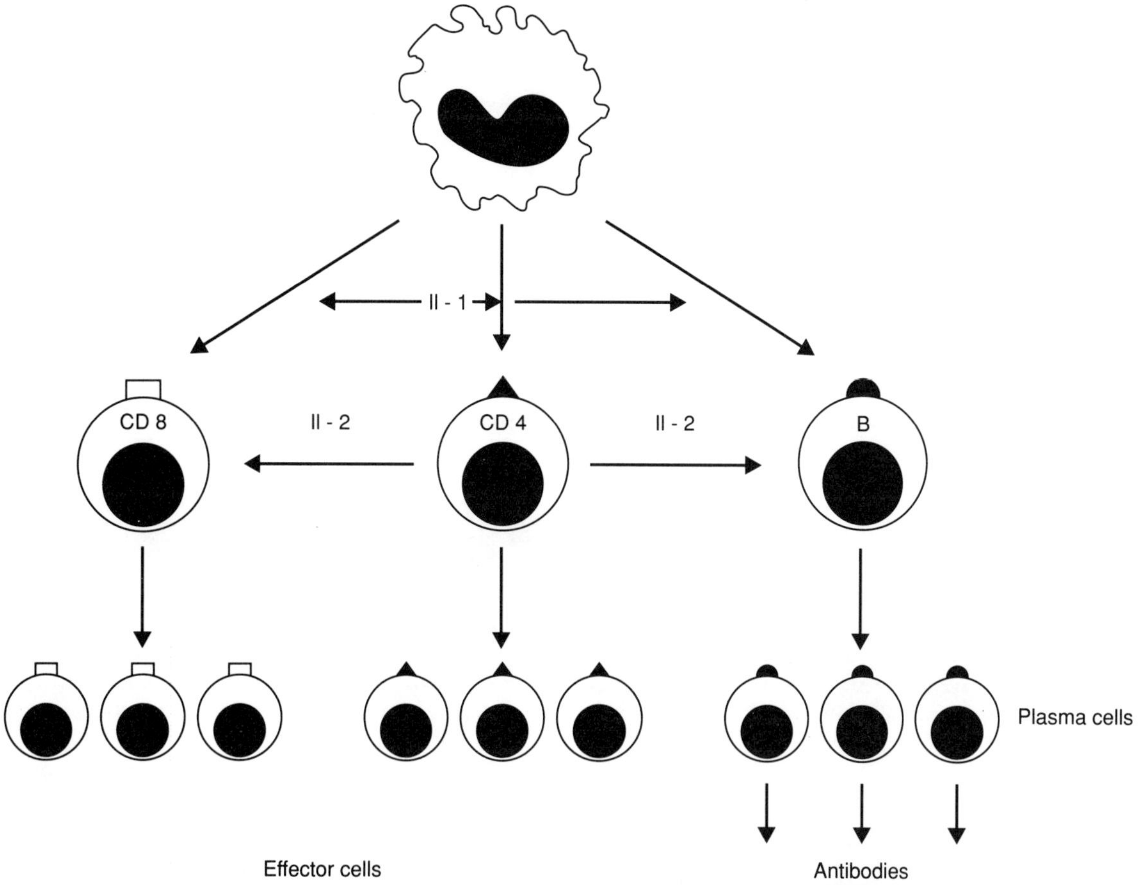

Figure 10.1 Immunology of rejection. The primary stimulus in the process of rejection is provided by the donor's antigen-presenting cells. They express both class I and class II HLA antigens and activate the CD8-positive cells (class I) and the CD4-positive cells (class II). As a consequence populations of CD4- and CD8-derived effector cells recognize the HLA antigens on the transplanted tissues and a classic cytotoxic reaction takes place. The activated B cells create a population of plasmocytes producing specific anti-HLA antibodies reacting with their respective antigens. Cytokines play an important role in the phase of activation (IL-1, IL-2). Further, other lymphokines (e.g. IFNγ) activate macrophages and are responsible for the inappropriate expression of class II antigens and thus for the enhancement of the rejection process.

Pericapillary interstitial oedema is present and increases progressively thereafter. Occasionally, intercapillary mononuclear cells with dense nuclear chromatin, resembling peripheral blood monocytes, closely attached to capillary endothelial cells and accompanied by other larger mononuclear cells, increase progressively in number. Capillary damage is soon followed by the exudation of fibrin, erythrocytes, serum (oedema) fluid and a rapidly increasing infiltration by host mononuclear cells (Fig. 10.2).

Immunological mechanisms described above operate in the acute and late-acute rejection episode, and lymphocytes and plasma cells positively stained with Methyl Green–Pyronin, frequently present in biopsies, can be used as markers. Occasionally, very large cells with large vesicular nuclei, prominent, often multiple nucleoli, and

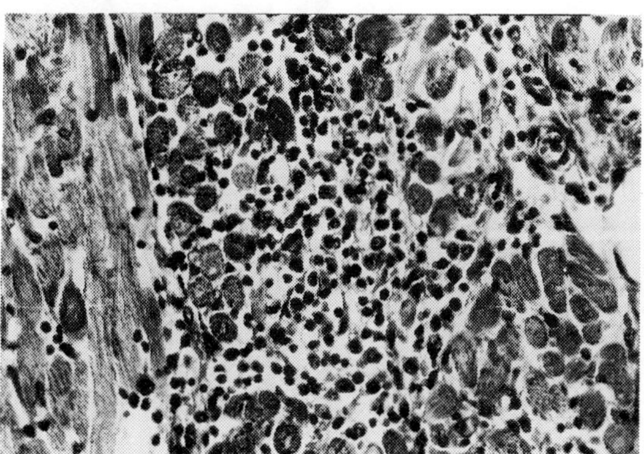

Figure 10.2 Transplanted dog heart in acute rejection with extensive mononuclear interstitium infiltration in and around vessels (Masson trichrom staining).

a rim of densely pyroninophilic cytoplasm are also present. This group of cells is regarded as "immunoblasts" (Dameshek, 1963).

Later, after 4–5 days, a well-developed cellular response composed of lymphocytes, plasma cells and numerous immature mononuclear cells in and around vessels of all sizes is present in all areas of the myocardium examined (Figs 10.2–10.5). The end result, usually within 7–10 days, is extensive myocardial necrosis and mononuclear cell infiltration, mainly of perivascular distribution (Fig. 10.3). The capillary walls become almost impossible to identify and they are disrupted in places. In the terminal stages some capillaries contain small fibrin thrombi (Fig. 10.3). Polymorphonuclear cells are seen in association with muscle cell necrosis but not elsewhere.

Ultrastructurally, a progressive injury to the myocardial fibres and endothelial cells of capillaries is observed in the first 24 h (Figs 10.7–10.9). The most pronounced

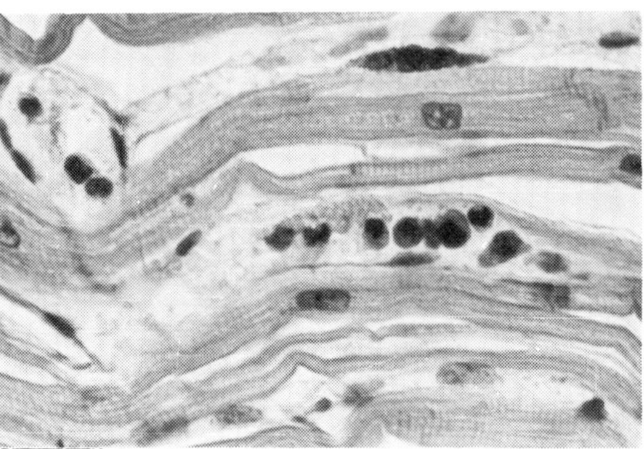

Figure 10.5 A small focus of myocytolysis invaded by monocytes in acute canine cardiac rejection (PTAH staining).

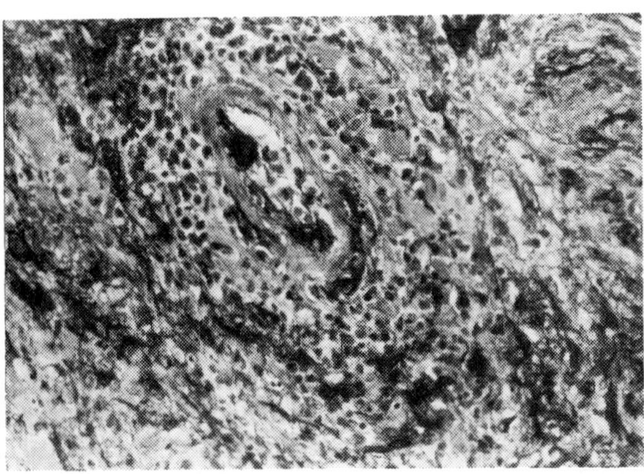

Figure 10.3 The same heart as in Fig. 10.2 showing fibrin deposition in and around vessels and massive perivascular infiltration with mononuclear cells. Some areas of focal necrosis are evident (PTAH staining).

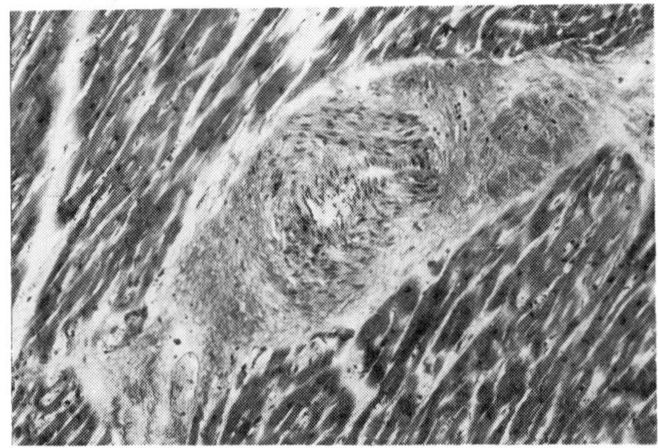

Figure 10.6 Chronic canine cardiac rejection after 6 months of treatment with immunosuppressive drugs. Intimal hyperplasia and luminal narrowing with fibrosis of media is the main finding (Masson trichrom staining).

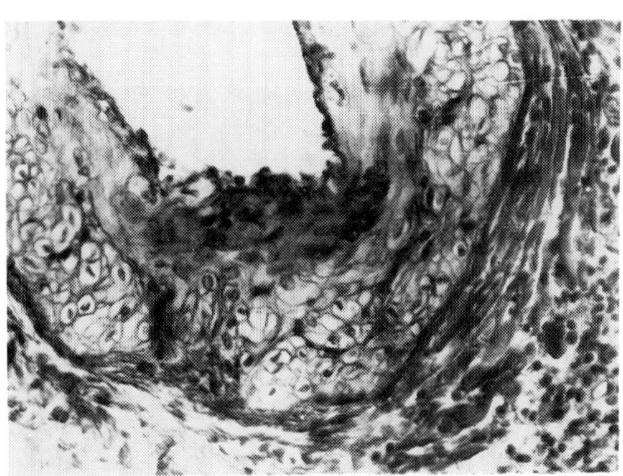

Figure 10.4 Artery of the dog heart in 16th day of acute rejection. Intimal hyperplasia, vacuolization of media and massive periarterial lymphocyte infiltration (Masson trichrom staining).

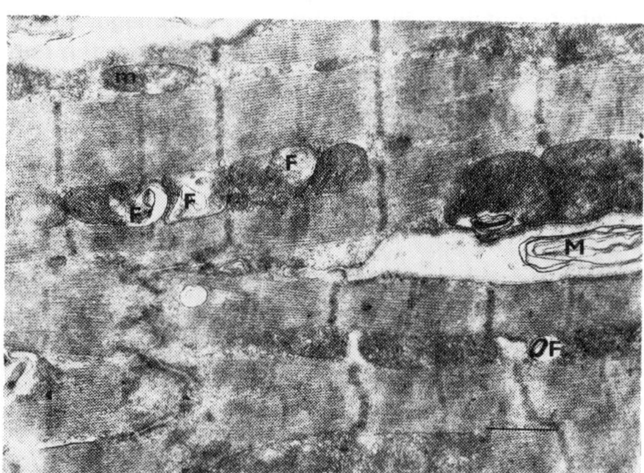

Figure 10.7 Electron micrograph of the dog heart 24 h after transplantation. Numerous myelin figures in mitochondria (F), decreased density of mitochondrial matrix and loss of cristae in mitochondria (m), loss of glycogen and presence of membranous material in extracellular space (M) represent the early stage of degeneration. Scale bar represents 1 μm.

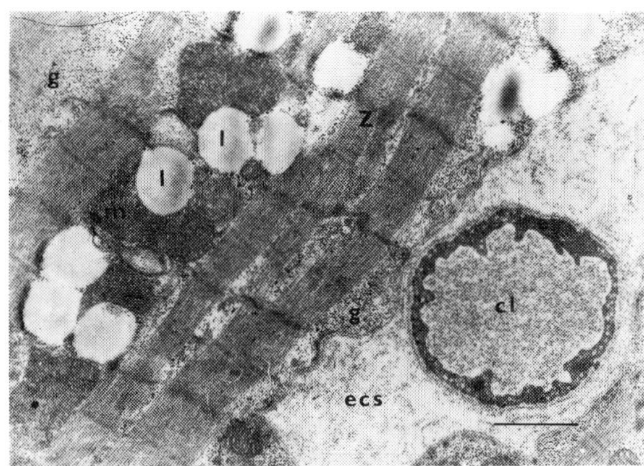

Figure 10.8 Different myocyte of the same heart as in Fig. 10.7 with a vast amount of glycogen granules (g) and numerous lipid droplets (l). Enlarged extracellular space (ecs) contains fibrinous material. Endothelial cells of capillaries (e) lose their fine structure. Mitochondria (m) contain sausage-like cristae in lucent matrix. cl, capillary lumen. Scale bar represents 1 μm.

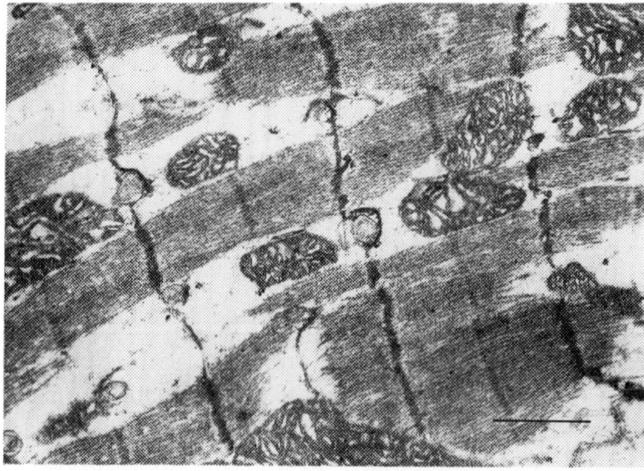

Figure 10.9 The same heart as in Figs 10.7 and 10.8. Myocyte showing extant interfibrillar oedema, absence of glycogen granules and a mitochondria with lucent matrix and decreased amount of cristae. Scale bar represents 1 μm.

changes are intercellular oedema, decreased density of the mitochondrial matrix, varying amounts of glycogen, occurrence of a great number of myelin figures, and other degenerative changes. Endothelial cells lose their fine structure, homogenize and necrotize.

2.1.2.1 Effect of Rejection on the Conducting System

The histopathological findings within the conducting system of rejected hearts point to a primary immunological response. In allografts undergoing rejection, these findings represent a diffuse infiltration of the A–V node and bundle of His by mononuclear cells with abundant basophilic cytoplasm displaying various degree

of pyroninophilia. This infiltrate, which includes macrophages, often surrounds the atrial nerve bundles adjacent to the A–V node and the A–V node itself.

2.1.2.2 Effect of Treatment

The same basic structural changes occur in cases treated by immunosuppressive therapy but the timing and severity are different (Leedham *et al.*, 1971). Cellular rejection, as indicated by mononuclear margination, presence of immunoblasts and microvascular disruption, develops later and is less severe. Ischaemic damage, as indicated by muscle cell necrosis and polymorph infiltration, is also reduced.

However, a reduced cellular response in treated cases may still produce enough microvascular disruption to cause severe ischaemic damage and this would suggest that a cellular rejection response above the critical level can initiate massive ischaemic damage to the myocardium (Leedham *et al.*, 1971).

2.2 HYPERACUTE REJECTION

Preformed ABO alloantibodies, and alloantibodies against class I HLA antigens produced by previous pregnancy, blood transfusion and/or transplantation, when present in sufficient quantity, will bind to the vascular endothelium and trigger a cascade of immunological events. Initially, complement fixation and activation occurs, followed by an activation of the clotting pathway. This series of events, if intensive enough, can result in microthrombi leading to severe ischaemia and necrosis (Colvin, 1991).

2.3 CHRONIC REJECTION

2.3.1 Immunology

Chronic rejection is characterized by the narrowing of the vascular arterial lumen due to the growth of endothelial cells lining the vascular bed (Johnson *et al.*, 1989). The actual control mechanisms for this response are unknown but may include immunological injury signals, release of IL-1 by monocytes and release of platelet-derived growth factor by platelets and endothelial cells. Initially, the proliferating endothelial cell lesion is reversible, but once it proceeds to fibrotic changes within the blood vessel wall itself, it becomes unresponsive to current modes of therapy and develops to graft ischaemia, extensive interstitial fibrosis and ultimate loss of graft function (Hammond *et al.*, 1991).

2.3.2 Pathology

A process of chronic rejection seems to be primarily related to repeated or continuous injury to the endothelium of the arterial system and is manifested by intimal hyperplasia with gradual luminal narrowing (Fig. 10.4) as well as by eventual destruction and fibrosis of the media, probably due to involvement of the vasa vasorum.

Microscopic examination has revealed a myocardial interstitial infiltrate consisting of lymphocytes, a small number of histiocytes and varying numbers of polymorphonuclear leukocytes. The infiltrate is concentrated mainly in the perivascular area, and a similar infiltrate is found in the thickened intima of the coronary arteries.

The degree of rejection can be quantified by the amount of mononuclear margination, the presence of immunoblasts, and the microvascular disruption and graft necrosis associated with polymorphonuclear infiltration. Fresh or fibrotic necrosis can be found scattered throughout the myocardial wall. The chronic, insidious process of intimal proliferation and fibrosis is usually difficult to detect until damage is far advanced (Fig. 10.6). It undoubtedly results from inadequate immunosuppression or poor histocompatibility or a combination of both. A chronic rejection process aggravates circulatory disturbances and gives rise to ECG changes.

2.4 IMMUNOGENETICS OF HEART TRANSPLANTATION

Allograft survival and function is a question of the balance between the level of the histocompatibility and the level of immunosuppressive therapy sufficient to combat it. Pretransplant immunogenetic evaluation consists of HLA typing, determination of ABO antigens and cross-match testing of the recipient's serum against the lymphocytes of the donor (Dausset and Rapaport, 1966; Rapaport and Dausset, 1987).

2.4.1 Major Histocompatibility Complex

The role of the MHC antigens in heart transplantation is uncertain (Nyulassy *et al.*, 1991; Zerbe *et al.*, 1991). Recent data have revealed a close relationship between the HLA match and the outcome of the heart transplantation (Opelz, 1992; Figure 10.11). However, this knowledge usually cannot influence the success of engraftment because, in emergency situations, the heart is transplanted regardless of the level of the histocompatibility. Therefore, the usual approach is the transplantation of an allograft compatible in the ABO system and with negative cross-match.

2.5 Diagnosis of Rejection

2.5.1 Clinical Assessment

Immunological methods are not sensitive enough to predict rejection episodes and the clinical diagnosis of acute rejection is usually based on evidence of graft dysfunction, i.e. by haemodynamic assessment. The

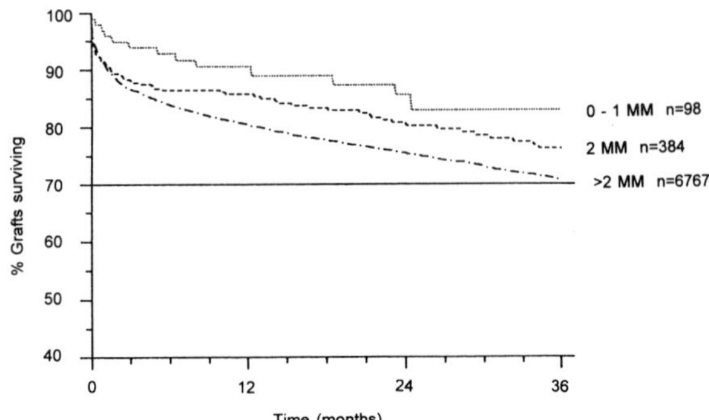

Figure 10.11 Heart transplantation and HLA mismatches, (Collaborative Transplant Study; Opelz, 1992)

same criteria are used in the diagnosis of late acute rejection (Salaman, 1991).

Acute and late acute rejection episodes may be characterized by mild symptoms of congestive heart failure or diminished cardiac output. Electrocardiographic findings of a decrease in the QRS voltage and/or arrhythmias frequently constitute the only indicators of acute rejection (Reader *et al.*, 1990).

Treatment with high doses of immunosuppressive drugs generally causes a prompt reversal of the reduction in ECG voltage coincident with marked clinical improvement (Lower, 1969).

2.5.2 Myocardial Biopsy

Endocardial biopsy is necessary in the assessment of the development of rejection. Morphological assessment of acute and chronic rejection allows gradation of the severity of the rejection process. This approach provides a basis for modifications in the immunosuppressive therapy (Dec *et al.*, 1991).

3. *Therapy of Rejection*

For the aforementioned reasons, a lower level of histocompatibility in cardiac transplantation requires a more aggressive therapeutic approach towards the management of rejection than that in renal transplantation (where graft selection can be done more carefully so allowing better matching of host and donor histocompatibility).

Over the past two decades considerable progress has been achieved in the development of compounds capable of providing a suppression of the immune response against the graft. Non-specific immunosuppression is still dominating at present. The essential goal, however, is specific inhibition of the immune response against the incompatible antigens of the donor. Development proceeded from the "classic" immunosuppressants, such as glucocorticosteroids and azathioprine, through to cyclosporin A (CyA), sometimes called a second generation drug, and the currently intensively studied FK 506, and to rapamycin and some newer compounds. Simultaneously, another line of development has led to use of anti-lympho(thymo)cyte globulins and, recently, monoclonal antibodies against certain cell markers and cytokines. Although the present situation is certainly far from ideal, the new therapeutic approaches substantially contribute to the increasing improvement of transplantation outcome (Briggs, 1991).

3.1 IMMUNOSUPPRESSIVE CHEMOTHERAPY

3.1.1 Classic Immunosuppression

3.1.1.1 *Glucocorticosteroids*
Glucocorticosteroids (GCS) are known to reduce the capacity of antigen-presenting cells to induce the expression of the HLA class II antigens on the surface of different cells and to release IL-1 (Larsson, 1980; Palacios, 1982). In addition, they inhibit the alloactivation of T cells and, consequently, the release of IL-2 (Kaplan *et al.*, 1983). The effect of GCS on the migration and function of the effector cells, as well as on their capacity to release IFNγ may explain their efficacy in reversing acute rejection (Arya *et al.*, 1984). GCS further produce lymphopaenia especially of CD4-positive T cells, by delaying the transit of lymphocytes through the bone marrow and the lymphoid tissue (Goodwin *et al.*, 1986).

3.1.1.2 *Azathioprine*
Azathioprine (AZA) is a nitroimidazol derivative of the purine antagonist 6-mercaptopurine which is rapidly converted to 6-mercaptopurine (Spreafico *et al.*, 1982). It blocks the synthesis of inosinic acid, the precursor of adenylic and guanylic acids. Therefore, the major effect of AZA is the impairment of DNA synthesis and consequently decrease of cell replication (Lee *et al.*, 1985). AZA preferentially inhibits B cell responses (Spreafico *et al.*, 1982). Nevertheless, both cell-mediated and humoral responses are suppressed. AZA further reduces the number of circulating NK and K cells (the K cells are responsible for the antibody-dependent cellular cytotoxicity) (Thiele and Lipsky, 1989).

3.1.2 Cyclosporin A
Cyclosporin A (CyA) is a cyclic undecapeptide isolated from *Tolypocladium inflatum* soil fungus. The compound has several *N*-methylated amino acids and one novel C$_9$-amino acid. CyA inhibits both humoral and cell-mediated immune responses (Borel *et al.*, 1976). Its effect is reversible and highly selective for T$_H$ lymphocytes with sparing of T$_S$ cells (Hess *et al.*, 1986).

Immunosuppression achieved by CyA does not lead to lympholysis. Most of the immunosuppressive effects of CyA are the consequence of inhibition of IL-2 (and other lymphokine) secretion by activated T cells (Kern *et al.*, 1981). At a cellular level, this implies that cytotoxic and other effector T cells that destroy the target cells by interaction with the antigen are inhibited. The lymphocytic subpopulation of T$_H$ cells no longer produces and secretes IL-2 or lymphokines necessary to activate B cells and their maturation into antibody-secreting plasma cells (Jenkins *et al.*, 1988). Inflammatory effector mechanisms cease to function in the absence of other delayed-type hypersensitivity T-lymphocyte-derived (T$_{DTH}$) lymphokines, such as macrophage-activating factor, interferons and colony stimulating factors (Sigal *et al.*, 1991). These effects are all rapid in onset, dose dependent and often quickly reversible upon withdrawal of treatment. Other CyA effects on the immune system are less well defined and more controversial, including: inhibition of IL-1 production; possible inhibition of receptor (IL-2R) expression on membranes; failure to affect T-suppressor cell function; and direct inhibition of activated cytotoxic T cells (Thompson *et al.*, 1983). It is noteworthy that CyA does not affect the main macrophage and granulocyte functions, such as phagocytosis, chemotaxis and the release of most monokines, thus suggesting that the drug has a greater selectivity than other immunosuppressant drugs currently in use (Sigal *et al.*, 1991).

3.1.3 New Drugs
FK 506 is a new compound derived from *Streptomyces tsukubaensis*. It is an antibiotic of the macrolide type. FK 506 inhibits the induction of T lymphocyte proliferation at a lower concentration than does CyA (Wasik *et al.*, 1991). Despite a completely different composition, the mechanism of its action is very much like that of CyA. It has substantially less side effects than CyA. It blocks IL-2 production and secondary proliferation of alloreactive T lymphocytes (Johansson and Möller, 1990). Finally FK 506 seems to inhibit the transcription of genes responsible for the early phase of T cell activation (Thomson, 1991).

Rapamycin is a molecule structurally related to FK 506. Similarly, it exerts an inhibition of the production of IL-2 and of other lymphokines required for T cell proliferation and differentiation (Kahan *et al.*, 1991). However, it was shown that the mechanism by which rapamycin exerts its immunosuppressive action seems to be different from that of CyA and FK 506 (Kay *et al.*, 1991). Results of Sigal and coworkers (1991) show that rapamycin acts as an antagonist of FK 506 bioactivity *in vitro*.

3.2 POLYCLONAL AND MONOCLONAL ANTIBODIES

3.2.1 Anti-lymphocyte Antibodies
Anti-lymphocyte and anti-thymocyte globulin (ALG,

ATG) are polyclonal antibodies prepared by immunizing animals with human lymphocytes or thymocytes. The immune serum gained is fractionated and the globulin portion is obtained (Najarian *et al.*, 1970). ALG and ATG are effective immunosuppressants (Cosimi, 1983). Their primary effect is the impairment of cell-mediated responses because of their specific cytolytic effect on T lymphocytes.

There are several problems associated with the use of these preparations. Their standardization is difficult, sometimes cross-reactions occur with other types of cells (e.g. thrombocytes – destructive thrombocytopaenia), the immunoglobulins are heterologous and they can elicit a humoral immune response and can cause serum sickness (Briggs, 1991).

3.2.2 Monoclonal Antibodies (mABs)

mABs have been investigated for their use in transplantation. mAB against the pan-T cell receptor CD3 (OKT3) is successfully used to prevent and/or reverse ongoing rejection episodes in clinical organ transplantation (Schulak *et al.*, 1991).

mABs against the interleukin-2 receptor CD25 28 (33B3.1, anti-Tac, YTH-906, BT 563) have been employed prophylactically as adjunctive therapy in several centers (Hancock *et al.*, 1990).

mABs against the CD4 molecule (BL4) are a recent development. Clinical trials in renal cadaver allograft recipients have begun and the preliminary results are promising (Morel *et al.*, 1990). However, data for cardiac transplantation are not yet available.

3.3 THERAPEUTIC STRATEGIES

Immunosuppressive therapy has been substantially modified over the last 20 years. During the 1970s the "classic" combination of steroids and azathioprine was the mainstay and ALG was added during the first month after transplantation. By 1983 CyA had replaced azathioprine for maintenance of immunosuppression leading to a marked improvement in the survival rate and a decrease in infectious complications (Dec *et al.*, 1991). Later, azathioprine was reintroduced in order to reduce the maintenance dose of CyA and to mitigate its potential side effects. The combination of the three agents is employed currently in the majority of heart transplant programmes as chronic maintenance therapy (Dec *et al.*, 1991).

There is a lack of uniformity among heart transplant programmes as regards therapeutic strategies. The usual scheme in an uncomplicated course is triple immunosuppression with prednisone, azathioprine and cyclosporine A, sometimes supported during the first week by prophylactic administration of ALG (Slapak *et al.*, 1991). There are controversies, particularly concerning the use of steroids as a maintenance therapy, as well as the prophylactic use of both ALG and ATG and OKT3. The

reason for uncertainty, with respect to steroids, stems from their well-known undesirable side effects when administered long term. Several groups have shown that a steroid-free maintenance therapy does not significantly change the prognosis of the graft survival and function (Briggs, 1991). Prophylactic cytolytic therapy decreases the risk of early rejection, but both ALG and OKT3 have a high incidence of side effects which may alter the character of infectious complications with a particular risk of cytomegalovirus (CMV) and Epstein–Barr virus infections, and they are probably responsible for the increased incidence of post-transplant lymphoproliferative disease (Cockfield *et al.*, 1991). In acute rejection, high-dose steroids are used as a bolus therapy with methylprednisolone i.v. or prednisone together with ALG, ATG or OKT3. For resistant rejection, methotrexate in low doses, total lymphoid irradiation or actinomycin C may be useful (Olsen *et al.*, 1990).

3.3.1 Other Agents

The most promising of all the new immunosuppressive agents are the naturally occurring biological molecules which can modify specific aspects of the immune reaction against the allograft. They include antibodies against cytokines and against relevant IL receptors (Simmons and Wang, 1991).

4. *Further Immunologically Relevant Problems*

4.1 IMMUNOSUPPRESSION AND INFECTION

Immunosuppression in a transplant recipient inevitably increases the risk of infection in the patient. The interaction of two factors plays a role: the net state of immunosuppression and the epidemiologic situation. If exposure to an infectious agent is great enough, even minimally immunosuppressed patients can develop severe infection and conversely, if the immunosuppression is great enough, minimal exposure to opportunistic infectious agents can result in life-threatening disease. Infections that occur in the transplant patient are dependent on the type, intensity, duration and sequence of the immunosuppressive agents administered. Immunomodulating viruses are of particular interest: the herpes group, particularly CMV, hepatitis viruses and HIV. When present, they contribute to the net state of immunosuppression and therefore, they exert a broader effect on the patient than in a normal host (Rubin and Tolkoff-Rubin, 1991).

4.2 GRAFT ATHEROSCLEROSIS

Accelerated atherosclerosis of the coronary vessels of the transplanted heart represents a major limitation to long-term graft survival (Hruban *et al.*, 1991). This is a particular nosological entity. Its development does not depend on risk factors usual in the non-transplanted heart such as smoking, hyperlipaemia, advanced hyperglycaemia, and preoperative ischaemic heart disease (Kaufman *et al.*, 1991). The histopathology is also different. In graft vessels less deposition of lipids and calcium as well as greater cellularity, comprising lymphocytes and plasma cells, can be usually observed. All findings point to an immune pathogenic mechanism. This mechanism involves T_H cells interacting with class II HLA antigens; this is different from accelerated atherosclerosis of the graft from typical acute rejection where T_C cells interact with HLA class I antigens (Libby *et al.*, 1988); a relationship with the frequency of rejection episodes was found. The production of anti-HLA antibodies is another major risk factor both for accelerated atherosclerosis and subsequent rejection (Rose *et al.*, 1989). Chronic immune reaction against the cell wall may be induced by the inappropriate expression of class II HLA antigens on the cell membrane of endothelial cells due to IFNγ and TNF production during the rejection episodes (Salomon *et al.*, 1991). Another triggering mechanism could be viral infection caused by herpes virus or CMV. Viruses can damage the endothelial cells and thus induce atherosclerosis (Bruggeman and van Dam-Mieras, 1991).

4.3 POST-TRANSPLANT LYMPHOPROLIFERATIVE DISEASE

Post-transplant lymphoproliferative disease (PLD) is an important complication of immunosuppressive treatment, which may frequently cause the death of these patients. PLD is much more common in patients with a heart transplant than in kidney transplant recipients. The proliferation of lymphocytes occurs primarily in the B subpopulation. In all cases, EBV is probably one of the aetiological agents (Cockfield *et al.*, 1991). The patient is immunocompromised because intensive immunosuppression represents a condition in which oncogenic factors can function well (Penn, 1991). There is a direct relationship between the level of immunosuppression and the incidence of PLD. The risk of PLD increases with the use of cytolytic therapy by ALG and OKT3 (Cockfield *et al.*, 1991). Therapy requires cessation of immunosuppression and the use of acyclovir as a specific inhibitor of EBV replication (Dec *et al.*, 1991).

5. *Perspectives*

Despite the recent advances in immunopharmacology, immunotherapy in heart transplantation still remains far from being ideal. Inadequate as well as excessive immunosuppression is a threat to the transplant recipient. Cardiac allograft failure due to rejection accounts for 25% of all deaths within 30 days and 35% of late deaths. Infectious complications account for an additional 40% of total mortality (Dec *et al.*, 1991). It is clear that more selective and efficient immunosuppressive agents are needed. Although new drugs such as FK 506, rapamycin and others are promising, the use of mABs which are directed against specific T cell subpopulations, cytokines and cytokine receptors may contribute to future advances in immunosuppressive therapy within this decade.

Immunotherapy aimed at control of deleterious humoral immune response needs to be developed (Rose *et al.*, 1989). It is to be expected that active induction of suppressor networks as well as anti-idiotypic regulation could be of value prior to transplantation. Monoclonal toxin-linked antibodies as well as toxin-linked cytokines may also be useful for the highly selective depletion of certain types of cells (Uckun *et al.*, 1989). Another promising therapeutic approach is the use of gene therapy. It will be possible within the next 15 years to sequence immunoregulatory genes and to transect them into somatic cells and thus to alter the ease of acceptance of a graft (Alexander, 1990).

The continuing research into heart transplantation will certainly result in new findings which will change substantially our approach and will make heart transplantation more feasible, safer and more cost effective.

6. *References*

Alexander, J.W. (1990). The cutting edge: A look to the future in transplantation. Transplantation 49, 237–240.

Arya, S.K., Wong-Staal, F. and Gallo, R.C. (1984). Dexamethasone mediated inhibition of T cell growth factor and gamma interferon messenger RNA. J. Immunol. 133, 273.

Borel, J. F., Feurer, C., Gubler, H.U. and St Ahelin, H. (1976). Biological effects of Cyclosporin A: a new anti-lymphocytic agent. Agents Actions 6, 468–475.

Bradley, J.A., Mowat, A. McI. and Bolton, E.M. (1992). Processed MHC class I alloantigen as the stimulus for $CD4^+$ T-cell-dependent antibody-mediated graft rejection. Immunol. Today 13, 434–438.

Briggs, J.D. (1991). A critical review of immunosuppressive therapy. Immunol. Lett. 29, 89–94.

Bruggeman, C.A. and van Dam-Mieras, C.E. (1991). The possible role of cytomegalovirus in atherogenesis. Prog. Med. Virol. 38, 1–26.

Cachera, J.P., Lacombe, M., Bui-Mong-Hung, Leandri, J., Harada, S. and Dubost, C. (1968). Azathioprine et survie après homotransplantation orthotopique du coeur conservé chez le jeune chien. Advances in Transplantation. Proceedings of the First International Congress of the Transplantation Society, 667.

Cockfield, S.M., Preiksaitis, J., Harvey, E., Jones, C., Hebert, D., Keown, P. and Halloran, P.F. (1991). Is sequential use of ALG and OKT3 in renal transplants associated with an increased incidence of fulminant post transplant lymphoproliferative disorder? Transplant. Proc. 1, 1106–1107.

Colvin, R.B. (1991). The pathogenesis of vascular rejection. Transplant. Proc. 23, 2052–2055.

Cosimi, A.B., (1983). The clinical usefulness of antilymphocyte antibodies. Transplant. Proc. 15, 583–589.

Dameshek, W. (1963). "Immunoblasts" and "Immunocates" – an attempt at a functional nomenclature. Blood 21, 243.

Dausset, J. and Rapaport, F.T. (1966). Role of ABO erythrocyte groups in human histocompatibility reactions. Nature (Lond.) 209.

Dec, G.W., Semigran, M.J. and Vlahakes, G.J. (1991). Cardiac transplantation: current indications and limitations. Transplant. Proc. 23, 2095–2106.

Garovoy, M.R., Melzer, J.S., Ascher, N., Magilligan, D. and Bozdech, M. (1990). In "Basic and Clinical Immunology" (ed D.P. Stites and A.I. Terr), pp. 747–765. Prentice-Hall International Inc., Los Angeles.

Goodwin, J.S., Alturu, D., Sierakowski, S. and Lianos, E.A. (1986). Mechanism of action of glucocorticosteroids: inhibition of T cell proliferation and interleukin 2 production is reversed by leukotriene B4. J. Clin. Invest. 77, 1244.

Hammond, E.H., Ensley, R.D., Yowell, R.L., Craven, C.M., Bristow, M.R., Renlund, D.G. and O'Connell, J.B. (1991). Vascular rejection of human cardiac allografts and the role of humoral immunity in chronic allograft. Transplant. Proc. 23 (Suppl. 2), 26–30.

Hancock, W.W., DiStefano, R., Braun, P., Schweizer, R.T., Tilney, N.L. and Kupiec-Weglinski, J.W. (1990). Cyclosporine and anti-interleukin 2 receptor monoclonal antibody therapy suppress accelerated rejection of rat cardiac allografts through different effector mechanisms. Transplantation 49, 416–421.

Hess, A.D., Colombani, P.M. and Esa, A.H. (1986) Cyclosporine and the immune response: basic aspects. Crit. Rev. Immunol. 6, 123.

Hruban, R.H., Beschorner, W.E., Baumgartner, W.A., Augustine, S.M., Reitz, B.A. and Hutchins, G.M. (1991). Accelerated arteriosclerosis in heart transplant recipients: an immunopathology study of transplanted hearts. Transplant. Proc. 23, 1230–1232.

Jenkins, M.K., Schwartz, R.N. and Pardoll, D.M. (1988). Effects of cyclosporine A on T cell development and clonal deletion. Science 241, 1655–1657.

Johansson, A. and Möller, E. (1990). Evidence that the immunosuppressive effects of FK 506 and cyclosporine are identical. Transplantation 50, 1001.

Johnson, D.E., Shao Zhou, G. and Schroeder, J.S. (1989). The spectrum of coronary artery pathologic findings in human cardiac allografts. J. Heart Transplant 8, 349–354.

Kahan, B.D., Chang, J.Y. and Sehgal, S.N. (1991). Preclinical evaluation of a new potent immunosuppressive agent Rapamycin. Transplantation 52, 185–191.

Kaplan, M.P., Lysz, K., Rosenberg, S.A. and Rosenberg, J.C. (1983). Suppression of interleukin-2 production by methylprednisolone. Transplant. Proc. 15, 407–410.

Kaufman, C., Zeevi, E., Zerbe, T., Keenan, R., Kormos, R., Griffith, B., Hardesty, R., Armitage, J., Uretsky, B. and Duquesnoy, R.J. (1991). In vitro culture of infiltrating lymphocytes from coronary arteries and endomyocardial biopsies: association with graft coronary disease. Transplant. Proc. 23, 1142–1143.

Kay, J.E., Kromwel, L., Sea, D. and Denyer, M. (1991). Inhibition of T and B cell proliferation by rapamycin. Immunology 72, 544–551.

Kern, D.E., Gillis, S., Okada, M. and Henney, C.S. (1981). The role of IL-2 in the differentiation of cytotoxic T cells: the effect of monoclonal anti-IL-2 antibody and adsorption with LI-2 dependent T cell lines. J. Immunol. 127, 1323–1327.

Kosek, J., Hurley, E.J. and Lower, R.R. (1968). Histopathology of orthotopic canine cardiac homografts. Lab. Invest. 19, 1968.

Larsson, E. (1980). Cyclosporine A and dexamethasone suppress T cell responses by selectively acting at distinct sites of the triggering process. J. Immunol. 124, 2828–2833.

Lee, H.J., Pawlak, K., Nguyen, B.T., Robins, R.K. and Sadee, W. (1985). Biochemical differences among four inosinate dehydrogenase inhibitors, mycophenolic acid, ribavirin, tiazofurin and selenazofurin studied in mouse lymphoma cell culture. Cancer Res. 45, 5512.

Leedham, P.W., Baum, M. and Cullum, P. (1971). Acute and modified rejection of heterotopic canine cardiac allotransplants studied by serial needle biopsy. Thorax 26, 534–542.

Libby, P., Salomon, R.N., Payne, D.D. and Schoen, F.J. (1988). Functions of vascular wall cells related to development of transplantation-associated coronary arteriosclerosis. Transplant Proc. 21, 3677–3684.

Lower, R.R. (1969). Rejection of the transplanted heart. Transplant. Proc. I, 733–737.

Morel, P., Vincent, C., Cordier, G., Panaye, G., Carosella, E. and Revillard, J.P. (1990). Anti-CD4 monoclonal antibody administration in renal transplanted patients. Clin. Immunol. Immunopathol. 56, 311–316.

Najarian, J.S., Simmons R.L. and Gewurz, H. (1970). Antihuman lymphoblast globulin. Fed. Proc. 29, 197–199.

Nyulassy, Š., Buc, M., Hašková, V., Ivašková, E. (1991). In "Imunologia 1991". Proceedings of the VIth Congress of the Czecho-Slovak Immunonologists (ed M. Ferenčík), pp. 57–63. Czecho-Slovak Immunological Society, Prague.

Olsen, S.L., O'Connell, J.B., Bristow, M.R. and Renlund, D.G. (1990). Methotrexate as an adjunct in the treatment of persistent mild cardiac allograft rejection. Transplantation 50, 773–775.

Opelz, G. (1992) (editor) "Collaborative Transplant Study, 10th Anniversary". CTS, Heidelberg.

Palacios, R. (1982) Mechanisms of T cell activation: role and functional relationship of HLA-DR antigens and interleukins. Immunol. Rev. 63, 73–110.

Penn, I. (1991). The changing pattern of posttransplant malignancies. Transplant. Proc. 23, 1101–1103.

Rapaport, F.T. and Dausset, J. (1987). Activity of the ABO system as a determinant of histocompatibility in human transplantation. Transplant. Proc. 19, 4487.

Reader, J.A., Burke, M.M., Counihan, P., Kirby, J.A., Adams, S., Davies, M.J. and Pepper, J.R. (1990). Noninvasive monitoring of human cardiac allograft rejection. Transplantation 50, 29–33.

Rose, E. A., Smith, C.R., Petrossian, G.A., Barr, M.L. and Reemtsma, K. (1989). Humoral immune responses after cardiac transplantation: correlation with fat rejection and graft atherosclerosis. Surgery 106, 203–208.

Rubin, R.H. and Tolkoff-Rubin, N.E. (1991). The impact of infection on the outcome of transplantation. Transplant. Proc. 23, 2068–2074.

Sablinski, T., Hancock, W.W., Tilney, N.L. and Kupiec-Weglinski, J. (1991). CD4 monoclonal antibodies in organ transplantation – a review of progress. Transplantation 52, 579–589.

Salaman, J.R. (1991). Monitoring of rejection in renal transplantation. Immunol. Lett. 29, 139–142.

Salomon, R.N., Hughes, C.C., Schoen, F.J., Payne, D.D., Pober, J.S. and Libby, P. (1991). Human coronary transplantation-associated arteriosclerosis. Evidence for a chronic immune reaction to activated graft endothelial cells. Am. J. Pathol. 138, 791–798.

Schulak, J.A., Mayes, J.T. and Hricik, E.E. (1991). Treatment with OKT3 and cyclosporin for acute allograft rejection. Transplant. Proc. 23, 2119–2122.

Sigal, N.H., Lin, C.S. and Siekierka, J.J. (1991). Inhibition of human T-cell activation by FK 506, rapamycin, and cyclosporin A. Transplant. Proc. 23(Suppl. 2), 1–5.

Simmons, R.L. and Wang, S.C. (1991). New horizons in immunosuppression. Transplant. Proc. 23, 2152–2156.

Slapak, M., Digard, N., Wise, M. and Setakis, N. (1991). Tripple therapy. Transplant. Proc. 23, 2186–2188.

Slezák, J. and Hubka, M. (1970). Histopathological examination of the myocardium after homotransplantation in dogs. J. Cardiovasc. Surg. II, 310–320.

Spreafico, F., Tagliabue, A. and Vecchi, A. (1982). In "Immunopharmacology" (ed P. Sirois and M. Rola-Pleszczinski), Elsevier Biomedical Press, Amsterdam, New York, Oxford.

Thiele, D.L. and Lipsky, P.E. (1989). The role of cell surface recognition structure in the initiation of MHC unrestricted "promiscuous" killing by T cells. Immunol. Today 10, 375–381.

Thompson, A.W., Moon, D.K., Geczy, C.L. and Nelson, D.S. (1983). Cyclosporin A inhibits lymphokine production but not the response of macrophages to lymphokines. Immunol. 48, 291–299.

Thomson, A.W. (1991). The immunosuppressive macrolides FK-506 and rapamycin. Immunol. Lett. 29, 105–112.

Uckun, F.M., Myers, D.E., Ledbetter, J.A., Wee, S.L. and Vallera, D.A. (1989). Cell type specific cytotoxicity of anti-CD4 and anti-CD8 rich immunotoxins against human alloreactive T cell clones. Blood 74, 2445.

Waldman, T.A. (1990). The multichain interleukin 2 receptor: a target for immunotherapy in lymphoma, autoimmune disorders, and organ allograft. JAMA 263, 272–278.

Wallach, D., Fellous, M. and Revel, M. (1982). Preferential effect of gamma-interferon on the synthesis of HLA antigens and their mRNAs in human cells. Nature 299, 833.

Wasik, M., Gorski, A., Stepien-Sopniewski, B. and Lagodzinski, Z. (1991). Effect of FK 506 versus cyclosporine on human natural and antibody dependent cytotoxicity reactions *in vitro*. Transplantation 51, 268.

Zerbe, T.R., Arena, V.C., Kormos, R.L., Griffith, B.P., Hardesty, R.L. and Duquesnoy, R.J. (1991). Histocompatibility and other risk factors for histological rejection of human cardiac allografts during the first three months following transplantations. Transplantation 52, 485–490.

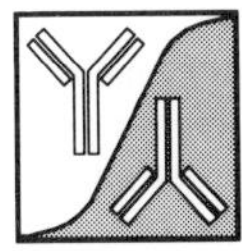

Glossary

A Absorbance
Å Angstrom
AA Arachidonic acid
AAb Autoantibody
Ab Antibody
Abcc Antibody dependent cytotoxic activity
ABA-L-GAT Arsanilicacid conjugated with the synthetic polypeptide L-GAT
AC Adenylate cyclase
ACAT Acyl co-enzyme A acyltransferase
ACE Angiotensin-converting enzyme
ACh Acetylcholine
α_1-ACT α_1-antichymotrypsin
ACTH Adrenocorticotrophin hormone
ADCC Antibody-dependent cell-mediated cytotoxicity
Ado Adenosine
ADP Adenosine diphosphate
AES Anti-eosinophil serum
Ag Antigen
AGE Advanced glycosylation end-product
AGEPC 1-*O*-Alkyl-2-acetyl-*sn*-glyceryl-3-phosphocholine
AI Angiotensin I
AII Angiotensin II
AID Autoimmune disease
AIDS Acquired immunodeficiency syndrome
A/J A Jackson inbred mouse strain
cAMP cyclic adenosine monophosphate (adenosine 3′,5′-phosphate)
AM Alveolar macrophage
AML Acute myelogenous leukaemia
AMP Adenosine monophosphate
ANAb Anti-nuclear antibodies
ANCA Anti-neutrophil cytoplasmic auto antibodies
cANCA Cytoplasmic ANCA
pANCA Perinuclear ANCA
AND Anaphylactic degranulation
ANF Atrial natriuretic factor
ANP Atrial natriuretic peptide
anti-Ig Antibody against an immunoglobulin
anti-RTE Anti-tubular epithelium
APA 13-azaprostanoic acid

APAS Antiplatelet antiserum
APC Antigen-presenting cell
APD Action potential duration
ApO-B Apolipoprotein B
ARDS Acute respiratory distress syndrome
AS Ankylosing spondylitis
4-ASA 4-aminosalicylic acid
5-ASA 5-aminosalicylic acid
ASA Acetylsalicylic acid (aspirin)
ATHERO-ELAM A monocyte adhesion molecule
ATL Adult T cell leukaemia
ATP Adenosine triphosphate
AUC Area under curve
AVP Arginine vasopressin

B₂ (CD18) A leukocyte integrin
β_2M β_2-microglobulin
BAF Basophil-activating factor
BAL Bronchoalveolar lavage
BALF Bronchoalveolar lavage fluid
BALT Bronchus-associated lymphoid tissue
B cell Bone marrow-derived lymphocyte
BCF Basophil chemotactic factor
BCG Bacillus Calmette–Guérin
bFGF Basic fibroblast growth factor
BG Birbeck granules
BHR Bronchial hyperresponsiveness
BI-CFC Blast colony-forming cells
Bk Bradykinin
BM Bone marrow
BMCMC Bone marrow cultured mast cell
BMMC Bone marrow mast cell
BOC-FMLP Butoxycarbonyl-FMLP
bp Base pair
BPB Para-bromophenacyl bromide
BPI Bacterial permeability-increasing protein
BSA Bovine serum albumin
β-TG β-thromboglobulin

CatG Cathepsin G
C1 The first component of complement
C1 inhibitor A serine protease inhibitor which inactivates C1r/C1s
C1q Complement fragment 1q (anaphylatoxin)

C1qR Receptor for C1w; facilitates attachment of immune complexes to mononuclear leucocytes and endothelium
C2 The second component of complement
C3 The third component of complement
C3a Complement fragment 3a (anaphylatoxin)
C3a₇₂₋₇₇ A synthetic carboxyterminal peptide C3a analogue
C3aR Receptor for anaphylatoxins, C3a, C4a, C5a
C4 The fourth component of complement
C4b Complement fragment 4b (anaphylatoxin)
C4BP C4 binding protein; plasma protein which acts as co-factor to factor I inactivate C3 convertase
C5 The fifth component of complement
C5a Complement fragment 5a (anaphylatoxin)
C5aR Receptor for anaphylatoxins C3a, C4a and C5a
C5b Complement fragment 5b (anaphylatoxin)
C6 The sixth component of complement
C7 The seventh component of complement
C8 The eighth component of complement
C9 The ninth component of complement
CAH Chronic active hepatitis
CALT Conjunctival associated lymphoid tissue
cAMP Cyclic adenosine monophosphate (adenosine 3′,5′-phosphate)
CAM Cell adhesion molecule
CBH Cutaneous basophil hypersensitivity
CBP Cromolyn binding protein
CCR Creatinine clearance rate
CD Cluster of differentiation (a system of nomenclature for surface molecules on cells of the immune system); cluster determinant

CD2 Present on T cells and involved in antigen non-specific cell activation

CD3 Present on T cells associated with the antigen receptor and involved in antigen-specific cell activation

CD4 Present primarily on helper T cells and involved in class II restricted interactions

CD8 Present primarily on cytotoxic T cells and involved in class I restricted interactions

CD11a α chain of LFA-1 (leucocyte function antigen-1) present on several types of leucocyte and which mediates adhesion

CD11b α chain of CR3 (complement receptor type 3) present on several types of leucocyte and which mediates adhesion

*__CD11c__ Granulocytes, monocytes, NK-cells, B-cell subset, T-cell subset

CD18 The common β chain of the CD11 family of molecules

*__CD20__ Pan B cell

*__CD31__ Platelets, monocytes, macrophages, granulocytes and B-cells

*__CD34−__ Stem cell marker

*__CD33+__ Monocyte and stem cell marker

*__CD45__ Pan leucocyte

*__CD59__ Low molecular weight HRf; many haemotopolatic and non-haemotopolatic cells

*__CD62__ Activated platelets, endothelial cells

CDC Complement-dependent cytotoxicity

C$_\varepsilon$2 Heavy chain of immunoglobulin E: domains 2

C$_\varepsilon$3 Heavy chain of immunoglobulin E: domain 3

C$_\varepsilon$4 Heavy chain of immunoglobulin E: domain 4

cDNA Complementary DNA

CDP Choline diphosphate

CDR Complementary-determining region

CD$_{xx}$ Common determinant xx

CEA Carcinoembryonic antigen

CETAF Corneal epithelial T cell activating factor

CF Cystic fibrosis

Cf Cationized ferritin

CFA Complete Freund's adjuvant

CFC Colony-forming cell

CFU Colony-forming unit

CFU-Eo/B Eosinophil/basophil colony-forming cell

CFU-GM Granulocyte-macrophage colony-forming cell

CFU-S Colony-forming unit, spleen

CGD Chronic granulomatous disease

cGMP Cyclic guanosine monophosphate (guanosine 3′,5′-phosphate)

CGRP Calcitonin gene-related peptide

CHO Chinese hamster ovary

CI Chemical ionization

CIBD Chronic inflammatory bowel disease

CK Creatine phosphokinase

CKMB The myocardial-specific isoenzyme of creatine phosphokinase

CL Chemiluminescent

CL18/6 Anti-ICAM-1 monoclonal antibody

CLC Charcot–Leyden crystal

CMC Critical micellar concentration

CMI Cell mediated immunity

CML Chronic myeloid leukaemia

CMV Cytomegalovirus

CNS Central nervous system

CO Cyclooxygenase

CoA Coenzyme A

CoA-IT Coenzyme A-independent Transacylase

Con A Concanavalin A

COPD Chronic obstructive pulmonary disease

CoVF Cobra venom

CP Creatine phosphate

CPJ Cartilage/pannus junction

CR Complement receptor

CR1 Complement receptor type 1

CR2 Complement receptor type 2

CR3 Complement receptor type 3

CR3-α Complement receptor type α

CRF Corticotrophin-releasing factor

CRI Cross-reactive idiotype

CRP C-reactive protein

CSA Cyclosporin A

CSF Colony-stimulating factor

CSS Churg–Strauss syndrome

CTAP-III Connective tissue-activating peptide

CTD Connective tissue diseases

CThp Cytotoxic T lymphocyte precursors

C terminus Carboxy terminus of peptide

CTL Cytotoxic T lymphocyte

CTMC Connective tissue mast cell

ct.min^{-1} Counts per minute

Da Dalton (the unit of relative molecular mass)

DAF Decay accelerating factor

DAG Diacylglycerol

DAO Diamine oxidase

D-Arg D-Arginine

DC Dendritic cell

DCF Oxidized DCFH

DCFH 2′,7′-dichlorofluorescin

DEC Diethylcarbamazine

DFMO α-Difluoromethyl ornithine

DFP Diisopropyl fluorophosphate

DGLA Dihomo-γ-linolenic acid

DH Delayed hypersensitivity

DHR Delayed hypersensitivity reaction

DIC Disseminated intravascular coagulation

DLE *Disciud lupus erythematosus*

DMARD Disease-modifying anti-rheumatic drug

DMF *N,N*-Dimethylformamide

DMSO Dimethylsulphoxide

DNA Deoxyribonucleic acid

D-NAME D-Nitroarginine methyl ester

DNase Deoxyribonuclease

DNCB Dinitrochlorobenzene

DNP Dinitrophenol

Dpt4 *Dermatophagoides pteronyssinus* allergen 4

DGW2 HLA phenotype

DR3 HLA phenotype

DR7 HLA phenotype

DREG-2 Murine IgG$_1$ monoclonal antibody against L-selectin

ds Double-stranded

DSCG Disodium cromoglycate

DST Donor-specific transfusion

DTH Delayed-type hypersensitivity

DTPA Diethylenetriamine pentaacetate

DTT Dithiothreitol

dv/dt Rate of change of voltage within time

ε Molar absorption coefficient

EA Egg albumin

EAE Experimental autoimmune encephalomyelitis

EAF Eosinophil-activating factor

EAR Early phase asthmatic reaction

EAT Experimental autoimmune thyroiditis

EBV Epstein–Barr virus

EC Electron capture

ECD Electron capture detector

ECE Endothelin converting enzyme

E-CEF Eosinophil cytotoxicity enhancing factor

ECF-A Eosinophil chemotactic factor of anaphylaxis

ECG Electrocardiogram

ECGF Endothelial cell growth factor

ECGS Endothelial cell growth supplement

E. coli *Escherichia coli*

ECP Eosinophil cationic protein

ED35 Effective dose producing 35% maximum response

ED50 Effective dose producing 50% maximum response

EDF Eosinophil differentiation factor

EDN Eosinophil-derived neurotoxin

EDRF Endothelium-derived relaxant factor

EDTA Ethylene diamine tetraacetic acid (etidronic acid)

EE Eosinophilic eosinophils

EEG Electroencephalogram

EET Epoxyeicosatrienoic acid

EFA Essential fatty acid
EFS Electrical field stimulation
EGF Epidermal growth factor
EGTA Ethylene glycol-bis(β-aminoethyl ether)$N,N,N'N'$-tetraacetic acid
EI Electron impact
ELAM Endothelial leucocyte adhesion molecule
ELAM-1 Endothelial leucocyte adhesion molecule-1
ELF Respiratory epithelium lung fluid
ELISA Enzyme-linked immunosorbent assay
EMS Eosinophilia-myalgia syndrome
ENS Enteric nervous system
EO Eosinophil
EOR Early onset reaction
EPA Eicosapentaenoic acid
EPO Eosinophil peroxidase
EpDIF Epithelial-derived inhibitory factor
EpDRF Epithelium-derived relaxant factor
EPX Eosinophil protein X
ER Endoplasmic reticulum
ESP Eosinophil stimulation promoter
ESR Erythrocyte sedimentation rate
ET Endothelin
ETYA Eicosatetraynoic acid

FA Fatty acid
FAB Fast-electron bombardment
factor B Serine protease in the C3 converting enzyme of the alternative pathway
factor D Serine protease which cleaves factor B
factor H Plasma protein which acts as a co-factor to factor I
factor I Hydrolyses C3 converting enzymes with the help of factor H
FBR Fluorescence photobleaching recovery
Fc Portion of immunoglobulin molecule
F$_c$R Receptor for F$_c$ region of antibody
Fc$_\varepsilon$RI High affinity receptor for IgE
Fc$_\varepsilon$RII Low affinity receptor for IgE
FCS Fetal calf (bovine) serum
FEV$_1$ Forced expiratory volume in 1 second
FGF Fibroblast growth factor
FID Flame ionization detector
FITC Fluorescein isothiocyanate
FKBP FK506-binding protein
FLAP 5-lipoxygenase-activating protein
FMLP N-Formyl-methionyl-leucyl-phenylalanine

FNLP Formyl-norleucyl-leucyl-phenylalanine
FSG Focal sequential glomerulosclerosis
FTS Facteur thymic serique
5-FU 5-Fluorouracil

G6PD Glucose 6-phosphate dehydrogenase
GABA γ-Aminobutyric acid
GAG Glycosaminoglycan
GALT Gut-associated lymphoid tissue
GBM Glomerular basement membrane
GC Guanylate cyclase
GC-MS Gas chromatography mass spectroscopy
G-CSF Granulocyte colony-stimulating factor
GDP Guanosine 5'-diphosphate
GEC Glomerular epithelial cells
GF-1 Insulin-like growth factor
GFR Glomerular filtration rate
GH Growth hormone
GH-RF Growth hormone releasing factor
GI Gastrointestinal
GIP Granulocyte inhibitory protein
GMC Gastric mast cell
GM-CSF Granulocyte-macrophage colony-stimulating factor
GMP Guanosine monophosphate (guanosine 5'-phosphate)
GMP-140 Granule-associated membrane protein-140
GP Glycoprotein
GPIIb-IIIa Glycoprotein IIb-IIIa, a platelet membrane antigen
GSH Glutathione (reduced)
GSSG Glutathione (oxidized)
GTP Guanosine triphosphate
GTP-γ-S Guanarine 5'O-(3-thiotriphosphate)
GTPase Guanidine triphosphatase
GVHD Graft versus host disease
GVHR Graft versus host reaction

H Haemagglutinin
H$_1$ Histamine receptor type 1
H$_2$ Histamine receptor type 2
H$_2$O$_2$ Chemical symbol for hydrogen peroxide
H$_3$ Histamine receptor type 3
HA Histamine
H & E Haematoxylin and eosin
hIL Human interleukin
Hb Haemoglobin
HBBS Hank's balanced salt solution
HDC Histidine decarboxylase
HDL High-density lipoprotein
HEL Hen egg white lysozyme
HEPE Hydroxyeicosapentanoic acid
HEPES N-2-hydroxylethylpiperazine-N'-2-ethane sulphonic acid

HES Hypereosinophilic syndrome
HETE 5, 8, 9, 11 and 15 Hydroxyeicosatetraenoic acid
HETrE Hydroxyeicosatrienoic acid
HEV High endothelial venule
HFN Human fibronectin
HGF Hepatocyte growth factor
HHT 12-hydroxy-5,8,10-heptadecatrienoic acid
HHTrE 12(S)-Hydroxy-5,8,10-heptadecatrienoic acid
HIV Human immunodeficiency virus
HLA Human leucocyte antigen
HLA-DMA HLA-DM α chain
HLA-DMB HLA-DM β chain
HLA-DNA HLA-DN α chain
HLA-DOB HLA-DO β chain
HLA-DPA1 HLA-DP α chain
HLA-DPA2 HLA-DP α chain-related pseudogene
HLA-DPB1 HLA-DP β chain
HLA-DPB2 HLA-DP β chain-related pseudogene
HLA-DQA1 HLA-DQ α chain
HLA-DQA2 HLA-DQ α chain-related sequence
HLA-DQB1 HLA-DQ β chain
HLA-DQB2 HLA-DQ β chain-related sequence
HLA-DQB3 HLA-DQ β chain-related sequence
HLA-DRA HLA-DR α chain
HLA-DRB1 HLA-DRB$_1$ chain
HLA-DRB2 Pseudogene with HLA-DR β-like sequence
HLA-DRB3 HLA-DRB β_3 chain
HLA-DRB4 HLA-DRB β_4 chain
HLA-DRB5 HLA-DRB β_5 chain
HLA-DRB6 HLA-DRB β pseudogene
HLA-DRB7 HLA-DRB β pseudogene
HLA-DRB8 HLA-DRB β pseudogene
HLA-DRB9 HLA-DRB β pseudogene
HLA-E Class I-like α chain
HLA-F Class I-like α chain
HLA-G Class I-like α chain
HLA-H Class I pseudogene
HLA-J Class I pseudogene
HMG CoA Hydroxyl methyl glutaryl Co-enzyme A
HMW High molecular weight
HMT Histidine methyltransferase
HMVEC Human microvascular endothelial cells
HNC Human neutrophil collagenase (MMP-8)
HNE Human neutrophil elastase
HNG Human neutrophil gelatinase (MMP-9)
HODE Hydroxyoctadecanoic acid
HPETE Hydroperoxyeicosatetraenoic acid

HPETrE Hydroperoxytrienoic acid
HPODE Hydroperoxyoctadecanoic acid
HPLC High-performance liquid chromatography
HRA Histamine-releasing activity
HRAN Neutrophil-derived histamine-releasing activity
HRf Homologous-restriction factor
HRF Histamine-releasing factor
HRP Horseradish peroxidase
HSA Human serum albumin
HSP Heat-shock protein
HS-PG Heparan sulphate proteoglycan
HSV Herpes simplex virus
^{3}HTdR Tritiated thymidine
5-HT 5-Hydroxytryptamine (Serotonin)
HUVEC Human umbilical vein endothelial cell

Ia Immune reaction-associated antigen
Ia+ Murine class II major histocompatibility complex antigen
I_{sc} Short-circuit current
IB$_4$ Anti-CD18 monoclonal antibody
IBD Inflammatory bowel disease
IBMX Isobutylmethylxanthine
IBS Inflammatory bowel syndrome
IC$_{50}$ Concentration producing 50% inhibition
ICAM Intercellular adhesion molecule
ICAM-1 Intercellular adhesion molecule-1
ICAM-2 Intercellular adhesion molecule-2
ICE IL-1β-converting enzyme
IDC Interdigitating cell
IDD Insulin-dependent (typ 1) diabetes
IEL Intraepithelial leucocytes
IFA Incomplete Freund's Adjuvant
IFN Interferon
IFNα Interferon α
IFNβ Interferon β
IFNγ Interferon γ
Ig Immunoglobulin
IgA Imunoglobulin A
IgE Immunoglobulin E
IgG Immunoglobulin G
IgG1 Immunoglobulin G class 1
IgG$_{2a}$ Immunoglobulin G class 2a
IgM Immunoglobulin M
IGF-1 Insulin-like growth factor
IHC Immunohistochemistry
IHES Idiopathic hypereosinophilic syndrome
IL Interleukin
IL-1 Interleukin-1
IL-2 Interleukin-2
IL-3 Interleukin-3
IL-5 Interleukin-5
IL-6 Interleukin-6

IL-5R Interleukin-5-receptor
IL-8 Interleukin-8
IL-1α Interleukin-1α
IL-1β Interleukin-1β
IL-1Ra Interleukin-1 receptor antagonist
IL-2R Interleukin-2 receptor
IL-3R Interleukin-3R
ILR Interleukin receptor
IMMC Intestinal mucosal mast cell
INCAM Inducible cell adhesion molecule
INCAM110 Inducible cell adhesion molecule 110
i.p. Intraperitoneally
IP$_3$ Inositol triphosphate
IP$_4$ Inositol tetrakisphosphate
IPO Intestinal peroxidase
IpOCOCq Isopropylidene OCOCq
I/R Ischaemia-reperfusion
IRAP IL-1 receptor antagonist protein
ISCOM Immune-stimulating complexes
ISGF3 Interferon-stimulated gene factor 3
iSRS Immunoreactive slow-reacting substance
IT Immunotherapy
ITP Idiopathic thrombocytopenic purpura
i.v. Intravenous

K_a Association constant
kb Kilobase
20KDHRF A homologous restriction factor; binds to C8
65KDHRF A homologous restriction factor, also known as C8 binding protein; interferes with cell membrane pore-formation by C5b-C8 complex
K_d Equilibrium dissociation constant
kD Kilodalton
K_D Dissociation constant
KD Kallidin
Ki Antagonist binding affinity
Ki67 Nuclear membrane antigen
KLH Keyhole limpet haemocyanin
KOS KOS strain of herpes simplex virus

λ_{max} Wavelength of maximum absorbance
LAD Leucocyte adhesion deficiency
LAK Lymphocyte-activated killer (cell)
LAM Leucocyte adhesion molecule
LAM-1 Leucocyte adhesion molecule-1
LAR Late-phase asthmatic reaction
L-Arg L-Arginine
LBP LPS binding protein
LC Langerhans cell

LCF Lymphocyte chemoattractant factor
LCR Locus control region
LDH Lactate dehydrogenase
LDL Low-density lipoprotein
LDV Laser Doppler velocimetry
LECAM Lectin adhesion molecule
LECAM-1 Lection adhesion molecule-1
LFA Leucocyte function-associated antigen
LFA-1 Leucocyte function-associated antigen-1
LG β-Lactoglobulin
LHRH Luteinizing hormone-releasing hormone
LI Labelling index
LIS Lateral intercellular spaces
LMP Low molecular mass polypeptide
LMW Low molecular weight
L-NOARG L-Nitroarginine
LP(a) Lipoprotein a
LPS Lipopolysaccharide
LT Leukotriene
LTA$_4$ Leukotriene A$_4$
LTB$_4$ Leukotriene B$_4$
LTC$_4$ Leukotriene C$_4$
LTD$_4$ Leukotriene D$_4$
LTE$_4$ Leukotriene E$_4$
L$_y$-1$^+$ (Cell line)
LXA$_4$ Lipoxin A$_4$
LXB$_4$ Lipoxin B$_4$
LXC$_4$ Lipoxin C$_4$
LXD$_4$ Lipoxin D$_4$
LXE$_4$ Lipoxin E$_4$

M-540 Merocyanine-540
α_2-M α_2-macroglobulin
mAb Monoclonal antibody
mAB IB4 Monoclonal antibody IB4
mAB PB1.3 Monoclonal antibody PB1.3
mAB R 3.1 Monoclonal antibody R 3.1
mAB R 3.3 Monoclonal antibody R 3.3
mAB 6.5 Monoclonal antibody 6.5
mAB 60.3 Monoclonal antibody 60.3
MAC Membrane attack molecule
Mac-1 Macrophage-1
MAF Macrophage-activating factor
MAO Monoamine oxidase
MAP(s) Monophasic action potential(s)
MBP Major basic protein
MBSA Methylated bovine serum albumin
MC Mesangial cells
M cell Microfold or membranous cell of Peyer's patch epithelium
MCP Membrane co-factor protein
M-CSF Monocyte colony-stimulating factor

MC$_T$ Tryptase-containing mast cell
MC$_{TC}$ Tryptase- and chymase-containing mast cell
MDA Malondialdehyde
MDGF Macrophage-derived growth factor
MDP Muramyl dipeptide
MEA Mast cell growth-enhancing activity
MEL Metabolic equivalent level
MEM Minimal essential medium
MG *Myasthenia gravis*
MHC Major histocompatibility complex
MI Myocardial ischaemia
MIF Migration inhibition factor
mIL Mouse interleukin
MI/R Myocardial ischaemia/reperfusion
MIRL Membrane inhibitor of reactive lysis
MLC Mixed lymphocyte culture
MLR Mixed lymphocyte reaction
MMC Mucosal mast cell
MMCP Mouse mast cell protease
MMP Matrix metalloproteinase
MMP1 Matrix metalloproteinase 1
mNA 6-methoxy-2-napthylacetic acid
MNC Mononuclear cells
MO Macrophage
MPO Myeloperoxidase
MRI Magnetic resonance imaging
mRNA Messenger ribonucleic acid
MS Mass spectrometry
MSS Methylprednisoline sodium succinate
MT Malignant tumour
MW Molecular weight

NA Noradrenaline
NAAb Natural autoantibody
NAb Natural antibody
NADH Reduced nicotinamide adenine dinucleotide
NADP Nicotinamide adenine diphosphate
NADPH Reduced nicotinamide adenine dinucleotide phosphate
L-NAME L-Nitroarginine methyl ester
NANC Nonadrenergic, noncholinergic
NAP Neutrophil-activating peptide
NAP-1 Neutrophil-activating peptide-1
NAP-2 Neutrophil-activating peptide-2
NC1 Non-collagen 1
N-CAM Neural cell adhesion molecule
NCEH Neutral cholesteryl ester hydrolase
NCF Neutrophil chemotactic factor
NDGA Nordihydroguaretic acid

Neca 5'-(N-ethyl carboxamido)-adenosine
NED Nedocromil sodium
NEP Neutral endopeptidase (EC 3.4.24.11)
NF-AT Nuclear factor of activated T lymphocytes
NF-κB Nuclear factor-κB
NGF Nerve growth factor
NGPS Normal guinea-pig serum
NIMA Non-inherited maternal antigens
Nk Neurokinin
NK Natural killer
NkA Neurokinin A
NkB Neurokinin B
L-NMMA L-Nitromonomethyl arginine
NMR Nuclear magnetic resonance
NO Chemical symbol for nitric oxide
NPY Neuropeptide Y
NRS Normal rabbit serum
NSAID Non-steroidal anti-inflammatory drug
NSE Nerve-specific enolase
NT Neurotensin
N terminus Amino terminus of peptide

*O$_2$ Oxygen free radical
OA Osteoarthritis
OD Optical density
ODC Ornithine decarboxylase
ODS Octadecylsilyl
OH Chemical symbol for hydroxyl radical
OVA Ovalbumin
ox-LDL Oxidized low-density lipoprotein

Ξ Probability
P Phosphate
P$_a$O$_2$ Arterial oxygen pressure
P$_i$ Inorganic phosphate
α$_1$-PI α$_1$-proteinase inhibitor
P150,95 A leucocyte integrin
PA Phosphatidic acid
pA_2 Negative logarithm of the antagonist dissociation constant
PADGEM Platelet-activation dependent granule external membrane
PAF Platelet-activating factor
PAGE Polyacrylamide gel electrophoresis
PAI Plasminogen activator inhibitor
PAM Pulmonary alvcolar macrophages
PAS Periodic acid–Schiff reagent
PBA Polyclonal B cell activators
PBC Primary biliary cirrhosis
PBL Peripheral blood lymphocytes
PBMC Peripheral blood mononuclear cells
PBS Phosphate-buffered saline

PC Phosphatidylcholine
PCA Passive cutaneous anaphylaxis
PCNA Proliferating cell nuclear antigen
PCR Polymerase chain reaction
p.d. Potential difference
PDBu 4a-phorbol 12,13-dibutyrate
PDE Phosphodiesterase
PDGF Platelet-derived growth factor
PE Phosphatidylethanolamine
PECAM Platelet endothelial cell adhesion molecule
PEG Polyethylene glycol
PET Positron emission tomography
PEt Phosphatidylethanol
PF$_4$ Platelet factor 4
PG Prostaglandin
PGA Polyglandular autoimmune syndrome
PGE$_2$ Prostaglandin E$_2$
PGF Prostaglandin F
PGI Prostacyclin
PGI$_2$ More standard abbreviation of above
PGD$_2$ Prostaglandin D$_2$
PGE$_1$ Prostaglandin E$_1$
PGF$_{2a}$ Prostaglandin F$_{2a}$
PGF$_2$ Prostaglandin F$_2$
PGG$_2$ Prostaglandin G$_2$
PGH Prostaglandin H
P$_a$O$_2$ Arterial oxygen pressure
PGH$_2$ Prostaglandin H$_2$
PGI$_2$ Prostaglandin I$_2$
PGP Protein gene-related peptide
Ph^1 Philadelphia (chromosome)
PHA Phytohaemagglutinin
PHI Peptide histidine isoleucine
PHM Peptide histidine-methionine
PI Phosphatidyl inositol
P$_i$ Inorganic phosphate
PIP Phosphatidylinositol monophosphate
PIP$_2$ Phosphatidylinositol biphosphate
PK Protein kinase
PKA Protein kinase A
PKC Protein kinase C
PL Phospholipase
PLA Phospholipase A
PLA$_2$ Phospholipase A$_2$
PLAP Putative phospholipase activating protein
PLC Phospholipase C
PLD Phospholipase D
PLP Proteolipid protein
PLT Primed lymphocyte typing
PMA Phorbol myristate acetate
PMC Peritoneal mast cell
PMD Piecemeal degranulation
PML Polymorphonuclear leucocyte
PMN Polymorphonuclear neutrophil
PMSF Phenylethylsulphonyl fluoride
PNU Protein nitrogen unit
p.o. *Per os* (by mouth)
PPD Purified protein derivative

PRA Percentage reactive activity
PRD Positive regulatory domain
PR3 Proteinase-3
proET-1 Proendothelin-1
PS Phosphatidylserine
PTA$_2$ Pinane thromboxne A$_2$
PTCA Percutaneous transluminal coronary angioplasty
PTCR Percutaneous transluminal coronary recanalization
PteH$_4$ Tetrahydropteridine
PtX Pertussis toxin
PUFA Polyunsaturated fatty acid
PUMP-1 Punctuated metalloproteinase
PWM Pokeweed mitogen
PYY Peptide YY

q.i.d. *Quater in die* (four times a day)
QRS Segment of electrocardiogram

·R Free radical
R15.7 Anti-CD18 monoclonal antibody
RA Rheumatoid arthritis
RANTES Regulated on activation, normal T expressed and secreted
RAST Radioallergosorbent test
RBC Red blood cell
RBF Renal blood flow
RBL Rat basophilic leukaemia
RE RE strain of herpes simplex virus type 1
REA Reactive arthritis
REM Relative electrophoretic mobility
RER Rough endoplasmic reticulum
RF Rheumatoid factor
RFL-6 Rat fetal lung-6
RFLP Restriction fragment length polymorphism
rh- (prefix) recombinant human (usually refers to peptide)
RIA Radioimmunoassay
RMCP Rat mast cell protease
RMCPII Rat mast cell protease II
RNA Ribonucleic acid
RNase Ribonuclease
RNHCl *N*-Chloramine
RNL Regional lymph nodes
ROM Reactive oxygen metabolites
ROS Reactive oxygen species
R-PIA R-NG-(1-methyl-1-phenyltheyl)-adenosine
RPMI 1640 Roswell Park Memorial Institute 1640 medium
rrSF Recombinant rat SCF
RS Reiter's syndrome
RSV Rous sarcoma virus
RTE Rabbit tubular epithelium
RW Ragweed

S Svedberg (unit of sedimentation density)

SALT Skin associated lymphoid tissue
SAZ Sulphasalazine
SC Secretory component
SCF Stem cell factor
SCFA Short chain fatty acid
SCG Sodium cromoglycate
SCID Severe combined immunodeficiency syndrome
sCR$_1$ Soluble type-1 complement receptors
SCW Streptococcal cell wall
SD Standard deviation
SDS Sodium dodecyl sulphate
SDS–PAGE Sodium dodecyl sulphate–polyacrylamide gel electrophoresis
SEM Standard error of the mean
SGAW Specific airway conductance
SIRS Soluble immune response suppressor
SK Streptokinase
Sl Murine Steel mutation
SLE Systemic lupus erythematosus
SLe^x Sialyl Lewis X antigen
SLO Streptolysin-O
SLPI Secretory leucocyte protease inhibitor
SM Sphingomyelin
SNAP *S*-Nitroso-*N*-acetylpenicillamine
SNP Sodium nitroprusside
SOD Superoxide dismutase
SOM Somatostatin
SOZ Serum-opsonized zymosan
SP Sulphapyridine
S Protein vitronectin
SR Systemic reaction
SRBC Sheep red blood cells
SRIF Somatotrophin release-inhibiting factor (somatostatin)
SRS Slow-reacting substance
SRS-A Slow-reacting substance of anaphylaxis
Sub P Substance P

t$_{1/2}$ Half-life
T84 Human intestinal epithelial cell line
TauNHCl Taurine monochloramine
TBM Tubular basement membrane
TCA Trichloroacetic acid
T cell Thymus-derived lymphocyte
TCR T cell receptor
TDI Toluene diisocyanate
TDID$_{50}$ Tissue culture infectious dose – 50%
TEC Tubular epithelial cells
TF Tissue factor
Tg Thyroglobulin
TGF Transforming growth factor
TGFβ Transforming growth factor β
TGFβ_1 Transforming growth factor-β_1
T$_H$ T helper cells
T$_{H0}$ T Helper o

T$_{HP}$ T helper precursor
T$_H$0, T$_H$1, T$_H$2 Subsets of helper T cells
Thy 1+ Murine T cell antigen
t.i.d. *Ter in die* (three times a day)
TIL Tumour-infiltrating lymphocytes
TIMP Tissue inhibitors of metalloproteinase
TIMP-1 Tissue inhibitor of metalloproteinases 1
Tla Thymus leukaemia antigen
TLC Thin-layer chromatography
TLP Tumour-like proliferation
Tm T memory
TNF Tumour necrosis factor
TNF-α Tumour necrosis factor-α
tPA Tissue-type plasminogen activator
r-tPA Recombinant tissue-type plasminogen activator
TPA 12-*o*-Tetradeconylphorbol-13-acetate
TPK Tyrosine protein kinases
TPP Transpulmonary pressure
Tris Tris(hydroxymethyl)amino-methane
TSH Thyroid-stimulating hormone
TTX Tetrodotoxin
***TWF-R** TWF receptor – ask author for definition of TWF
TX Thromboxane
TXA$_2$ Thromboxane A$_2$
TXB$_2$ Thromboxane B$_2$
Tyk2 Tyrosine kinase

UC Ulcerative colitis
UDP Uridine diphosphate
UPA Urokinase-type plasminogen activator
UV Ultraviolet

VC Veiled cells
VCAM Vascular cell adhesion molecule
VCAM-1 Vascular cell adhesion molecule-1
VF Ventricular fibrillation
VIP Vasoactive intestinal peptide
VLA Very late antigen
VLA-4 Very late activating antigen-4
VLDL Very low-density lipoprotein
V$_{max}$ minimal velocity
VP Viral protein
VPB Ventricular premature beat
VT Ventricular tachycardia

W Murine dominant white spotting mutation
WBC White blood cell
WGA Wheat germ agglutinin

XO Xanthine oxidase

ZA Zonulae adherens
ZAS Zymosan-activated serum
ZO Zonulae occludentes

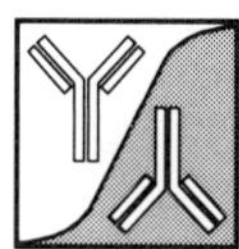

Key to Illustrations

Helper lymphocyte

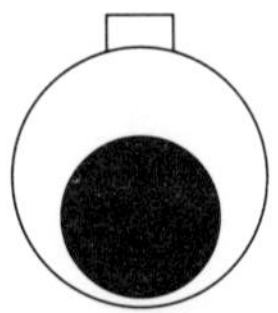

Suppressor lymphocyte

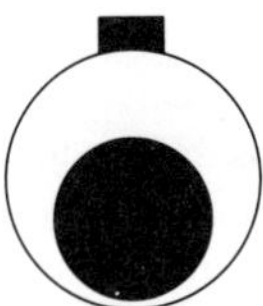

Killer lymphocyte

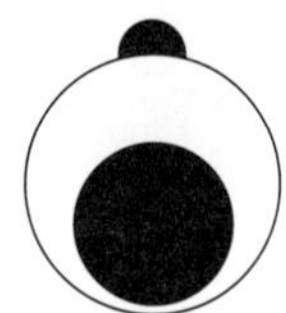

Plasma cell

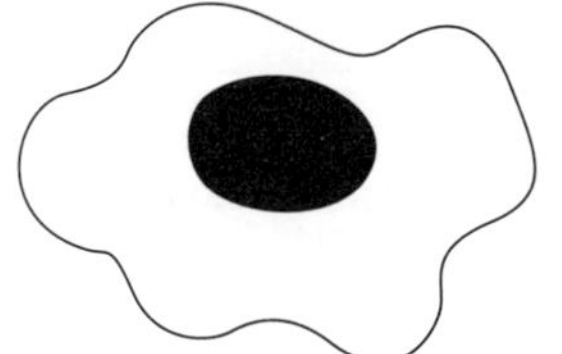

Bacterial or Tumour cell

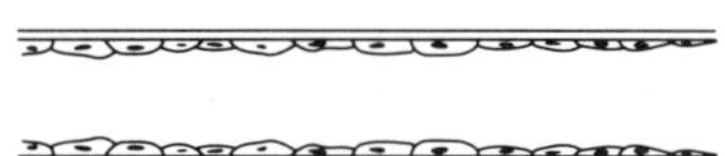

Blood vessel lumen

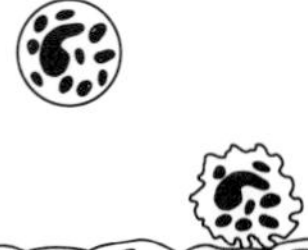
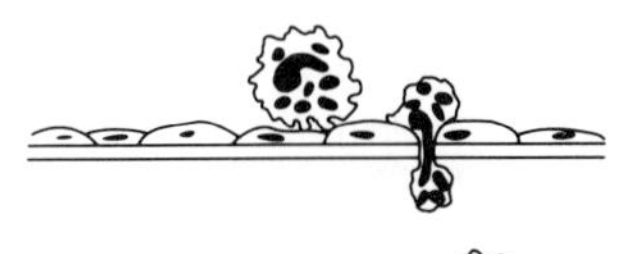

Eosinophil passing through vessel wall

Neutrophil passing through vessel wall

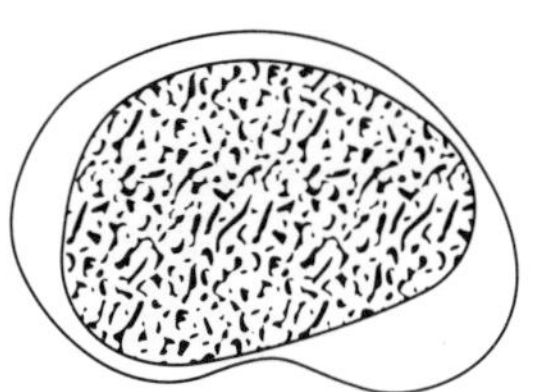

Fibroblast

Resting neutrophil

Activated neutrophil

Resting eosinophil

Activated eosinophil

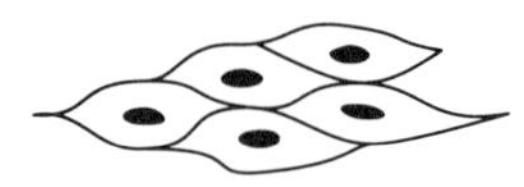

Smooth muscle

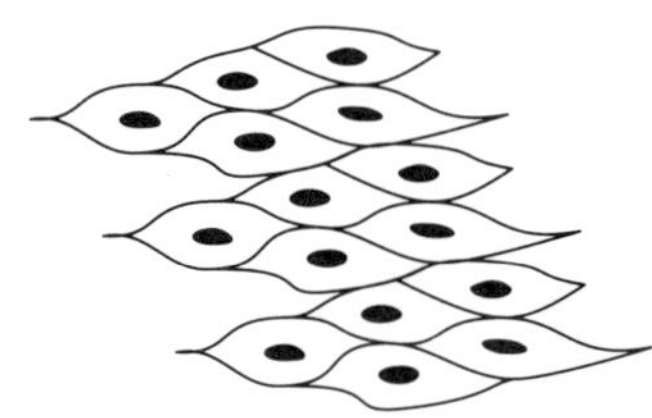

Smooth muscle thickening

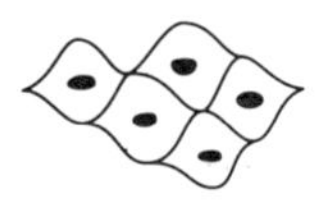

Smooth muscle contraction

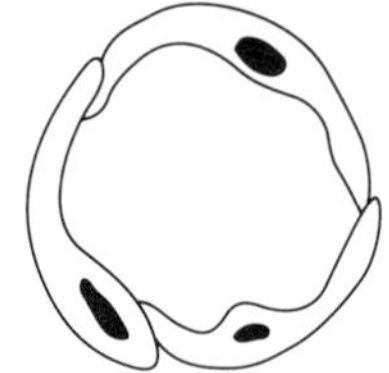

Normal blood vessel

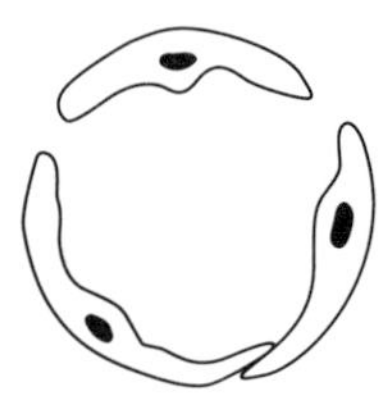

Endothelial cell permiability

Resting macrophage

Activated macrophage

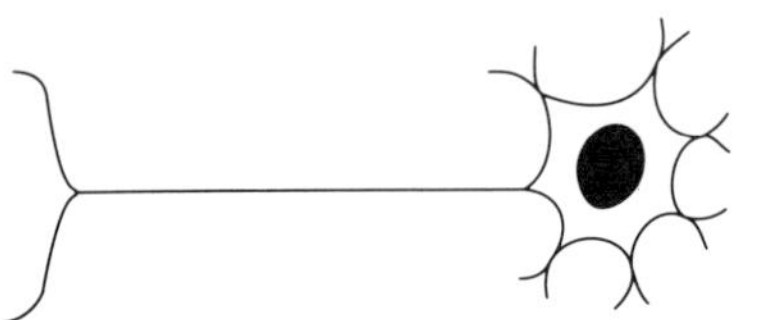

Nerve

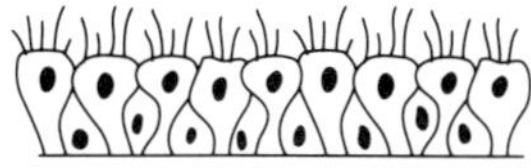

Intact epithelium

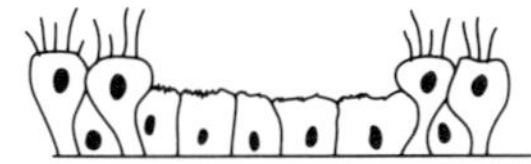

Damaged epithelium

Intact epithelium with submucosal gland

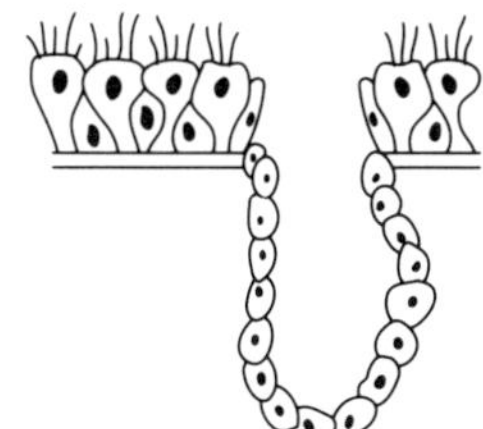

Normal submucosal gland

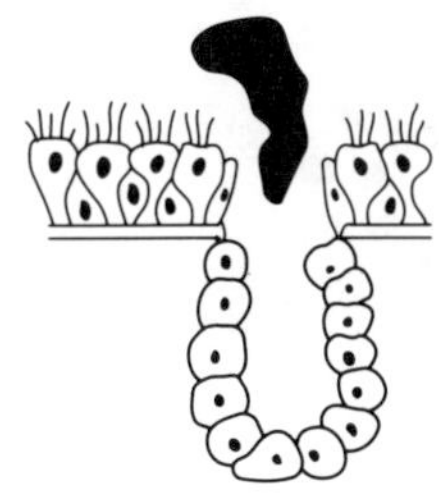

Hypersecreting submucosal gland

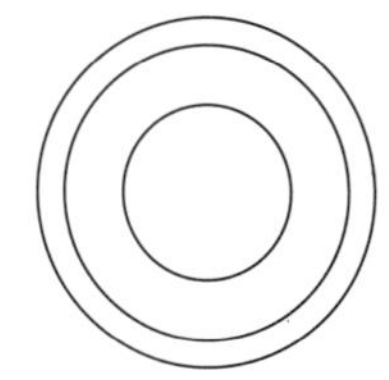

Normal airway

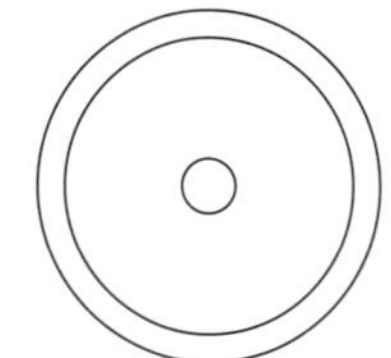

Oedema

Bronchospasm

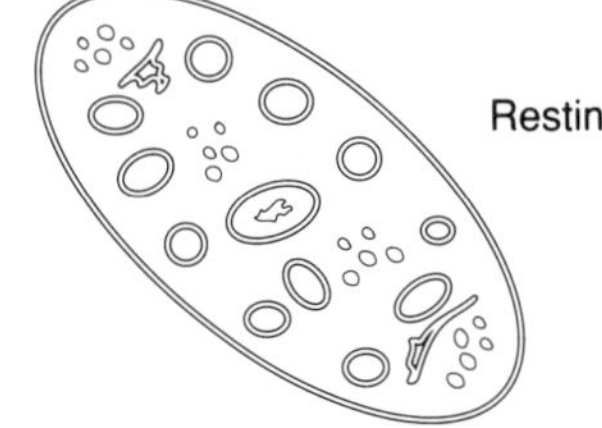

Resting platelet

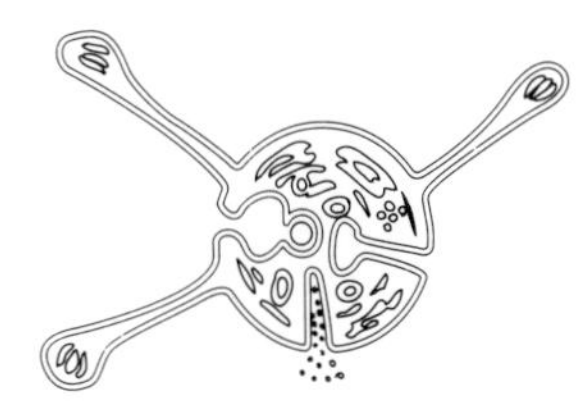

Activated platelet

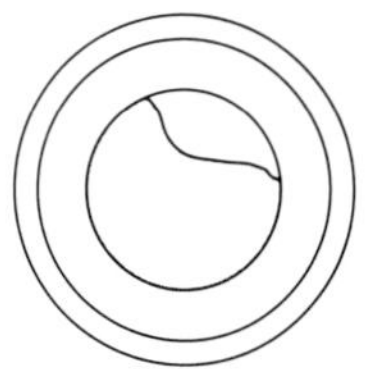

Airway hypersecreting mucus

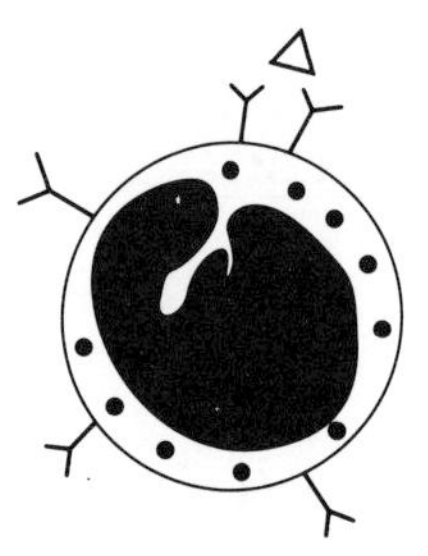

Resting
basophil

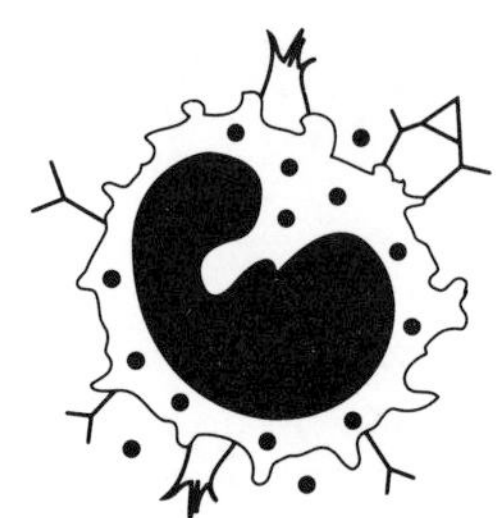

Activated
basophil

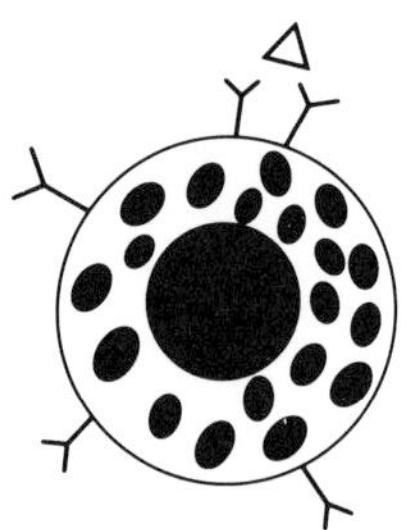

Resting
mast cell

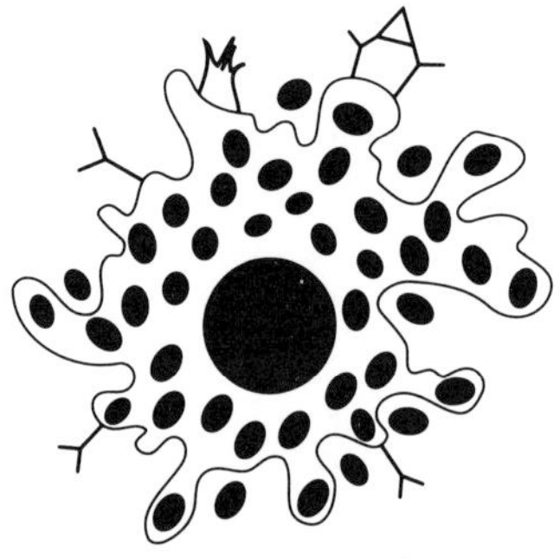

Activated
mast cell

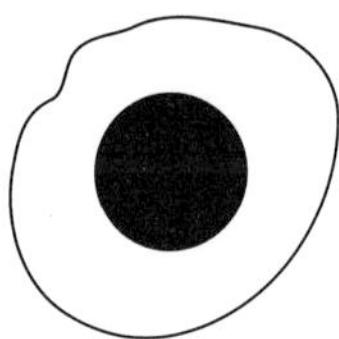

Resting
chondrocyte

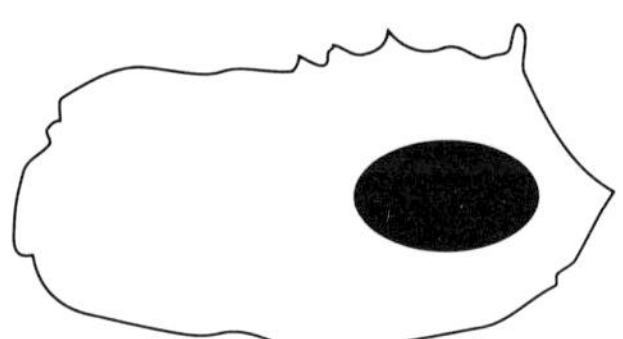

Activated
chondrocyte

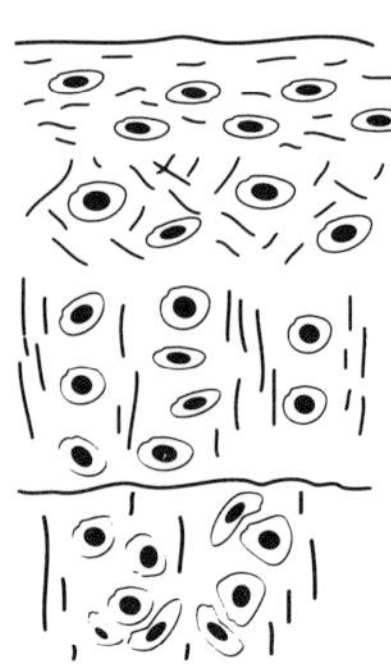

Cartilage

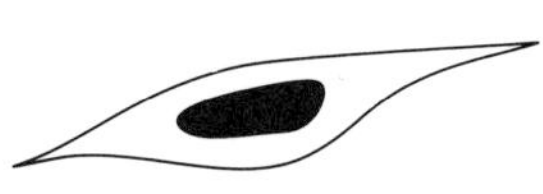

Fibroblast

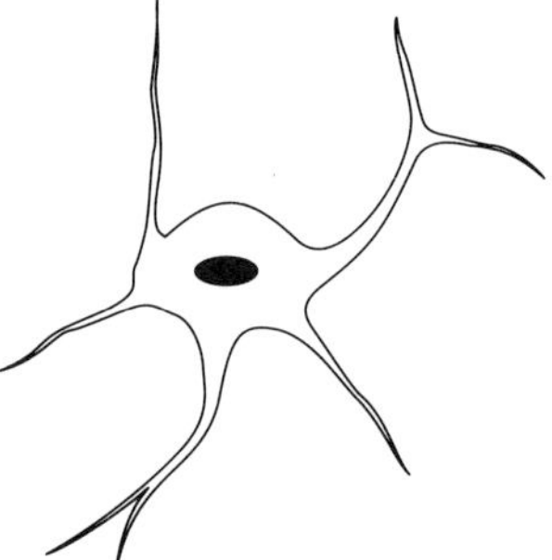

Dendritic cell/
Langerhans cell

Arteriole

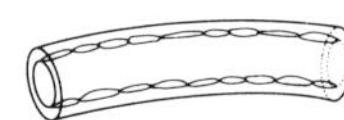

Venule

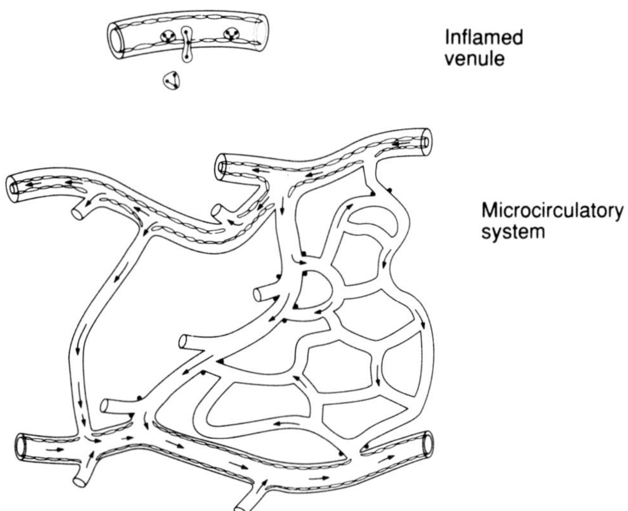

Inflamed
venule

Microcirculatory
system

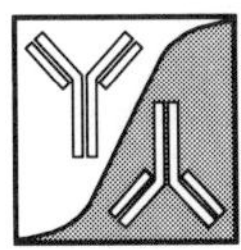

Index